T0139699

Partial Differential Equations

The subject of partial differential equations (PDE) has undergone great change during the last 70 years or so, after the development of modern functional analysis; in particular, distribution theory and Sobolev spaces. In the modern concept, the PDE is visualized in a more general setup of functional analysis, where we look for solutions in a sense weaker than the usual classical sense to address the more physically relevant solutions. Though the aim of the present book is to introduce the fundamental topics in a classical way as in any first book on PDE, the authors have demonstrated the basic topics in a way that opens the doors to the modern theory. Readers can immediately and naturally sense the importance of studying these topics in a modern approach. As a lead, after introducing method of characteristics for first order equations, the authors have immediately discussed the importance of introducing the notion of weak solutions to two important classes of first order equations, namely conservation laws and Hamilton–Jacobi equations. The implication is that physically relevant solutions cannot be obtained within the realm of classical solutions. Almost all the chapters cover something about modern topics. Also included are many exercises in most chapters, which help students get better insight. Hints or answers are provided to some selected exercises.

A. K. Nandakumaran is Professor, Department of Mathematics, Indian Institute of Science, Bangalore. His areas of interest are analysis of partial differential equations, in particular homogenization of PDE, control and controllability problems, inverse problems and computations. He received the Sir C. V. Raman Young Scientist State Award in Mathematics in 2003. He and P. S. Datti are also co-authors of the book 'Ordinary Differential Equations: Principles and Applications', published by the Cambridge University Press in 2017.

P. S. Datti is former faculty at Tata Institute of Fundamental Research (TIFR) Centre for Applicable Mathematics, Bangalore. His main areas of research include non-linear hyperbolic equations, hyperbolic conservation laws, ordinary differential equations, evolution equations and boundary layer phenomena. He has written TIFR Lecture Notes for the lectures delivered by G. B. Whitham (CalTech) and Cathleen Morawetz (Courant Institute).

CAMBRIDGE-IISc SERIES

Cambridge–IISc Series aims to publish the best research and scholarly work on different areas of science and technology with emphasis on cutting-edge research.

The books will be aimed at a wide audience including students, researchers, academicians and professionals and will be published under three categories: research monographs, centenary lectures and lecture notes.

The editorial board has been constituted with experts from a range of disciplines in diverse fields of engineering, science and technology from the Indian Institute of Science, Bangalore.

Titles in print in this series:

- *Continuum Mechanics: Foundations and Applications of Mechanics* by C. S. Jog
- *Fluid Mechanics: Foundations and Applications of Mechanics* by C. S. Jog
- *Noncommutative Mathematics for Quantum Systems* by Uwe Franz and Adam Skalski
- *Mechanics, Waves and Thermodynamics* by Sudhir Ranjan Jain
- *Finite Elements: Theory and Algorithms* by Sashikumaar Ganesan and Lutz Tobiska
- *Ordinary Differential Equations: Principles and Applications* by A. K. Nandakumaran, P. S. Datti and Raju K. George
- *Lectures on von Neumann Algebras, 2nd Edition* by Serban Valentin Strătilă and László Zsidó
- *Biomaterials Science and Tissue Engineering: Principles and Methods* by Bikramjit Basu
- *Knowledge Driven Development: Bridging Waterfall and Agile Methodologies* by Manoj Kumar Lal

Cambridge-IISc Series

Partial Differential Equations

Classical Theory with a Modern Touch

A. K. Nandakumaran
P. S. Datti

CAMBRIDGE
UNIVERSITY PRESS

University Printing House, Cambridge CB2 8BS, United Kingdom

One Liberty Plaza, 20th Floor, New York, NY 10006, USA

477 Williamstown Road, Port Melbourne, vic 3207, Australia

314 to 321, 3rd Floor, Plot No.3, Splendor Forum, Jasola District Centre, New Delhi 110025, India

79 Anson Road, #06-04/06, Singapore 079906

Cambridge University Press is part of the University of Cambridge.

It furthers the University's mission by disseminating knowledge in the pursuit of
education, learning and research at the highest international levels of excellence.

www.cambridge.org
Information on this title: www.cambridge.org/9781108839808

First published 2020

Printed in India by Nutech Print Services, New Delhi 110020

A catalogue record for this publication is available from the British Library

ISBN 978-1-108-83980-8 Hardback

To Our Teachers

Contents

Illustrations

Preface

We ventured into writing this book *Partial Differential Equations* knowing very well that writing a textbook on a very old discipline, that too for beginners, is indeed a formidable task. This exercise was partly due to the good response we received for our first book, *Ordinary Differential Equations*, co-authored with Raju K. George, whose contents were also classical. The venture was also partly due to the suggestions we have received during our interactions with students and teachers from various institutions in the country. The choice of the contents for this book are largely based on such interactions and also on our training in the subject. It is our wish that such a course on partial differential equations (PDE) should seriously be taught at senior undergraduate or beginning graduate level at various institutions in the country, so as to prepare a student for a more serious study of the advanced topics.

This book should be accessible to anyone with sound knowledge in several variable calculus, save for a couple of chapters where the reader is expected to have knowledge of the modern integration theory. The book essentially deals with first-order equations, the classical Laplace and Poisson equations, heat or diffusion equation and the wave equation. The full generality was never on our minds. Numerical analysis and computations are not considered here. Nevertheless, students and researchers working on these aspects of the subject can also gain something from the book. Almost all the topics considered here, of course, arise from the real-world applications in physics, engineering, biology, and so on. Though there is no discussion on the applications in the book, the community of students and researchers from these applied fields can also benefit from the book. We have also presented a detailed description of the classification of PDE, including a motivation behind classification.

A few words about the title. The subject of PDE has undergone great change during the last 70 years or so after the development of modern functional analysis, in particular *distribution theory* and *Sobolev spaces*. In the modern concept, the PDE is visualized in a more general setup of functional analysis, where we look for solutions in a sense *weaker* than the usual classical sense to address the more physically relevant solutions. Though the aim of the present book is to introduce the fundamental topics in a classical way as in any first book on PDE, we have demonstrated the basic topics in such a way that the doors of the modern theory are open to interested readers. They can, immediately and naturally, sense the importance of studying these topics in a modern approach. For example, after introducing the *method of characteristics* for the first-order equations, we have immediately discussed the importance of introducing the notion of weak solutions to two important class of first-order equations, namely *conservation laws* and *Hamilton–Jacobi* equations. These examples suggest that the physically relevant solutions cannot be obtained within the realm of classical solutions. Also in almost all the chapters, we have written something about the modern topics. This is the modern touch we have envisaged and decided to put in the title.

We have included many exercises in most of the chapters. Students should work on them to get better insight into the subject. Hints or answers are provided to some selected exercises.

Bangalore, India
November 2019

A. K. Nandakumaran
P. S. Datti

Acknowledgments

We wish to express our sincere appreciation to the IISc Press for suggesting to publish our second book through the joint venture of IISc Press and Cambridge University Press. Our first book, *Ordinary Differential Equations* was also published through the same joint venture. We thank them for the continued encouragement. We wish to acknowledge the encouragement we received from our respective institutions and for the moral support we received from our colleagues during the preparation of the manuscript. We also wish to thank the anonymous referees for their constructive criticism and suggestions, improving the overall presentation of the book. We thank our academic fraternity for making valuable suggestions while reading parts of the manuscript. Our sincere thanks to the staff of the IISc Press and Cambridge University Press for their coordination from the very beginning of this project, and timely communications. In particular, our special appreciation to Ms Kavitha Harish of IISc Press and Ms Taranpreet Kaur of Cambridge University Press who have coordinated with us during the preparation of the manuscript.

All illustrations in the book have been drawn using the freely available *tikz package*. This package has indeed made our lives simple by reproducing exactly what we had in mind.

Last but not least, we wish to thank our family members for their patience and understanding during the preparation of this book.

Notations

The notations used in the book are the standard ones extensively used in the literature. Below is a list of these notations.

- Abbreviation PDE means partial differential equation(s) and PDO means partial differential operator(s).
- Points in the Euclidean space $\mathbb{R}^n$ are denoted by $x = (x_1, \dots, x_n)$.
- For $x, y \in \mathbb{R}^n$, their *dot product* or *scalar product* or *inner product* is defined by $x \cdot y = \sum_{i=1}^n x_i y_i$. We also write $(x, y) = x \cdot y$. The *standard norm* in $\mathbb{R}^n$ is denoted by $|x| = \sqrt{(x, x)}$.
- If A and B are subsets of $\mathbb{R}^n$, we write $A \subset\subset B$ if $\bar{A} \subset B$, where $\bar{A}$ denotes the *closure* of A.
- For $x \in \mathbb{R}^n$ and $r > 0$, the *open ball* with centre at x and radius r is denoted by $B_r(x)$; the *sphere* with centre at x and radius r is denoted by $S_r(x)$ or $\partial B_r(x)$, which is the boundary of $B_r(x)$. Thus,

  $$B_r(x) = \{y \in \mathbb{R}^n : |x - y| < r\}$$

  and

  $$S_r(x) = \{y \in \mathbb{R}^n : |x - y| = r\}.$$
- The *volume* of the unit ball $B_1(0)$ in $\mathbb{R}^n$ is denoted by ω_n and the *surface area* of the unit sphere $S_1(0)$ in $\mathbb{R}^n$ is denoted by σ_n. Thus, $\sigma_n = \frac{2\pi^{n/2}}{\Gamma(n/2)}$ and $\omega_n = \frac{\sigma_n}{n}$, where Γ is the Euler *gamma function*. The *volume* of $B_r(x)$ and the *surface area* of $S_r(x)$ are denoted by $|B_r(x)|$ and $|S_r(x)|$ respectively. Thus, $|B_r(x)| = \omega_n r^n$ and $|S_r(x)| = \sigma_n r^{n-1}$.
- The closure of the open ball $B_r(x)$ is denoted by $\bar{B}_r(x)$ (also $\overline{B_r(x)}$).
- The partial derivatives are denoted by

  $$D_j = \partial_j = \frac{\partial}{\partial x_j}, \; j = 1, 2, \dots, n.$$
- A multi-index is an n-tuple $\alpha = (\alpha_1, \dots, \alpha_n)$ with α_j all non-negative integers. The *order* of α is the non-negative integer $|\alpha| = \alpha_1 + \cdots + \alpha_n$.

- If α is a multi-index, we write

$$D^\alpha = D_1^{\alpha_1} \cdots D_n^{\alpha_n} = \frac{\partial^{|\alpha|}}{\partial x_1^{\alpha_1} \cdots \partial x_n^{\alpha_n}}.$$

- If α is a multi-index and $x \in \mathbb{R}^n$, we write $x^\alpha = x_1^{\alpha_1} \cdots x_n^{\alpha_n}$.
- Unless otherwise stated, Ω denotes a bounded open set in $\mathbb{R}^n$ with smooth boundary.
- The space $C^k(\Omega)$ denotes the set of all real valued functions defined on Ω which have continuous partial derivatives of order up to k, for a non-negative integer k. We also write $C = C^0$ when $k = 0$.
- The set $C^k(\overline{\Omega}) \subset C^k(\Omega)$ denotes the set of all real valued functions defined on Ω whose partial derivatives up to order k, are continuous in $\overline{\Omega}$.
- For $0 < \alpha \le 1$, The set $C^{k,\alpha}(\Omega)$ denotes the subset of $C^k(\Omega)$ consisting of all those u such that $D^\beta u$, $|\beta| = k$ are Hölder continuous of order α in Ω (Lipschitz continuous if $\alpha = 1$).
- For a function $f : \Omega \to \mathbb{R}$, its *support* is defined as the closure of the set $\{x \in \Omega : f(x) \neq 0\}$ and is denoted by supp f.
- The set of functions in $C^k(\Omega)$ having compact support in Ω is denoted by $C_c^k(\Omega)$.
- The set of all infinitely differentiable functions defined on Ω and having compact support in Ω is denoted by $C_c^\infty(\Omega)$. The space of test functions $\mathcal{D}(\Omega)$ is the set $C_c^\infty(\Omega)$, with a specified topology.
- If Ω is an open subset (or more generally, a Lebesgue measurable subset) of $\mathbb{R}^n$, the class of all the real or complex valued measurable functions f defined on Ω satisfying the condition

$$\int_\Omega |f(x)|^p \, dx < \infty,$$

where $1 \le p < \infty$ is denoted by $L^p(\Omega)$. When $p = \infty$, the class of all the real or complex valued measurable functions f defined on Ω which are *essentially bounded* is denoted by $L^\infty(\Omega)$. The norm in $L^p(\Omega)$, $1 \le p \le \infty$ is denoted by $\|\cdot\|_{L^p(\Omega)}$:

$$\|f\|_{L^p(\Omega)}^p = \int_\Omega |f(x)|^p \, dx \, (p < \infty) \text{ and } \|f\|_{L^\infty(\Omega)} = \text{esssup}_\Omega |f|.$$

- If $u : \Omega \to \mathbb{R}$, we write $u_{ij} = D_i D_j u$ for $i, j = 1, 2, \ldots, n$. The *Hessian* of u is the real symmetric matrix $[u_{ij}]$ and is denoted by $D^2 u$.
- For functions of $(x, t) \in \mathbb{R} \times \mathbb{R}$, we use the notation $D_x = \partial_x = \frac{\partial}{\partial x}$ and $D_t = \partial_t = \frac{\partial}{\partial t}$ and similarly for mixed derivatives.

- The operator ∇ denotes the *grad* operator: $\nabla u = (\partial_1 u, \dots, \partial_n u)$. When needed, we also stress the variable, as in ∇_x, ∇_y.
- For a vector valued function $u = (u_1, \dots, u_n)$, its *divergence* is defined by $\mathrm{div} u = \nabla \cdot u = \partial_1 u_1 + \cdots + \partial_n u_n$.
- The *Laplace operator* or the *Laplacian* in $\mathbb{R}^n$ is the PDO defined by

$$\Delta = \nabla \cdot \nabla = D_1^2 + \cdots + D_n^2 = \frac{\partial^2}{\partial x_1^2} + \cdots \frac{\partial^2}{\partial x_n^2}.$$

- The *wave operator* or the *D'Alembertian* in $\mathbb{R}^{n+1}$ is the PDO defined by

$$\Box_c = \partial_t^2 - c^2 \Delta_x = \partial_{tt} - c^2 \Delta_x$$

  where $c > 0$ is a constant, $t \in \mathbb{R}$ and $x \in \mathbb{R}^n$; Δ_x denotes the Laplacian with respect to the x variables.
- The *heat operator* is $\partial_t - a^2 \Delta_x$.
- For an $m \times n$ real matrix A, its *transpose* is denoted by A^T, which is an $n \times m$ matrix.

CHAPTER 1
Introduction

1.1 GENERAL NATURE OF PDE

It is no exaggeration to state that partial differential equations (PDE) have played a vital role in the development of science and technology, primarily since the beginning of the twentieth century. In the earlier stage, PDE were mainly used to describe physical phenomena, like vibrations of strings, heat conduction in solids, transport phenomena, to mention a few. Later, with the advantage of mathematical modelling, the scope of using PDE for the description of phenomena occurring in biology, economics and even sociology became prominent.

Since the days of Newton or even earlier, many have attempted to describe physical processes using mathematics.[1] Such a mathematical description often leads to linear differential, integral and even integro-differential equations. Thus, a large number of PDE naturally come from mathematical physics. The initial developments in PDE, though, were mainly geared towards obtaining solutions to a particular physical or engineering problem, it was soon realized that many of the problems will have common features and similarities. This naturally led to the grouping of PDE that can be tackled in a single framework. This automatically leads to the abstraction of the subject and the theoretical analysis that follows, hence, becomes more important. This is one of the features we try to follow in the present book. Indeed, unlike ordinary differential equations (ODE), all PDE including the linear ones cannot be treated in a single theoretical framework, leading to the necessity of a classification. In fact, due to the diverse nature of physical phenomena, we remark that we cannot classify all the PDE. Nevertheless, a fairly good classification is available for the second-order equations and interestingly a large number of physical and other problems lead to second-order equations. Also, for the three important classes of equations, namely elliptic, hyperbolic and parabolic, general theories have been developed.

As mentioned above, a wide class of physical problems is described by second-order linear differential equations of the form

$$\sum_{i,j=1}^{n} a_{ij}(x)u_{x_ix_j} + \sum_{i=1}^{n} b_i(x)u_{x_i} + c(x)u = f(x). \tag{1.1}$$

[1]It is a historical fact that the *calculus* was born during such a process.

Here the variable x varies in an open set in the physical space $\mathbb{R}^n$, $n = 1, 2, 3$ and the coefficients a_{ij}, b_i and c are known from the physical process; u is the unknown function and f denotes an external quantity, if any, influencing the physical process.

We only mention a few real-world situations where PDE occur. For more examples and their detailed discussion, the reader is referred to Barták et al. (1991), Markowitz (2005), Murray (2003), Rhee et al. (1986), and Vladimirov (1984).

Many problems in mechanics like vibrations of strings, rods, membranes and three-dimensional objects and also the mathematical description of electromagnetic waves lead to the equation of vibrations, which is the wave equation in one more space dimension. If the mean free path of the particles is much larger than their dimensions, then the propagation of a particle may be more accurately described by an equation, in comparison with the diffusion equation, called the *transport* or *kinetic equation*. This is also called the *Boltzmann equation*. This is an integro-differential equation.

The Heisenberg principle states that the position of a particle and its momentum cannot be *simultaneously* described, according to the laws of quantum mechanics. Thus, for example, the position of a quantum particle can be confirmed only with certain probability. The Schrödinger's equation is an attempt to describe the dynamics of a quantum particle of a given mass moving in an external force field with a given potential. Reaction–Diffusion equations describe the interaction of two or more chemical concentrations of distinct diffusivity coefficients, in a chemical or biological process. These equations are also used in the modelling of pattern formation and form an important part of *Mathematical Biology* and constitute a system of non-linear diffusion equations.

The equation of heat diffusion in a medium and the diffusion of a chemical species are described by the heat or diffusion equation. Euler's equations of gas dynamics describe the dynamics of an ideal fluid, that is, a fluid with no or negligible viscosity. These equations form a system of first-order hyperbolic equations. In a particular situation where liquid is incompressible and has a potential, these equations reduce to the Poisson's equation for the potential function. The system of Maxwell's equations describe the dynamics of a charged particle in a medium with varying electromagnetic field, invoking *Ampere's law* and *Faraday's law*. In some particular cases, each component of the electric and magnetic fields satisfies the *telegraph equation*.

1.2 TWO EXAMPLES

The following two situations perhaps describe a general nature in the analysis of solutions to PDE. These are quite simple to state and involve second-order equations in two variables. The equations are the Laplace equation, the heat or diffusion equation and the wave equation: $L_i u = 0$, $i = 1, 2, 3$, where

$$L_1 = \partial_t^2 + \partial_x^2,\ L_2 = \partial_t - \partial_x^2 \text{ and } L_3 = \partial_t^2 - \partial_x^2.$$

The first situation involves the determination of solutions of $L_i u = 0$, $i = 1,2,3$ with prescribed data on the boundary of a rectangle $ABCD$ with the side AB situated on the x-axis in the $x-t$ plane. Without much concern whether to prescribe u or its first derivatives on the sides of $ABCD$, let us dwell on the number of conditions required for each of the operators L_i, in order to determine a solution of $L_i u = 0$. It turns out that L_1 requires *four* conditions one each on the four sides of the rectangle $ABCD$; L_2 requires *three* conditions one each on the sides AB, BC and AD of the rectangle $ABCD$; L_3 requires *four* conditions – two on AB and one each on BC and AD of the rectangle $ABCD$.

Note that all the three operators are linear and of second order. Yet, the number of data to be prescribed and the part of the boundary where to be placed become important in order to determine a solution. Apparently, there is no simple explanation for this anomaly. Perhaps the reader will find an answer after studying the relevant chapters in the book. This is quite different from the analysis of an initial value problem (IVP) of a system of linear ODE; here the problem can be studied for a system of *any order* in a single framework. However, in the case of PDE, as the above examples exhibit, it is not possible to do an analysis even for second-order linear equations in two dimensions, in a single framework. This leads to the notion of a classification of PDE, and a particular condition on the data like initial values or boundary values depends on the type of PDE under consideration.

The second situation also concerns the operators L_i, but now with regard to *weak solutions* of them. A continuous or a locally integrable function u defined in an open set Ω in $\mathbb{R}^2$ is said to be a *weak solution* of $L_i u = 0$ for $i = 1,2,3$, if $\iint u(x,t) L_i \varphi(x,t)\, dxdt = 0$ for all $\varphi \in C_c^\infty(\Omega)$.

It is shown in Chapter 9 that any continuous or locally integrable function u of the form $v(x \pm t)$ is a weak solution for L_3 and thus it can admit discontinuous (weak) solutions. For the operators L_1 and L_2, it turns out that any weak solution is in fact a C^∞ function, may be after making corrections in a *set of measure zero*. The apparently strange behavior of the operators L_1 and L_2 cannot be explained in simple terms and the reader will not find a complete answer in this book! The operators L_1 and L_3 are quite different, but the operator L_2 may share some properties with L_1 (regularity) and some other properties with L_3 (energy estimates).

The above two situations describe, we hope, the complexities that are involved in the analysis of PDE. There is indeed constant evolution of the subject as and when some peculiar phenomenon is observed through an example or otherwise. In this connection, it is an interesting fact that a somewhat *true* picture of linear operators started emerging only after the work of Peetre (1960), even though there were already quite many advancements in the modern theory of PDE which had emerged through the works of Leray, Petrowski, Schwartz and others. With appropriate domain and range of the operator, what Peetre showed was that the linear operators are precisely the *local operators*. This means that $\text{supp}Pu \subset \text{supp}u$, where P is linear and u is in its domain. This then led to the discovery of *pseudo-differential operators* and *Fourier integral operators*. Roughly speaking, the *inverse* of an elliptic operator is a pseudo-differential operator and the *inverse* of a wave operator is a Fourier integral operator (see Nirenberg, 1976).

1.3 DESCRIPTION OF THE CONTENTS

This then sets the stage for the present book, with a modest list of contents.

- The first chapter briefly discusses certain general notions of PDE, their occurrence in physical and other sciences and engineering. It also describes the contents of the book, chapter-wise.
- The theory[2] of modern PDE is quite vast and demands a great amount of prerequisites such as Lebesgue integration theory, functional analysis, distributions and Sobolev spaces. Since we are discussing mostly classical theory in the present book, the prerequisites are minimal – a good understanding of multivariable calculus should suffice for studying this book. Exceptions do occur in Chapters 4 and 5, where the reader is expected to have a good knowledge of the modern theory of integration, especially in the proofs of uniqueness of solutions. In Chapter 2, we collect a good number of results from multivariable calculus, ODE and related topics that are used in the book. To make the book as self-contained as possible, we have also provided the proofs when they are not too lengthy.
- Chapter 3 is about the first-order equations. Here we study the general Cauchy problem (IVP) for such equations. The (local) theory is fairly complete as the problem is reduced to an IVP for a system of ODE. The geometry, however, does get complicated as we move from linear to quasilinear to general first-order equations. Because of their importance in applications, we mention two classes of first-order equations, namely the conservations laws and the Hamilton–Jacobi equations. These two classes are studied in detail in further chapters.
- In Chapters 4 and 5, we consider certain important class of first-order equations – Hamilton–Jacobi Equation (HJE) and Conservation Laws (CL) – which have been topics of great interest among researchers owing to their importance in many applications. Though these equations have been mentioned in Chapter 3, the emphasis here is on a new concept of a solution of these equations. A beginner perhaps encounters for the first time the concept of a *weak* solution to a PDE, which is in general a non-differentiable function! Furthermore, to obtain uniqueness of a solution, additional condition(s) need to be imposed. Since the theory of modern PDE largely deals with *weak* solutions, we thought it is a good idea to introduce this concept of solution to a beginner in the context of HJE and CL. However, these chapters may be skipped for the first reading as the uniqueness results require a good knowledge of modern theory of integration.
- In the context of ODE, the theory dealing with the Cauchy problem of a single equation or that of a system of first-order equations is essentially the same. In particular, the

[2]This is not to suggest there is a *single* theory of PDE, like theory of ODE or theory of functions of real or complex variable. In fact, we see in the literature different *theories* of PDE owing to the sheer vastness of the subject.

analysis is the same for both the first-order equations and higher-order equations, in the study of ODE. In contrast, such is not the situation about PDE. This makes the subject of PDE more complicated and also interesting. In Chapter 6, we explain how the data in a Cauchy problem for a second-order equation cannot in general be arbitrary. This naturally leads to the concept of classification of second- and higher-order equations. The main discussion in this chapter is about second-order equations and their solutions.

It should, however, be noted that some important developments in science in nineteenth and twentieth centuries, especially quantum mechanics and fluid dynamics, resulted in new types of PDE – the Schrödinger equation, Navier–Stokes equations and Kortweg-de Vries (K-dV) equation, for example. These equations and many more equations do not fall in the ambit of the above-mentioned classification. Thus, there were attempts to make the subject of PDE a unified subject without mentioning the class to which a PDE belongs. However, such attempts have not been that successful. This is one of the reasons we see a great number of books written on a particular equation or on a particular class of equations.

- Undoubtedly, the three major equations of mathematical physics – the Laplace equation (Poisson equation), the heat or diffusion equation, and the wave equation – have had great impact on the development of much of the modern theory of PDE. These equations are the topics of discussion in Chapters 7 through 10, respectively.
- The Laplace operator is a prototype of uniformly elliptic operators. Some important properties – mean value property, maximum (minimum) principle, Harnack's inequalities – enjoyed by a solution of the Laplace's equation are discussed at length in Chapter 7. We have also indicated that the maximum (minimum) principle is also enjoyed by a solution of a general uniformly elliptic equation. The existence and uniqueness of the solutions are also discussed via Perron's method and Newtonian potential.
- In Chapter 8, the heat equation and its solutions are studied in great detail. This equation is a prototype of parabolic equations. In a way this equation *sits* between the Laplace's equation and the wave equation. Therefore, its solution enjoys certain properties from both sides. For example, maximum (minimum) principle from Laplace's equation and energy estimate from the wave equation. Its solution also enjoys a mean value property and backward uniqueness property.
- The study of Laplace's equation and the heat equation largely does not depend on the dimension. However, the analysis of the wave equation does depend on the dimension and this is the reason to consider the study of the wave equation in one dimension and higher dimensions separately. These are dealt with in Chapters 9 and 10, respectively. The wave equation is a prototype of hyperbolic equations.
- The Cauchy–Kovalevsky theorem is, historically, an important result in the subject field of PDE. It is one of the first results proving the existence and uniqueness of solution to a Cauchy problem for a general equation, though in a restricted class of

equations with analytic coefficients. Nevertheless, the contents of its proof are full of a priori estimates, a hallmark of the modern theory of PDE. In Chapter 11, we present the details of this theorem and a generalization. We also briefly discuss the Holmgren's uniqueness result.

- We also briefly mention some aspects of the modern theory without going into details in Chapter 12. An existence result of L^2 weak solution is discussed here, to give a general flavor of a modern theory.

CHAPTER 2

Preliminaries

2.1 MULTIVARIABLE CALCULUS

2.1.1 Introduction

We plan to briefly introduce the calculus on $\mathbb{R}^n$, namely the concept of total derivative of multivalued function, $f = (f_1, \cdots, f_m) : \mathbb{R}^n \to \mathbb{R}^m$. We are indeed familiar with the notion of partial derivatives $\partial_i f_j = \frac{\partial f_j}{\partial x_i}$, $1 \leq i \leq n, 1 \leq j \leq m$. In the sequel, we will introduce the important concept of *total derivative* and discuss its connection to the partial derivatives. We remark that the total derivative (known also as *Frechét derivative*) can be extended to infinite dimensional normed linear spaces, which is used in the analysis of more complicated problems especially arising from optimal control problems, calculus of variations, partial differential equations, and so on.

Motivation: One of the fundamental problems in mathematics (and hence in applications as well) is the following: Let $f : \mathbb{R}^n \to \mathbb{R}^n$. Given $y \in \mathbb{R}^n$, solve the system of equations

$$f(x) = y \tag{2.1}$$

and represent the solution as $x = g(y)$ and if possible find good properties of g, namely its smoothness. More generally, if $f : \mathbb{R}^{n+m} \to \mathbb{R}^n$, $x \in \mathbb{R}^n, y \in \mathbb{R}^m$, solve the implicit system of equations

$$f(x, y) = 0 \tag{2.2}$$

and represent the solution as $x = g(y)$. Consider the one-dimensional case, where $f : \mathbb{R} \to \mathbb{R}$ which is C^1. Suppose that $f'(a) \neq 0$ for some a. Then, by the continuity of f', we see that $f'(x) \neq 0$ in a neighborhood interval I of a. Hence f' preserves the sign in I, f is monotonic in I and $f(I)$ is an interval. Thus, if $f(a) = b$, then the above argument shows that $f(x) = y$ is solvable for all y in $f(I)$, a neighborhood of b. This is the local solvability that is obtained by the non-vanishing property of the derivative of f at a. This immediately shows the importance of understanding the derivatives in the solvability of algebraic equations. We remark that the mere existence of all partial derivatives does not guarantee the local solvability. We need the stronger concept of total derivative.

Linear Systems: Let us look at the well-known linear system

$$Ax = y, \tag{2.3}$$

where $A = [a_{ij}]$ is a given $n \times n$ matrix. That is $f(x) = Ax$. The system (2.3) can be rewritten as

$$\sum_{j=1}^{n} a_{ij}x_j = y_i, \quad 1 \le i \le n. \tag{2.4}$$

The system (2.3) or (2.4) is uniquely solvable for x in terms of y if and only if $\det A \neq 0$ (global solvability). In this case

$$x = A^{-1}y$$

and A^{-1} is also an $n \times n$ matrix. We would like to address the solvability of (2.1) and (2.2) giving appropriate conditions similar to non-vanishing determinant as in the case of a linear system.

Example 2.1. Define $f : \mathbb{R} \to \mathbb{R}$ by $f(x) = x^2$. Clearly $f(0) = 0$. For $y > 0$, the equation $x^2 = y$ has two solutions $x_1 = +\sqrt{y}$ and $x_2 = -\sqrt{y}$ (non-uniqueness) and $y < 0$, the equation has no solution. Thus, we sense a difficulty around $x = 0$. Note that $\left.\frac{\partial f}{\partial x}\right|_{x=0} = 2x|_{x=0} = 0$. This shows that we cannot decide the sign of $\frac{\partial f}{\partial x}$ around 0. If we take any $a \neq 0$, and $b = a^2 = f(a)$, then, for any $y \in (b - \varepsilon, b + \varepsilon)$, ε small, there exists unique $x \in (a - \delta, a + \delta)$ for some δ such that $f(x) = y$. That is, the equation is solvable in a neighborhood of the point $b = f(a)$. Here, observe that $\left.\frac{\partial f}{\partial x}\right|_{x=a} = 2x|_{x=a} = 2a \neq 0$ and thus the sign of $\frac{\partial f}{\partial x}(a)$ is known. □

Example 2.2. Consider the function $f : \mathbb{R}\times\mathbb{R} \to \mathbb{R}$ defined by $f(x, y) = x^2 + y^2 - 1$. Indeed, the solutions (x, y) of the equation $f(x, y) = 0$ are points on the unit circle. Consider the solvability of x in terms of y near the solution $(0, 1)$ of $x^2 + y^2 - 1 = 0$, that is $x^2 = 1 - y^2$.

For y near $1, y < 1$, we have two solutions $x_1 = +\sqrt{1 - y^2}$, $x_2 = -\sqrt{1 - y^2}$. Similarly the case near the point $(0, -1)$. Again observe that $\frac{\partial f}{\partial x}|_{(0,\pm 1)} = 2x|_{(0,\pm 1)} = 0$.

On the other hand, consider the point $(+1, 0)$. For y near 0, there exists unique solution $x = +\sqrt{1 - y^2}$; and for the point $(-1, 0)$ and y near 0, there exists unique solution $x = -\sqrt{1 - y^2}$. In fact, for any (a, b) with $a^2 + b^2 - 1 = 0$ and $a \neq 0$, we get $\left.\frac{\partial f}{\partial x}\right|_{(a,b)} \neq 0$ and the system is uniquely solvable for x in terms of y in a neighborhood of b. The situation is reversed if we look at the possibility of solving y in terms of x. □

Thus, we see the impact of non-vanishing of the derivative on the solvability as in the linear systems. In higher dimensions, we have many partial derivatives and we need a systematic procedure to deal with such a complicated case. In other words, we would like to understand the solvability of a system of non-linear equations in several unknowns. This is given via inverse and implicit function theorems. We also remark that in general, it is only possible to obtain a local solvability result and not a global result as in linear systems.

2.1.2 Partial, Directional and Frechét Derivatives

Let $f : \mathbb{R} \to \mathbb{R}$ and $x_0 \in \mathbb{R}$. Then $f'(x_0)$ is normally defined as

$$f'(x_0) = \lim_{h\to 0} \frac{f(x_0 + h) - f(x_0)}{h}, \tag{2.5}$$

when the limit exists. We are also aware of the fact that $f'(x_0)$ is the slope of the tangent to the curve $y = f(x)$ at the point $(x_0, f(x_0))$. This allows for another interpretation of the derivative via linear transformation, which is at the heart of the concept of Frechét derivative. Let U be an open subset of $\mathbb{R}^n$ and $f : U \to \mathbb{R}^m$ be a vector-valued map represented by $f = (f_1, \cdots, f_m)^T$, where $f_i : U \to \mathbb{R}$ are real-valued maps. The limit definition can easily be used to define the directional derivatives in any direction and in particular partial derivatives are nothing but the directional derivatives along the co-ordinate axes.

Directional and Partial Derivatives: Recall that the derivative in (2.5) is the instantaneous rate of change of the output $f(x)$ with respect to the input x. Thus, if we consider $f(x)$ at $x_0 \in \mathbb{R}^n$, there are infinitely many radial directions emanating from x_0. Any given vector $v \in \mathbb{R}^n$ determines a direction given by its position vector. Thus, for $x_0 \in \mathbb{R}^n, f(x_0+hv)-f(x_0), h \in \mathbb{R}$ is the change in f in the direction v. This motivates us to define the derivative of f at $x_0 \in \mathbb{R}^n$ in the direction v, denoted by $D_v f(x_0)$, as

$$D_v f(x_0) = \lim_{h\to 0} \frac{f(x_0 + hv) - f(x_0)}{h} \tag{2.6}$$

whenever the limit exists. Note that if $f = (f_1, \cdots, f_m)^T$, then

$$D_v f(x_0) = (D_v f_1(x_0), \ldots, D_v f_m(x_0))^T.$$

If v is a unit vector, then $D_v f(x_0)$ is called the *directional derivative* of f at x_0 in the direction v. If $v = e_i = (0, \cdots, 0, 1, 0, \cdots, 0)$ is the co-ordinate axis vector, then clearly

$$D_{e_i} f(x_0) = \frac{\partial f}{\partial x_i}(x_0) = \left(\frac{\partial f_1}{\partial x_i}(x_0), \cdots, \frac{\partial f_m}{\partial x_i}\right)^T.$$

Example 2.3. Define $f : \mathbb{R}^n \to \mathbb{R}$ by $f(x) = |x|^2$. Then $\dfrac{\partial f}{\partial x_i}(x_0) = 2x_{0i}$. Now, for $v \in \mathbb{R}^n$,

$$
\begin{aligned}
f(x_0 + hv) &= \sum_{i=1}^{n} (x_{0i} + hv_i)^2 \\
&= f(x_0) + 2h(x_0, v) + h^2|v|^2.
\end{aligned}
$$

It follows that

$$D_v f(x_0) = 2(x_0, v).$$

□

As seen earlier the existence of all directional derivatives implies the existence of partial derivatives. But, the converse is not true.

Example 2.4. Let $f : \mathbb{R}^2 \to \mathbb{R}$ be defined by

$$f(x, y) = \begin{cases} x + y \ \text{if } x = 0 \text{ or } y = 0 \\ 1 \ \ \text{otherwise.} \end{cases}$$

Then, $D_{(1,0)}f(0, 0) = D_{(0,1)}f(0, 0) = 1$, but $D_{(a,b)}f(0, 0)$, $a \neq 0, b \neq 0$ does not exist. □

Normally, we expect differentiable functions to be continuous, which is true in one dimension. But the existence of all directional derivatives at a point does not imply the continuity at that point. This is a serious drawback and prompts us to look for a stronger concept of derivative, namely the notion of *total derivative*.

Example 2.5. Consider the function $f : \mathbb{R}^2 \to \mathbb{R}$ defined by

$$f(x, y) = \begin{cases} \frac{xy^2}{x^2+y^4}, \ \text{if } x \neq 0 \\ 0 \text{ if } x = 0. \end{cases}$$

It is easily seen that $D_v f(0, 0)$ exists for all $v \in \mathbb{R}^2$, but f is not continuous at $(0, 0)$. □

Example 2.6. Let the function $f : \mathbb{R}^2 \to \mathbb{R}$ be defined by

$$f(x, y) = \begin{cases} \dfrac{xy(x^2 - y^2)}{x^2 + y^2}, \ \text{if } (x, y) \neq (0, 0) \\ 0 \ \text{if } (x, y) = (0, 0). \end{cases}$$

Again, it is easily checked that $\frac{\partial^2 f}{\partial x \partial y}(0,0) = 1$ and $\frac{\partial^2 f}{\partial y \partial x}(0,0) = -1$. This shows that, in general, the order of the mixed partial derivatives cannot be interchanged. □

Total (Frechét) Derivative: Recall that if $f : \mathbb{R} \to \mathbb{R}$, then $f'(x_0) = \alpha$ represents the slope of the tangent to the curve $y = f(x)$ at the point $(x_0, f(x_0))$. In this case, a linear equation is associated, namely the line $y - f(x_0) = \alpha(x - x_0) = f'(x_0)(x - x_0)$. In other words, the derivative can be viewed as a linear mapping, $T_\alpha : \mathbb{R} \to \mathbb{R}$ defined by

$$T_\alpha x = \alpha x = f'(x_0)x.$$

Thus, interpreting any differentiation concept as a linearization is the crux of the matter not only in finite dimension, but in infinite dimensions as well. Once again recall $f'(x_0)$ in dimension one can be recast as

$$f(x_0 + h) = f(x_0) + f'(x_0)h + r(h)$$

where the reminder (or error) term satisfies

$$\lim_{h \to 0} \frac{r(h)}{h} = 0.$$

Definition 2.7 (Frechét Derivative). Let U be an open subset of $\mathbb{R}^n$ and $x_0 \in U$. We say that a function $f : U \to \mathbb{R}^m$ is differentiable (or Frechét differentiable) at x_0, if there is a linear operator $T = T(x_0) : \mathbb{R}^n \to \mathbb{R}^m$ such that[1]

$$\lim_{h \to 0} \frac{|f(x_0 + h) - f(x_0) - Th|}{|h|} = 0. \tag{2.7}$$

□

Of course, it is understood that the norm in the numerator of (2.7) is the norm in $\mathbb{R}^m$ whereas the one in the denominator is the norm in $\mathbb{R}^n$. It is easy to see that T is unique, if it exists. We denote it by $T = f'(x_0)$ and call it the *Frechét derivative* or the *total derivative* of f at x_0. Furthermore, it is also easy to see that if f is differentiable at x_0, then f is continuous at x_0. If f is differentiable at all points in U, we say f is differentiable in U.

[1]This notion is extended to a mapping $f : X \to Y$, where X, Y are Banach spaces. Here the requirement is that $T : X \to Y$ is a bounded linear operator.

Equivalently, (2.7) can be written as

$$f(x_0 + h) = f(x_0) + f'(x_0)h + r(h)$$

with $r(h) = o(h)$.

Example 2.8. Suppose $A \in L(\mathbb{R}^n, \mathbb{R}^m)$ be an $m \times n$ matrix. Define $f : \mathbb{R}^n \to \mathbb{R}^m$ by $f(x) = Ax$. Then, clearly $f(x_0+h) - f(x_0) = A(x_0+h) - Ax_0 = Ah$ by linearity. Therefore $r(h) = 0$ and $f'(x_0) = A$ for any x_0. □

Example 2.9. Let $f : \mathbb{R}^n \to \mathbb{R}$ by $f(x) = |x|^2 = (x, x)$. Then, $f'(x_0)h = 2(x_0, h)$ or $f'(x_0) = 2x_0$. □

Proposition 2.10 (Chain Rule). Suppose U, V are open subsets of $\mathbb{R}^n$, $\mathbb{R}^m$, respectively. Let $f : U \to \mathbb{R}^m$ and $g : V \to \mathbb{R}^k$ be mappings such that $f(U) \subset V$. Assume f is differentiable at x_0 and g is differentiable at $y_0 = f(x_0)$. Then, the composite function $F(x) = g \circ f(x) = g(f(x))$, defined on U, is differentiable at x_0 and

$$F'(x_0) = g'(f(x_0))f'(x_0). \tag{2.8}$$

□

We have to interpret the product $g'(f(x_0))f'(x_0)$ as the product or composition of linear operators.

The example given earlier indicates that the existence of all directional derivatives is not sufficient to guarantee the existence of the total derivative. But if the total derivative exists, then all the directional derivatives exist and in fact we can compute the total derivative using the partial derivatives.

Let $\{e_1, \dots, e_n\}$ and $\{\tilde{e}_1, \dots, \tilde{e}_m\}$ be the standard bases of $\mathbb{R}^n$ and $\mathbb{R}^m$ respectively and $f : \mathbb{R}^n \to \mathbb{R}^m$. If $f'(x_0)$ exists, then for $1 \leq j \leq n$, we get

$$\begin{aligned} f(x_0 + he_j) - f(x_0) &= f'(x_0)(he_j) + r(he_j), \\ &= hf'(x_0)e_j + r(he_j) \end{aligned}$$

where $h \in \mathbb{R}$ and $\dfrac{|r(he_j)|}{h} \to 0$ as $h \to 0$. Dividing by h and taking the limit as $h \to 0$, we get

$$\frac{\partial f}{\partial x_j} = f'(x_0)e_j = \left(\frac{\partial f_1}{\partial x_j}, \dots, \frac{\partial f_m}{\partial x_j}\right)^T. \tag{2.9}$$

Thus, each partial derivative $\frac{\partial f_i}{\partial x_j}(x_0)$ exists for $1 \leq i \leq m$ and $1 \leq j \leq n$, and $\frac{\partial f}{\partial x_j}(x_0) = f'(x_0)e_j$, $1 \leq j \leq n$. More generally, if $v \in \mathbb{R}^n$, $v = \sum_{i=1}^{n} v_i e_i$, then, $D_v f(x_0)$ exists and equals $\sum_{i=1}^{n} v_i \frac{\partial f}{\partial x_i}$.

Thus, the matrix representation of $f'(x_0)$ is given by

$$f'(x_0) = \left[\frac{\partial f_i}{\partial x_j}\right]_{\substack{1\leq i\leq m \\ 1\leq j\leq n}}. \tag{2.10}$$

The above results can be consolidated in the following theorem:

Theorem 2.11. Let $f : U \subset \mathbb{R}^n \to \mathbb{R}^m$ be differentiable at $x_0 \in U$. Then, $\frac{\partial f_i}{\partial x_j}$, exists for all $1 \leq i \leq m, 1 \leq j \leq n$ and $f'(x_0)$ is given as in (2.10). That is,

$$f'(x_0)e_i = \sum_{j=1}^{m} \frac{\partial f_j}{\partial x_i}\tilde{e}_j.$$

In other words the matrix representation of $f'(x_0)$ in the standard bases of $\mathbb{R}^n$ and $\mathbb{R}^m$ is given by the $m \times n$ matrix $\left[\frac{\partial f_j}{\partial x_i}\right]$ which is the Jacobian of f. □

2.1.3 Inverse Function Theorem

In this section, we address the solvability of a system of non-linear algebraic equations in explicit form:

$$f(x) = y, \tag{2.11}$$

where $f : U \subset \mathbb{R}^n \to \mathbb{R}^n$ and $y \in \mathbb{R}^n$ is given and U is open. These are a set of n non-linear equations in n unknowns:

$$\begin{cases} f_1(x_1, \cdots x_n) = y_1 \\ \quad\vdots \\ f_n(x_1, \cdots x_n) = y_n. \end{cases}$$

Given $a \in U$, let $b = f(a)$, then (a, b) is a solution to (2.11). We want to give conditions under which (2.11) can be solved for x for all y in a neighborhood of b.

Theorem 2.12 (Inverse Function Theorem). Suppose $E \subset \mathbb{R}^n$ is open and $f : E \to \mathbb{R}^n$ is a C^1 map, that is, $f'(x)$ exists for all $x \in E$ and the mapping $x \mapsto f'(x)$ from E into $L(\mathbb{R}^n, \mathbb{R}^n)$ is continuous. Furthermore, assume that the matrix $f'(a)$ is invertible. Then, there exist open sets U and V in $\mathbb{R}^n$ containing a and b, respectively, such that

(1) $f : U \to V$ is 1-1 and onto.
(2) $g = f^{-1} : V \to U$ given by $g(f(x)) = x$ for all $x \in U$ is also a C^1 map. □

The above theorem tells us that $y = f(x)$ can be uniquely solved for x in terms of y in a neighborhood of b. Furthermore, the inverse map is also smooth. We will not present a proof of the above theorem, but it is based on the contraction mapping theorem from functional analysis. Also observe that the space $L(\mathbb{R}^n, \mathbb{R}^n)$ is equipped with the operator norm and it is a Banach space.

Theorem 2.13 (Banach Contraction Mapping Theorem). Let (X, d) be a complete metric space. Assume $\phi : X \to X$ is a contraction map, that is there exists $0 \le \alpha < 1$ such that $d(\phi(x), \phi(y)) \le \alpha d(x, y)$ for all $x, y \in X$. Then, there exists a unique solution to the problem

$$\phi(x) = x.$$

The proof is easy and constructive. Take any arbitrary point $x_0 \in X$. Construct x_n inductively: $x_{n+1} = \phi(x_n), n = 0, 1, 2, \cdots$. Then, for $n > m \ge 1$, using the definition of x_n and contraction inequality, we get

$$d(x_n, x_m) \le \left(\alpha^{n-1} + \cdots + \alpha^{m-1}\right) d(x_1, x_0) = \alpha^{m-1}\frac{1-\alpha^{n-m}}{1-\alpha} d(x_1, x_0) \to 0$$

as $m, n \to \infty$ as $\alpha < 1$. This shows that $\{x_n\}$ is a Cauchy sequence and hence converges, say, to $x \in X$. Since ϕ is continuous, it follows that $\phi(x_n) \to \phi(x)$. But, $\phi(x_{n+1}) = x_{n+1} \to x$. This proves $\phi(x) = x$. Uniqueness follows again from contraction. □

Proof (of Inverse Function Theorem): We only sketch the proof. Given that $A = f'(a)$ is invertible. Since f' is continuous, given $\varepsilon > 0$, there is an open set $U \subset E$ such that $\|f'(x) - A\|_{L(\mathbb{R}^n,\mathbb{R}^n)} \le \varepsilon$ for all $x \in U$. Now, for $y \in \mathbb{R}^n$, define $\phi(x) = x + A^{-1}(y - f(x))$. Then $f(x) = y$ if and only if $\phi(x) = x$ has a solution. The proof follows by establishing ϕ is a contraction and the inverse map thus obtained is a C^1 function. □

Corollary 2.14 (Open Mapping Theorem). Let $f : U \subset \mathbb{R}^n \to \mathbb{R}^n$ be C^1 and $\det f'(x) \neq 0$ for all $x \in U$. Then f is an open map. □

The matrix of $f'(x)$ is also known as the Jacobian matrix.

2.1.4 Implicit Function Theorem

Quite often, we do not expect to get equations in explicit form $y = f(x)$ like in $x^2 + y^2 - 1 = 0$. Rather, we may get a relation connecting the variables x and y. To consider such a general situation, let $f : E \subset \mathbb{R}^n \times \mathbb{R}^m \to \mathbb{R}^n$ be a C^1 map. We wish to solve for x in terms of y of the system of equations

$$f(x, y) = 0. \tag{2.12}$$

This is a system of n equations in $n + m$ variables:

$$\begin{cases} f_1(x_1, \ldots x_n, y_1, \ldots y_m) = 0 \\ \quad\vdots \\ f_n(x_1, \ldots x_n, y_1 \ldots y_m) = 0. \end{cases} \tag{2.13}$$

First consider the linear system

$$Ax + By = 0, \tag{2.14}$$

where A is an $n \times n$ matrix and B is an $n \times m$ matrix. If A is invertible, then x can be solved as

$$x = -A^{-1}By.$$

Let $T \in L(\mathbb{R}^{n+m}, \mathbb{R}^n)$ be a linear transformation from $\mathbb{R}^{n+m}$ into $\mathbb{R}^n$. Indeed T can be represented as an $n \times (n + m)$ matrix like $[A\ B]$, where A is an $n \times n$ matrix and B is an $n \times m$ matrix. For $(h, k) \in \mathbb{R}^{n+m}$, we write $(h, k) = (h, 0) + (0, k)$, $h \in \mathbb{R}^n, k \in \mathbb{R}^m$ and by linearity of T, we get

$$T(h, k) = T(h, 0) + T(0, k).$$

Define $T_x : \mathbb{R}^n \to \mathbb{R}^n, T_y : \mathbb{R}^m \to \mathbb{R}^n$ by

$$T_x h = T(h, 0),\ T_y k = T(0, k)$$

and thus

$$T(h, k) = T_x h + T_y k$$

We will write this as

$$T = T_x + T_y$$

with $T_x \in L(\mathbb{R}^n, \mathbb{R}^n)$ and $T_y \in L(\mathbb{R}^m, \mathbb{R}^n)$.

Theorem 2.15 (Implicit Function Theorem: Linear Version). Assume $T \in L(\mathbb{R}^{n+m}, \mathbb{R}^n)$ and T_x is invertible. Then, for any $k \in \mathbb{R}^m$, there exists a unique $h \in \mathbb{R}^n$ such that $T(h, k) = 0$ and the solution is given by $h = -T_x^{-1} T_y(k)$. □

Theorem 2.16 (Implicit Function Theorem: Non-linear Version). Let E be open in $\mathbb{R}^{n+m}$ and $f : E \to \mathbb{R}^n$ be a C^1 map such that $f(a,b) = 0$ for some $(a,b) \in E$. Put $T = f'(a,b) \in L(\mathbb{R}^{n+m}, \mathbb{R}^n)$ and write $T = T_x + T_y$ as above and assume T_x is invertible.[2] Then, there are open sets $U \subset E$, $W \subset \mathbb{R}^m$ with $b \in W$, $(a,b) \in U$ satisfying

(1) for every $y \in W$, there exists unique x such that $(x,y) \in U$ and $f(x,y) = 0$.
(2) define $g : W \to \mathbb{R}^n$ by $g(y) = x$, then g is a C^1 map such that $g(b) = a$, and $f(g(y),y) = 0$. Further

$$g'(b) = -T_x^{-1}T_y. \qquad \square$$

The proof follows by an application of inverse function theorem applied to the function $F : E \to \mathbb{R}^{n+m}$ defined by

$$F(x,y) = (f(x,y), y).$$

We will not go into the details. We also remark that the implicit function theorem can also be proved directly and the inverse function theorem can be deduced from it. For the system (2.13), we have $T = [T_x \ T_y]$, where

$$T_x = \begin{bmatrix} D_{x_1}f_1 \cdots D_{x_n}f_1 \\ \cdots \cdots \cdots \\ \cdots \cdots \cdots \\ D_{x_1}f_n \cdots D_{x_n}f_n \end{bmatrix} \text{ and } T_y = \begin{bmatrix} D_{y_1}f_1 \cdots D_{y_m}f_1 \\ \cdots \cdots \cdots \\ \cdots \cdots \cdots \\ D_{y_1}f_n \cdots D_{y_m}f_n \end{bmatrix}.$$

Example 2.17. Define $f : \mathbb{R}^5 \to \mathbb{R}^2$, $n = 2, m = 3$, by

$$f_1(x_1, x_2, y_1, y_2, y_3) = 2e^{x_1} + x_2 y_1 - 4y_2 + 3$$

$$f_2(x_1, x_2, y_1, y_2, y_3) = x_2 \cos x_1 - 6x_1 + 2y_1 - y_3.$$

Let $a = (0,1)$, $b = (3,2,7)$. Then $f(a,b) = 0$. Compute $T = (T_x, T_y)$ where

$$T_x = \begin{bmatrix} 2 & 3 \\ -6 & 1 \end{bmatrix} \text{ and } T_y = \begin{bmatrix} 1 & -4 & 0 \\ 2 & 0 & -1 \end{bmatrix}.$$

[2]This amounts to assuming that T has maximum rank n. We can then write $T = T_x + T_y$ with T_x invertible, perhaps after multiplying T by a permutation matrix.

Clearly the matrix T_x is invertible and

$$T_x^{-1} = \frac{1}{20}\begin{bmatrix} 1 & -3 \\ 6 & 2 \end{bmatrix}.$$

Hence we can solve for $x = g(y)$ in a neighborhood of (a, b). □

For an extensive discussion of the topics presented here, we refer to Apostol (2011), Munkres (1991), Rudin (1976), and Spivak (1965).

2.2 MULTIPLE INTEGRALS AND DIVERGENCE THEOREM

In this section, we state two important results, namely, the *Green's theorem* and the *divergence theorem*. These are extensively used in the analysis of PDE. In a sense these theorems are the higher dimensional versions of the familiar fundamental theorem of calculus and integration by parts formula in one dimension. Hence they are also important by themselves and are particular cases of a general result called *Stokes theorem* on smooth manifolds. However, even for the statements of these theorems, we need a lot of machinery from differential geometry; some of these are the concept of a smooth domain, smooth surface (curve in two dimensions), integration over these objects, tangent space and smoothly varying unit normal to a surface. In this brief introduction to the subject, it will not be possible to present all these concepts in a rigorous manner. The reader should refer to the cited references for much deeper treatment of these topics. Our discussion here is just explaining these important theorems in some simple geometric situations.

2.2.1 Multiple Integrals

In this sub-section, we briefly discuss the notion of multiple integrals, iterated integrals and integration by parts in multiple integrals leading to *Green's theorem* in two dimensions and *divergence theorem* (called Gauss-Ostrogadskii formula) in more than two variables. The general references for multiple integrals and their applications are Rund (1973), Spivak (1965), Apostol (2002), Taylor and Mann (1983), and Widder (1961).

The Riemann integral of a bounded function in a rectangular domain in $\mathbb{R}^n$ is very similar to the one-dimensional integral. However, for a general domain on which we wish to define the concept of integral, certain restrictions apply, especially in relation to the boundary of the domain. A very general description of these restrictions is indeed a difficult task, requiring the tools from differential geometry. Our presentation here will be simple, requiring the domains we consider have some specific geometric properties.

If $a_i < b_i$ for $i = 1, 2, \dots, n$, then the set $\mathcal{R}$ defined by

$$\mathcal{R} = [a_1, b_1] \times \cdots \times [a_n, b_n]$$

is termed as a *rectangle* in $\mathbb{R}^n$ or an *n-dimensional rectangle*. The *volume* of $\mathcal{R}$, denoted by $|\mathcal{R}|$, is defined by

$$|\mathcal{R}| = \prod_{i=1}^{n}(b_i - a_i).$$

In one dimension, it is simply the length of the interval; it is the area of the rectangle in two dimensions and volume in three and more dimensions. If P_i is a partition of $[a_i, b_i]$, that is

$$P_i : a_i = a_i^0 < a_i^1 < \cdots < a_i^{k_i} = b_i$$

for $i = 1, 2, \ldots, n$, then the Cartesian product $P = P_1 \times \cdots \times P_n$ is termed as a partition of $\mathcal{R}$. These partitions divide the rectangle $\mathcal{R}$ into finitely many sub-rectangles, denoted by I_j, $j = 1, 2 \ldots, k$, where $k = k_1 \cdots k_n$. The largest of all $|I_j|$ is called the *norm* of the partition P and is denoted by $\|P\|$. A partition P is said to be *finer* than another partition $\tilde{P}$ if $\tilde{P} \subset P$.

Definition 2.18. Let $f : \mathcal{R} \to \mathbb{R}$ be a bounded function and P be a partition of $\mathcal{R}$ with sub-rectangles I_j. A *Riemann sum* of f with respect to the partition P is defined by

$$S(P, f) = \sum_{j=1}^{k} f(t_j)|I_j|, \tag{2.15}$$

where $t_j \in I_j$. The function f is said to be *Riemann integrable* or simply *integrable* over $\mathcal{R}$ if a number A can be found with the property: given any $\varepsilon > 0$, there exists a partition P_ε such that for all partitions P finer than P_ε, there holds the inequality $|A - S(P, f)| < \varepsilon$. □

We may also write the condition in the definition as

$$\lim_{\|P\| \to 0} S(P, f) = A$$

for all possible choices of $t_j \in I_j$. This number A is unique when exists, and is denoted by $\int_{\mathcal{R}} f(x)\, dx$. The notations $\iint_{\mathcal{R}} f(x, y)\, dxdy$ and $\iiint_{\mathcal{R}} f(x, y, z)\, dxdydz$ are often used in two and three dimensions respectively. Similar to the one-dimensional integral, we may also define the notion of multiple integrals using the upper and lower Riemann sums and integrals.

Definition 2.19. Let f, $\mathcal{R}$ and P be as in Definition 2.18. The numbers

$$U(P, f) = \sum_{j=1}^{k} M_j|I_j| \quad \text{and} \quad L(P, f) = \sum_{j=1}^{k} m_j|I_j|$$

are respectively called the *upper Riemann sum* and the *lower Riemann sum* of f with respect to the partition P; here

$$m_j = \inf\{f(t_j) : t_j \in I_j\} \text{ and } M_j = \sup\{f(t_j) : t_j \in I_j\}.$$

The *upper Riemann integral* and the *lower Riemann integral* of f are respectively defined by

$$\overline{\int_{\mathcal{R}}} f(x)\, dx = \inf_P U(P,f) \text{ and } \underline{\int_{\mathcal{R}}} f(x)\, dx = \sup_P L(P,f)$$

where inf and sup are taken over all the partitions of $\mathcal{R}$.

The function f is said to satisfy the *Riemann condition* if, for any given $\varepsilon > 0$ there exists a partition P_ε such that for any partition P finer than P_ε there holds the inequality

$$U(P,f) - L(P,f) < \varepsilon.$$

□

Theorem 2.20. Let $f : \mathcal{R} \to \mathbb{R}$ be a bounded function defined on a rectangle in $\mathbb{R}^n$. Then, the following conditions are equivalent:

1. f is Riemann integrable over $\mathcal{R}$.
2. f satisfies the Riemann's condition in $\mathcal{R}$.
3. $\underline{\int_{\mathcal{R}}} f(x)\, dx = \overline{\int_{\mathcal{R}}} f(x)\, dx.$

□

As in the case of one dimension, we can easily deduce that f is Riemann integrable over $\mathcal{R}$ if f is continuous. In one dimension, it is known that a bounded function f defined on an interval $[a, b]$ is Riemann integrable if and only if f is continuous *almost everywhere* (a.e.), that is f is continuous in $[a, b]$ except on a subset of *measure zero*. This result has a straightforward extension to multi-dimensions.

A subset $S \subset \mathbb{R}^n$ is said to be of *measure zero* if, given any $\varepsilon > 0$, the set S can be covered by a countable number of n-dimensional rectangles, the sum of whose volumes is less than ε. Any countable set, for example, is a set of measure zero. We can now state the following result analogous to the one-dimensional case.

Theorem 2.21. Let $f : \mathcal{R} \to \mathbb{R}$ be a bounded function defined on a rectangle $\mathcal{R}$ in $\mathbb{R}^n$. Then, f is Riemann integrable over $\mathcal{R}$ if and only if f is continuous a.e. in $\mathcal{R}$. □

Iterated Integrals: We will consider the case of $n = 2$, as it is much simpler to explain the underlying procedure. Let $\mathcal{R} = [a, b]\times[c, d]$ be a two-dimensional rectangle and we use x, y

to denote the variables in $\mathbb{R}^2$. The one-dimensional intervals $[a, b]$ and $[c, d]$ are respectively the projections of $\mathcal{R}$ onto the lines $y = 0$ and $x = 0$. For any partitions P_1 of $[a, b]$ and P_2 of $[c, d]$, the Riemann sum defined in (2.15) can be written as an *iterated sum*:

$$S(P,f) = \sum_{j=1}^{k} f(t_j)|I_j| = \sum_{P_1} |I_j^1| \sum_{P_2} f(t_j)|I_j^2| = \sum_{P_2} |I_j^2| \sum_{P_1} f(t_j)|I_j^1|,$$

where $I_j = I_j^1 \times I_j^2$. Thus, if f is continuous in $\mathcal{R}$, we easily deduce the following result:

Theorem 2.22. Suppose $\mathcal{R} = [a, b] \times [c, d]$ be a two-dimensional rectangle and $f : \mathcal{R} \to \mathbb{R}$ be a continuous function. Then,

$$\int_{\mathcal{R}} f(x, y)\, dxdy = \int_a^b dx \int_c^d f(x, y)\, dy = \int_c^d dy \int_a^b f(x, y)\, dx. \tag{2.16}$$

□

The left most integral in (2.16) is called the *double integral* of f and is also written as $\iint_{\mathcal{R}} f(x, y)\, dxdy$. The other two integrals are called *iterated integrals*, which are one-dimensional integrals in this case. The above formula is called *Fubini theorem* and is very useful in the computation of multiple integrals. Analogous result can be written down in higher dimensions. Thus, the computation of a multiple integral of a continuous function over an n-dimensional rectangle may be reduced to those of n one-dimensional integrals.

Integral over an Arbitrary Set: Our next task is to extend the definition of the multiple integrals over suitable subsets of $\mathbb{R}^n$. We have in mind the examples of circular disk in $\mathbb{R}^2$, balls and solid cylinders in $\mathbb{R}^3$. Any bounded set S in $\mathbb{R}^n$ can be enclosed in a rectangle and, by suitably extending a given continuous function f on S, we may define the integral over S, without much difficulty. However, when it comes to the computation of the integral using the iterated integrals or while deriving formula for integration by parts, a more closer look at the set S is essential. This leads to the analysis of the *boundary* of S and certain restrictions need be put in order to derive some useful formulas.

Let S be a bounded set in $\mathbb{R}^n$. Choose a rectangle $\mathcal{R}$ in $\mathbb{R}^n$ containing S. The (topological) *boundary* of the set S is the closed set $\bar{S} \cap \bar{S^c}$, where S^c is the complement of S in $\mathbb{R}^n$. The boundary of S is denoted by ∂S. The elements of ∂S are called the *boundary points* of S and the points of $S \setminus \partial S$ are called the *interior points* of S. For a partition P of $\mathcal{R}$, define $\bar{J}(P, S)$ to be the sum of all the volumes of sub-rectangles arising from the partition P, which contain the points of $S \cup \partial S$ and $\underline{J}(P, S)$ to be the sum of all the volumes of sub-rectangles arising from the partition P, which contain only the interior points of S. Next, define the numbers

$$\underline{c}(S) = \sup \underline{J}(P, S) \text{ and } \bar{c}(S) = \inf \bar{J}(P, S),$$

where inf and sup are taken over all the partitions of $\mathcal{R}$ as described above. These numbers are respectively called the *inner* and *outer Jordan content* of S. It is further evident that $0 \le \underline{c}(S) \le \bar{c}(S)$ and that the definitions do not depend on $\mathcal{R}$. The set S is said to be *Jordan measurable* if $\underline{c}(S) = \bar{c}(S)$ and the common number is called the *Jordan content* of S. In general, we may not be able to include any rectangle inside a set and hence may not be able to define its inner content. In such a situation, the outer content is termed as content.

Lemma 2.23. If S is a bounded subset of $\mathbb{R}^n$, then $\bar{c}(\partial S) = \bar{c}(S) - \underline{c}(S)$. Thus, the set S is Jordan measurable if and only if its boundary ∂S has zero content. □

Apparently, only the outer Jordan content of ∂S makes sense.

Definition 2.24. Let S be a bounded Jordan measurable set in $\mathbb{R}^n$ and $f : S \to \mathbb{R}$ be a bounded function. For any n-dimensional rectangle $\mathcal{R}$ containing S, define $\tilde{f} : \mathcal{R} \to \mathbb{R}$ by

$$\tilde{f}(x) = \begin{cases} f(x) \text{ if } x \in S \\ 0 \text{ if } x \in \mathcal{R} \setminus S. \end{cases}$$

We say that f is *(Riemann) integrable* if $\int_{\mathcal{R}} \tilde{f}(x)\,dx$ exists. In this case, we write

$$\int_S f(x)\,dx = \int_{\mathcal{R}} \tilde{f}(x)\,dx.$$

□

It is not difficult to see that the definition is independent of the rectangle $\mathcal{R}$. We have the following result:

Theorem 2.25. Let S and f be as in Definition 2.24. Then, f is integrable over S if and only if the discontinuities of f form a set of measure zero. In particular, f is integrable over S if it is continuous on S. □

For unbounded domains and/or unbounded functions, we may use *partition of unity* to define the integral. For details, we refer to Spivak (1965). Though the above discussion regarding the multiple integrals over a bounded Jordan measurable set is satisfactory from a theoretical point of view, it is not so from a computational point of view. Ultimately, we wish to reduce the computation of a multiple integral to that of one-dimensional integral, as we did in the case of a rectangle. This forces us to consider a set with some special properties. We begin with a set in two dimensions.

Consider the set S in $\mathbb{R}^2$ defined by

$$S = \{(x, y) \in \mathbb{R}^2 : x \in [a, b],\ \varphi(x) \le y \le \psi(x)\},$$

where $\varphi, \psi : [a, b] \to \mathbb{R}$ are continuous functions, such that $\varphi(x) \leq \psi(x)$ for all $x \in [a, b]$. The set S is contained in the rectangle $[a, b] \times [c, d]$ with $c = \min \varphi$ and $d = \max \psi$ and is Jordan measurable. The set S is called a *x-simple* set. The boundary ∂S is part of the lines $x = a, x = b$ and the curves $y = \varphi(x), y = \psi(x)$. We have the following result:

Theorem 2.26. Suppose S is an x-simple set, as above, and $f : S \to \mathbb{R}$ is continuous. Then, f is integrable over S and

$$\iint_S f(x, y)\, dxdy = \int_a^b dx \int_{\varphi(x)}^{\psi(x)} f(x, y)\, dy. \tag{2.17}$$

□

The formula (2.17) is quite satisfactory from a computational point of view and is a version of Fubini theorem. The extension of this result to y-simple sets (defined in a similar manner) and to higher dimensions is obvious. In general, a x-simple set need not be y-simple and vice-versa. Also there are many subsets that are neither x-simple nor y-simple. In practice, we encounter many sets that can be decomposed into a finite number of subsets that are either x-simple or y-simple or both. For such sets, we can easily extend the results discussed above.

Change of Variables in Multiple Integrals: Often a change of variables is desired in the computation of integrals, including those in several variables. We begin with the following definition:

Definition 2.27 (Co-ordinate transformation). Let Ω be an open set in $\mathbb{R}^n$. A vector-valued function $g : \Omega \to \mathbb{R}^n$ is called a *co-ordinate transformation* if the following conditions are satisfied:

1. g one-one.
2. If we write $g = (g_1, \ldots, g_n)$, then each $g_i \in C^1(\Omega)$.
3. The Jacobian $\det J(x) \neq 0$ for all $x \in \Omega$, where $J(x) = J_g(x) = \left[\frac{\partial g_i}{\partial x_j}(x)\right]$ is the Jacobian matrix of the function g.

□

It follows that g^{-1} defined in $g(\Omega)$ is also a C^1 function.

Theorem 2.28 (Change of Variables). Let Ω be a bounded open set and $g : \Omega \to \mathbb{R}^n$ be a co-ordinate transformation. Let $\tilde{\Omega} = g(\Omega)$. If $f \in C(\overline{g(\Omega)})$, then[3]

$$\int_{\tilde{\Omega}} f(y)\, dy = \int_{\Omega} f(g(x))|J(x)|\, dx. \tag{2.18}$$

[3]This formula also holds for Lebesgue integral.

Using the inverse g^{-1} in place of g, we can also write the formula (2.18) as

$$\int_{\Omega} f(x)\,dx = \int_{\tilde{\Omega}} f(g^{-1}(y))|J(g^{-1}(y))|^{-1}\,dy \tag{2.19}$$

for $f \in C(\overline{\Omega})$. □

A familiar and quite useful co-ordinate transformation is that of *polar co-ordinates* (also called *spherical co-ordinates* in three and higher dimensions). Using familiar notations, this transformation in $\mathbb{R}^n$ is described by

$$\begin{aligned}
x_1 &= r\cos\theta_1 \sin\theta_2 \cdots \sin\theta_{n-1}\\
x_2 &= r\sin\theta_1 \sin\theta_2 \cdots \sin\theta_{n-1}\\
x_3 &= r\cos\theta_2 \sin\theta_3 \cdots \sin\theta_{n-1}\\
&\cdots\cdots\\
x_n &= r\cos\theta_{n-1}.
\end{aligned}$$

Here $r > 0$ and $\theta_1 \in (0, 2\pi)$, $\theta_j \in (0, \pi)$ for $j > 1$. We have

$$r^2 = x_1^2 + \cdots + x_n^2.$$

A lengthy calculation shows that the Jacobian (determinant) of this transformation is given by

$$r^{n-1}\sin^{n-2}\theta_2 \cdots \sin\theta_{n-1}.$$

Thus, by taking $\Omega = B_R(a)$ for any fixed $a \in \mathbb{R}^n$ and $R > 0$, we can write (2.18) for $f \in C(\overline{B_R(a)})$,

$$\int_{B_R(a)} f(x)\,dx = \int_0^R dr \int_{\partial B_r(a)} f\,dS$$

where the integral $\int_{\partial B_r(a)} f\,dS$ on the right-hand side is a surface integral (see the discussion below). If, as $R \to \infty$, the above integrals remain finite, then we have

$$\int_{\mathbb{R}^n} f(x)\,dx = \int_0^\infty dr \int_{\partial B_r(a)} f\,dS$$

for any fixed $a \in \mathbb{R}^n$.

In $\mathbb{R}^2$, we write the polar co-ordinates as $x = r\cos\theta,\ y = r\sin\theta$, with the Jacobian r and in $\mathbb{R}^3$ the transformation is given by

$$x = r\cos\theta\sin\varphi,\ y = r\sin\theta\sin\varphi,\ z = r\cos\varphi$$

with the Jacobian $r^2\sin\phi$.

Curves in the Space: A *curve* is a one-dimensional object in space. We are all familiar with conic curves – circle, ellipse, parabola and hyperbola – studied in elementary geometry. A most efficient way of describing a curve is through its parametric representation. A curve C in $\mathbb{R}^n$ is described by n continuous functions f_i, $1 \le i \le n$ defined on some (finite or infinite) interval $I \subset \mathbb{R}$. Thus,

$$C = \{(x_1, \dots, x_n) \in \mathbb{R}^n : x_i = f_i(t),\ t \in I,\ 1 \le i \le n\}. \tag{2.20}$$

Without further restrictions on f_i, a curve can be chaotic; think of Brownian motion of a particle. Also, the same curve can have different parametric representations.

The curve C is called *simple* if the mapping $t \mapsto (f_1(t), \dots, f_n(t))$ from I into $\mathbb{R}^n$ is one-one. If, in addition, $I = [a, b]$ and $(f_1(a), \dots, f_n(a)) = (f_1(b), \dots, f_n(b))$, the curve is said to be a simple closed curve or a *Jordan curve*. A curve is called a *smooth curve* if it is simple and has a tangent at every point. The latter condition is satisfied if we assume that all f_i are C^1 functions such that the derivatives $f_i'(t)$ do not vanish simultaneously, for all $t \in I$. In this case, the direction of the tangent is specified by the direction cosines:

$$\cos\gamma_i = \frac{f_i'(t)}{\sqrt{\sum_{j=1}^{n} |f_j'(t)|^2}}$$

with γ_i being the angle the tangent makes with the ith co-ordinate axis.

In what follows we consider either smooth curves or piecewise smooth curves, that is, a finite number of smooth curves joined end to end and call them, for simplicity, *smooth curves*. Such a curve may have corners at junction points as in the case of a rectangle or a triangle. We also encounter at times smooth curves that are intersections of two surfaces. For example, a circle in $\mathbb{R}^2$ is the intersection of a sphere and a hyperplane in $\mathbb{R}^3$. Though a parametric representation is not readily evident, we can obtain local parametric representation by the implicit function arguments.

Definition 2.29 (Regular Domains or Regions in Two Dimensions). Let Ω be a bounded open arc-wise connected subset of $\mathbb{R}^2$ such that its boundary $\partial\Omega$ consists of finitely many simple closed smooth curves that do not intersect each other. Such an Ω will be called a *regular domain* or *region*. □

In the above definition, arc-wise connected means the following: any two points in Ω are connected by a finite number of line segments *lying in* Ω.

Examples of regular domains include an open ball, an annulus among others. In practice, many regular domains are described as the interior of their boundaries. We will discuss a similar domain in higher dimensions after introducing the concept of a surface.

Line Integral: Suppose C is a simple curve having a parametric representation (2.20) with $I = [a, b]$. The *length* or the *arc-length* L of the curve C is given by

$$L = L(C) = \int_a^b \left(\sum_{j=1}^n |f_j'(t)|^2 \right)^{1/2} dt. \tag{2.21}$$

The above formula is derived as follows: if P is a partition of $[a, b]$ and $t_1, t_2 \in P$, then the distance between the corresponding points on the curve is given by

$$\left(\sum_{j=1}^n (f_j(t_1) - f_j(t_2))^2 \right)^{1/2} \approx \left(\sum_{j=1}^n |f_j'(\tau)|^2 \right)^{1/2} |t_1 - t_2|.$$

Then, forming the Riemann sums, taking the appropriate limit, the formula (2.21) results.[4] If the curve C has a different parametric representation

$$\tau \mapsto (\tilde{f}_1(\tau), \dots, \tilde{f}_n(\tau)),$$

for $\tau \in [c, d]$ and the mapping $t \mapsto \tau$ from $[a, b]$ to $[c, d]$ satisfies the condition $d\tau/dt > 0$ for all $t \in [a, b]$, it is easy to verify that L does not depend on a particular parametric representation. If we replace b by $t \in [a, b]$ in (2.21), then the variable $s = s(t)$ defined by

$$s = \int_a^t \left(\sum_{j=1}^n |f_j'(t)|^2 \right)^{1/2} dt \tag{2.22}$$

is called the *arc-length variable*; it is symbolically written as

$$ds^2 = \sum_{j=1}^n |f_j'(t)|^2 \, dt^2 = \sum_{j=1}^n dx_i^2$$

[4]The length of a curve can be defined as the appropriate Riemann integral, even if the functions f_i are not C^1 functions. In this situation, a curve is said to be *rectifiable* if $L < \infty$.

in the differential geometric notation. If $h : C \to \mathbb{R}$ is a continuous function, then the *line integral* of h is defined by

$$\int_C h\, ds = \int_a^b (h \circ f)(t) \left(\sum_{i=1}^n |f_i'(t)|^2 \right)^{1/2} dt \tag{2.23}$$

where $f(t) = (f_1(t), \ldots, f_n(t))$. We call ds the *line measure* along C. A few comments regarding the definition of the line integral. In the definition given by (2.23), we have inherently fixed a *direction* of the curve, namely the parameter t defining the curve C increases from a to b. In the evaluation of a line integral, it is necessary that a direction of the curve C is mentioned. Generally, there are two directions along a curve; we may choose one of them and call it the *positive direction*, and the other one the *negative direction*. When this choice is made, the curve is said to be an *oriented curve*. An oriented curve has an initial point and a terminal point. If a simple closed curve is oriented, it has neither an initial point nor a terminal point. However, it is often convenient to identify some point on such a curve and call it initial point and terminal point. As we will see in the Green's theorem later, orientation of the curve is important. If we reverse the orientation, the sign will change; remember the convention: $\int_a^b = -\int_b^a$.

In two dimensions, a line integral is commonly written as

$$\int_C P\, dx + Q\, dy, \tag{2.24}$$

where P, Q are continuous functions defined on C. The interpretation of (2.24) is the following: there is a parametric representation of the curve C given by $x = f(t)$, $y = g(t)$ with C^1 functions $f, g : [a, b] \to \mathbb{R}$ and the line integral in (2.24) is given by

$$\int_a^b \left[P(f(t), g(t)) f'(t) + Q(f(t), g(t)) g'(t) \right] dt.$$

Similar remarks apply in three and higher dimensions.

Hypersurface: A hypersurface or simply a surface is an $(n-1)$-dimensional object in $\mathbb{R}^n$. A sphere, the surface of a cylinder are examples of surfaces in $\mathbb{R}^3$. There are three common ways of describing a surface.

1. A surface can be described by an equation $x_n = F(x_1, \ldots, x_{n-1})$, where $F : \widetilde{\Omega} \to \mathbb{R}$ is a smooth function, with $\widetilde{\Omega} \subset \mathbb{R}^{n-1}$ (precise description of $\widetilde{\Omega}$ will

be done as and when required). The surface itself is described by the set of points $(x_1, \dots, x_{n-1}, F(x_1, \dots, x_{n-1}))$ in $\mathbb{R}^n$. Here a surface is viewed as the graph of a function.

2. A surface in $\mathbb{R}^n$ may also be described by an equation of the form $F(x_1, \dots, x_{n-1}, x_n) = 0$. Locally this representation may be reduced to a similar representation as in (1) by invoking implicit function theorem, under appropriate assumptions on F. More precisely, every point in the zero set of F has a neighborhood, which is the graph of a function.
3. From more of a theoretical point of view, a surface is defined through a parametric representation as:

$$x_i = f_i(s_1, \dots, s_{n-1}), \ 1 \le i \le n, \tag{2.25}$$

where $s = (s_1, \dots, s_{n-1})$ vary in a domain in $\mathbb{R}^{n-1}$ and f_i are smooth functions.

A surface is called a *smooth surface* if it possesses a tangent plane at each of its points, and if the direction of the normal to this tangent plane varies continuously. See a discussion below for the case of $n = 3$.

Surface Integral in $\mathbb{R}^3$: In analogy with a line integral, we now consider the surface integral in $\mathbb{R}^3$; the extension to higher dimensions can be carried out on similar lines. This is similar to a double integral. However, as we will see, so many restrictions need be imposed even to have a workable notion of the area of a surface. The analogy is similar to that of a curve in $\mathbb{R}^2$; merely assuming that a curve is described parametrically by continuous functions, does not assure that the curve is rectifiable.

Let S be a smooth surface in $\mathbb{R}^3$ described parametrically by

$$x = f(u, v), \ y = g(u, v), \ z = h(u, v), \tag{2.26}$$

where f, g, h are C^1 functions defined in a domain V in $\mathbb{R}^2$; here we have used (x, y, z) to denote the points in $\mathbb{R}^3$. To make the surface S a *genuine* two-dimensional object,[5] certain restrictions on the functions f, g, h apply. Here is a brief discussion. Fix $(u_0, v_0) \in V$ and let

$$x_0 = f(u_0, v_0), \ y_0 = g(u_0, v_0), \ z_0 = h(u_0, v_0).$$

The mapping $(u, v_0) \mapsto (f(u, v_0), g(u, v_0), h(u, v_0))$ defines a space curve in S as (u, v_0) varies in V. Its tangent vector at (u_0, v_0) is the vector $(f_u(u_0, v_0), g_u(u_0, v_0), h_u(u_0, v_0))$. Similarly, the tangent vector for the curve $(u_0, v) \mapsto (f(u_0, v), g(u_0, v), h(u_0, v))$ at (u_0, v_0) is the vector $(f_v(u_0, v_0), g_v(u_0, v_0), h_v(u_0, v_0))$. The two-dimensional nature of the surface S is indicated by the requirement that these two tangent vectors are linearly independent for all the points $(u_0, v_0) \in V$, and hence they span a two-dimensional subspace. This is the *tangent space*

[5] Otherwise S may degenerate to lines or points or combination of a genuine surface with lines or points.

at (u_0, v_0) or more precisely at (x_0, y_0, z_0) to the surface S. Since f, g, h are assumed to be C^1, we have a continuously varying family of tangent planes to the surface S. The linear independence of the tangent vectors at (u_0, v_0) is equivalent to saying that the rank of the 3×2 Jacobian matrix

$$\begin{bmatrix} f_u(u_0, v_0) & f_v(u_0, v_0) \\ g_u(u_0, v_0) & g_v(u_0, v_0) \\ h_u(u_0, v_0) & h_v(u_0, v_0) \end{bmatrix}$$

is 2. This means that the Jacobians j_1, j_2, j_3 do not vanish simultaneously, where j_1, j_2, j_3 are given by

$$j_1 = g_u h_v - g_v h_u,\ j_2 = f_v h_u - f_u h_v,\ j_3 = f_u g_v - f_v g_u.$$

In three dimensions, it is easy to describe the normal to this tangent plane: the normal is in the direction of the cross product

$$(f_u(u_0, v_0), g_u(u_0, v_0), h_u(u_0, v_0)) \times (f_v(u_0, v_0), g_v(u_0, v_0), h_v(u_0, v_0)).$$

If we denote by $\nu = \nu(u_0, v_0)$, the unit normal to the tangent plane, we can assign a *direction* to this normal with regard to the domain that the surface S *bounds*. In the statement of the divergence theorem, we speak of *outward normal*. For example, if $\Omega = \{|x| < 1\}$, the open unit ball centered at the origin, then its boundary $\partial\Omega$ is the unit sphere $\{|x| = 1\}$. If $|x_0| = 1$, then the outward unit normal (with respect to Ω) is in the direction of x_0, as for $t > 0$, $|x_0 + tx_0| = 1 + t > 1$. On the other hand, if Ω is the annulus $\{1 < |x - a| < 2\}$, where a is fixed, its boundary consists of two disjoint spheres, namely, $\{|x - a| = 1\}$ and $\{|x - a| = 2\}$. The outward unit normal to the outer sphere is $(x - a)/2$ and for the inner sphere it is $a - x$.

Let α, β and γ be the angles the normal to S makes with the positive axes, with designating the positive direction (or outward direction) of the normal by choosing γ acute. Our first task is to obtain a formula for the area of S. The arguments below are very intuitive in nature.

Consider a small rectangle in V with sides δu and δv. The image of this rectangle on S will be intuitively a parallelogram whose area is approximately given by[6]

$$\sqrt{j_1^2 + j_2^2 + j_3^2}\, \delta u \delta v.$$

It follows that the area of S is given by the double integral

$$\text{area}(S) = \iint_V \sqrt{j_1^2 + j_2^2 + j_3^2}\, du dv. \tag{2.27}$$

[6]This computation requires arguments from vector algebra.

When the surface S is described by the equation $z = f(x, y)$, where f is a C^1 function defined in a domain V in $\mathbb{R}^2$, formula (2.27) reduces to

$$\text{area}(S) = \iint_R \sqrt{1 + f_x^2 + f_y^2}\, dxdy = \iint_V \cos\gamma\, dxdy. \tag{2.28}$$

(Depending on the choice of the direction of the normal, we may replace $\cos\gamma$ by $|\cos\gamma|$.) Equation (2.27) will be used to define the surface integral on S. Let $F : S \to \mathbb{R}$ be a continuous function, where the surface S has the parametric representation given by (2.26). The *surface integral* of F over S is defined by

$$\iint_S F\, dS = \iint_V F(u, v)\sqrt{j_1^2 + j_2^2 + j_3^2}\, dudv, \tag{2.29}$$

where $F(u, v) = F(f(u, v), g(u, v), h(u, v))$. The measure dS is called the *surface measure* on S. Using the formula for change of variables in a double integral, it is not difficult to see that dS does not depend on a particular choice of parameterization used to define the surface S. Again, when the surface is described by a single equation $z = f(x, y)$, we may write (2.29) as

$$\iint_S F\, dS = \iint_V F(x, y, f(x, y))\sqrt{1 + f_x^2 + f_y^2}\, dxdy. \tag{2.30}$$

Let α, β, γ be the angles the normal makes to the positive x, y, z axes, respectively, the positive direction of the normal is so chosen that γ is acute. Since $\cos\alpha$, $\cos\beta$ and $\cos\gamma$ are the direction cosines of the normal, we have

$$\cos\alpha : \cos\beta : \cos\gamma = f_x : f_y : -1.$$

Hence, $\sec^2\gamma = 1 + f_x^2 + f_y^2$ and the formula (2.30) may also be written as

$$\iint_S F\, dS = \iint_V F(x, y, f(x, y)) \sec\gamma\, dxdy. \tag{2.31}$$

This formula is also applicable when the surface S is obtained by joining a finite number of surfaces, each of which is described by a function, using the usual properties of a double integral.

2.2.2 Green's Theorem

This theorem deals with a version of integration by parts in two dimensions. Recall the integration by parts in one dimension. If $u, v \in C^1([a, b])$, then

$$\int_a^b uv' = u(b)v(b) - u(a)v(b) - \int_a^b u'v.$$

In this case, the boundary of the interval consists of just two points, namely a and b. We assign unit normal at these boundary points, *pointing outward* to the interval. Thus, at b we assign $+1$ and at a we assign -1 as the unit normals. We can then rewrite the above integration by parts formula as

$$\int_a^b uv' = \int_{\{a,b\}} uv\nu - \int_a^b u'v$$

where ν is the outward unit normal to the boundary of $[a, b]$ and we interpret the *boundary integral* $\int_{\{a,b\}} uv\nu$ as the integral on the boundary with *counting measure.*

Now move to higher dimensions. Let Ω be an open bounded set in $\mathbb{R}^n$ and $u \in C^1(\overline{\Omega})$. Similar to one-dimensional case, the task now is to express the integral $\int_\Omega \frac{\partial u}{\partial x_i}\, dx$, $1 \leq i \leq n$, as a boundary integral over the boundary $\partial\Omega$. This is where we need to put sufficient smooth assumptions on the boundary so that a boundary integral is well-defined. This analysis leads to the Green's theorem in two dimensions and divergence theorem in more than two dimensions.

Theorem 2.30 (Green's Theorem). Let Ω be a regular domain in $\mathbb{R}^2$ (see Definition 2.29). Let C denote the union of curves forming $\partial\Omega$, each oriented in such way that the interior of Ω lies on the left as one advances along the curve in the positive direction. If $P, Q \in C^1(\Omega)$, then

$$\int_C P\,dx + Q\,dy = \iint_\Omega \left(\frac{\partial Q}{\partial x} - \frac{\partial P}{\partial y}\right) dxdy. \tag{2.32}$$

The left-hand side in (2.32) is the line integral over the curve C taken in the positive direction as described in the theorem. The formula (2.32) can also be written as

$$\oint_C F \cdot \nu\, ds = \iint_\Omega \mathrm{div} F(x, y)\, dxdy$$

in tune with the statement of the divergence theorem, discussed as in (2.35) below. Here $F = (Q, -P)$ is the two-dimensional vector field, ν is the outward unit normal to the curve C and div is the two-dimensional divergence operator, namely $\text{div}F(x, y) = \frac{\partial Q}{\partial x} - \frac{\partial P}{\partial y}$.

A word about the positive direction of the boundary curve. If the boundary consists of a single closed Jordan curve, like a circle, ellipse or rectangle, then the positive direction is the counter-clockwise direction. If on the other hand, the boundary consists of two disjoint closed Jordan curves (e.g. an annulus), then the positive direction means the counter-clockwise direction for the outer curve and the clockwise direction for the inner curve.

It is indeed a big task to prove the Green's theorem in its full generality. We will only sketch a proof assuming that Ω is x-simple. Suppose that Ω is bounded by the lines $x = a, x = b$ and the curves $y = \varphi(x), y = \psi(x)$, where $\varphi, \psi : [a, b] \to \mathbb{R}$ are continuous functions such that $\varphi(x) \leq \psi(x)$ for all $x \in [a, b]$. Then, using Theorem 2.26, we have

$$\begin{aligned}
\iint\limits_{\Omega} \frac{\partial P}{\partial y}(x, y)\, dxdy &= \int\limits_{a}^{b} dx \int\limits_{\varphi(x)}^{\psi(x)} \frac{\partial P}{\partial y}(x, y)\, dy \\
&= \int\limits_{a}^{b} P(x, \psi(x))\, dx - \int\limits_{a}^{b} P(x, \varphi(x))\, dx \\
&= - \int\limits_{C} P\, dx, \qquad (2.33)
\end{aligned}$$

as the positive direction of the line integral is counter-clockwise.

Similarly, if Ω is also y-simple, we obtain

$$\iint\limits_{\Omega} \frac{\partial Q}{\partial x}(x, y)\, dxdy = \int\limits_{C} Q\, dy. \qquad (2.34)$$

Combining equations (2.33) and (2.34), we arrive at (2.32). □

The domain Ω or rather its boundary C is called *simple* if it is both x-simple and y-simple. In general, a domain Ω can usually be divided into a finite number of smaller domains Ω_i, by introducing some new boundary curves, which are simple. The Green's theorem then follows by summing over all Ω_i; the line integrals over new curves that are created by the division occur twice and cancel each other as their positive directions have opposite signs.

Divergence Theorem[7]**:** This is the analogue of the Green's theorem in more than two dimensions. The assumptions on the domain become more technical here and it is indeed a difficult task to consider very general domains.

Let $\widehat{\Omega}_j$ be an open bounded set in the hyperplane $x_j = 0$ in $\mathbb{R}^n$ for some j, $1 \le j \le n$. The points in $\widehat{\Omega}_j$ will be denoted by $\hat{x}_j = (x_1, \dots, x_{j-1}, 0, x_{j+1}, \dots, x_n)$. Let S be the surface in $\mathbb{R}^n$, whose points are prescribed by the points

$$(x_1, \dots, x_{j-1}, \varphi(\hat{x}_j), x_{j+1}, \dots, x_n) \in \mathbb{R}^n$$

as the point $\hat{x}_j$ varies over $\widehat{\Omega}_j$, where $\varphi : \widehat{\Omega}_j \to \mathbb{R}$ is a C^1 function. Such a surface will be called a *simple piece*.

Let Ω be an open bounded set such that its boundary $\partial\Omega$ is a finite union of non-overlapping simple pieces as described above. For such an Ω, we are now going to state the divergence theorem.

Theorem 2.31 (Divergence Theorem)**.** Let Ω be as above and $u : \Omega \to \mathbb{R}^n$ be a vector-valued function, whose components $u_i \in C^1(\overline{\Omega})$, $1 \le i \le n$. Then,

$$\int_{\Omega} \operatorname{div} u \, dx = \int_{\partial\Omega} u \cdot \nu \, dS, \tag{2.35}$$

where ν is the outward unit normal to $\partial\Omega$. □

The formula (2.35) is known as *Gauss–Ostrogadskii formula*. In (2.35) $\operatorname{div} u = \sum_{i=1}^{n} \frac{\partial u_i}{\partial x_i}$ denotes the divergence of the vector field u, $\nu(x)$ denotes the outward unit normal to the boundary at $x \in \partial\Omega$ and dS is the surface measure on $\partial\Omega$. We are not going to present a proof here with this generality. Instead, we sketch a proof in three dimensions.

Let V be a regular domain in the xy–plane in $\mathbb{R}^3$ and $\varphi, \psi \in C^1(\overline{V})$ such that $\varphi(x, y) \le \psi(x, y)$ for all $(x, y) \in \overline{V}$. Consider the cylindrical surface (in $\mathbb{R}^3$) formed by drawing lines parallel to the z-axis at all the points of ∂V. Let S_3 be the part of this surface that is cut-off between the surfaces S_1, represented by the equation $z = \varphi(x, y)$ and the surface S_2 represented by $z = \psi(x, y)$ (S_3 may be empty). Consider now the region Ω bounded above by S_2, bounded below by S_1 and bounded laterally by S_3. The region Ω is called a z-simple region.

[7]This theorem or rather the formula is attributed to two mathematicians – Gauss and Ostrogadskii – in the literature.

Lemma 2.32. Let Ω be a z-simple region and $F \in C^1(\overline{\Omega})$. Then,

$$\iiint_{\Omega} \frac{\partial F}{\partial z}\, dxdydz = \iint_{\partial\Omega} F \cos\gamma \, dS \tag{2.36}$$

where γ is the angle between the positive z-axis and the outward unit normal to $\partial\Omega$. □

Proof We may write $F\cos\gamma = F\nu_z$, where ν_z is the z-component of the outward unit normal ν. Using iterated integrals, we can write

$$\begin{aligned}\iiint_{\Omega} \frac{\partial F}{\partial z}\, dxdydz &= \iint_{V} dxdy \int_{\varphi(x,y)}^{\psi(x,y)} \frac{\partial F}{\partial z}\, dz \\ &= \iint_{V} (F(x,y,\psi(x,y)) - F(x,y,\varphi(x,y))\, dxdy.\end{aligned} \tag{2.37}$$

Now as the boundary $\partial\Omega$ is the union of S_i, $i = 1, 2, 3$, the surface integral on the right-hand side of (2.36) can be written as the sum of three surface integrals. There is, however, no contribution from the surface integral over S_3 as $\gamma = \pi/2$ here and hence $\cos\gamma = 0$. At a point on S_1, the outward unit normal extends downward so that γ is obtuse and hence $\cos\gamma < 0$. On the other hand, γ is acute on S_2 and $\cos\gamma > 0$. Now the surface integrals over S_1 and S_2 can be written as double integral over V, using the formula (2.31). In this formula, γ represents the *acute* angle between the positive z-axis and the *undirected* unit normal to the surface. Hence, for the surface S_1 we should replace γ by $\pi - \gamma$ while using formula (2.31). Therefore, we have

$$\begin{aligned}\iint_{S_1} F(x,y,z)\, dS &= \iint_{V} F(x,y,\varphi(x,y)) \cos\gamma \sec(\pi-\gamma)\, dxdy \\ &= -\iint_{V} F(x,y,\varphi(x,y))\, dxdy.\end{aligned} \tag{2.38}$$

Similarly, we have

$$\begin{aligned}\iint_{S_2} F(x,y,z)\, dS &= \iint_{V} F(x,y,\psi(x,y)) \cos\gamma \sec\gamma\, dxdy \\ &= \iint_{V} F(x,y,\psi(x,y))\, dxdy.\end{aligned} \tag{2.39}$$

Finally, adding (2.38) and (2.39), the formula (2.36) follows from (2.37). The proof of the lemma is complete. □

Similar formulas hold for x-simple and y-simple regions. Thus, the proof of the divergence theorem, in three dimensions, follows if a region is simultaneously x-simple, y-simple and z-simple. For a domain, which is not simple, the remarks made at the end of the Green's theorem apply. The divergence theorem has several important consequences, which are collectively known as *Green's identities*. We now state them, whose proofs are left to the reader as exercises (see Exercise 1).

Theorem 2.33 (Green's Identities). Let $u, v \in C^2(\overline{\Omega})$. The following formulae hold:

1. $$\int_\Omega \Delta u \, dx = \int_{\partial\Omega} \frac{\partial u}{\partial \nu} dS(x).$$
2. $$\int_\Omega v\Delta u \, dx = -\int_\Omega \nabla u \cdot \nabla v \, dx + \int_{\partial\Omega} \frac{\partial u}{\partial \nu} v \, dS(x).$$
3. $$\int_\Omega (v\Delta u - u\Delta v) dx = \int_{\partial\Omega} \left(v\frac{\partial u}{\partial \nu} - u\frac{\partial v}{\partial \nu} \right) dS(x).$$

Here, $\frac{\partial u}{\partial \nu} = \nabla u \cdot \nu$ is the normal derivative and $\nabla = \left(\frac{\partial}{\partial x_1}, \cdots \frac{\partial}{\partial x_n} \right)$ is the *grad* operator. □

2.3 SYSTEMS OF FIRST-ORDER ORDINARY DIFFERENTIAL EQUATIONS: EXISTENCE AND UNIQUENESS RESULTS

In Chapter 3 on first-order equations, the questions related to existence and uniqueness of the solution to a Cauchy problem are tackled by the method of characteristics, which turn out to be the solutions of a system of first-order ordinary differential equations. We collect here some essential existence and uniqueness results for the system of ODE from the book Nandakumaran et al. (2017). For more details the reader can look into Nandakumaran et al. (2017) and the references therein.

Consider a system of n first-order equations:

$$\dot{y} = f(t, y), \tag{2.40}$$

or, explicitly written

$$\dot{y}_j = f_j(t, y_1, y_2, \cdots y_n), \quad j = 1, 2, \cdots, n.$$

Here y_j's, the unknowns, which are real-valued functions and f_j are real-valued functions defined on $\mathbb{R} \times \mathbb{R}^n$. The positive integer n is referred to as the *dimension* of the system. Since an ODE of any given order can be written in the form of a system of first-order equations, the above consideration is more general. System (2.40) is called an *autonomous* system if the right-hand side function f does not depend on t explicitly. When f depends on t explicitly as well, the system is referred to as *non-autonomous*. For example, the equation $\dot{y} = y + t$ (1D or one-dimensional equation) is non-autonomous; the equation $\dot{y} = \sin y$ is autonomous. A non-autonomous system may be converted into an autonomous system by increasing the dimension of the system by 1 and by introducing a new unknown variable τ satisfying $\dot{\tau} = 1$. However, even after this reduction, the study of a non-autonomous system is not easier. One reason for this is that since $\dot{\tau} = 1$, the reduced system does not have any equilibrium points; a point $y_0 \in \mathbb{R}^n$ is called an equilibrium point for (2.40) if $f(t, y_0) = 0$ for all t. For example, if we consider the linear system $\dot{x} = A(t)x(t)$, then $x = 0 \in \mathbb{R}^n$ is an equilibrium point for the unreduced system, but $0 \in \mathbb{R}^{n+1}$ is not an equilibrium point for the enlarged system, namely $\dot{x} = A(t)x(t)$, $\dot{\tau} = 1$.

We introduce the following *norm* in $\mathbb{R}^n$. For $x \in \mathbb{R}^n$, define

$$\|x\| = |x_1| + |x_2| + \cdots |x_n|,$$

where $x = (x_1, x_2, \cdots, x_n)$. This norm is referred to as ℓ^1 norm. It is easy to verify the following properties of the norm:

1. $\|x + y\| \leq \|x\| + \|y\|$,
2. $\|ax\| = |a|\|x\|$,
3. $\|x\| \geq 0$ and $= 0$ if and only if $x = 0$,

for all $x, y \in \mathbb{R}^n$ and $a \in \mathbb{R}$.

Suppose Ω is an open subset of $\mathbb{R}^n$ and $\phi : \Omega \to \mathbb{R}^n$. Then, ϕ is said to be *Lipschitz continuous* on Ω if there is a positive constant L such that $\|\phi(x) - \phi(y)\| \leq L\|x - y\|$, for all $x, y \in \Omega$. The smallest such an L is called the *Lipschitz constant* of ϕ. Equivalently, we can define the Lipschitz continuity through each component of ϕ. The class of Lipschitz continuous functions is huge. It includes in particular all the differentiable functions whose first derivatives are all bounded. We will now state the following theorem for the existence and uniqueness of solutions to the IVP for n-dimensional systems. As remarked earlier, since an nth-order equation may be written as an n-dimensional first-order system, we also have existence and uniqueness result for the solutions of an IVP for nth-order equation.

Let Ω be an open set in $\mathbb{R}^{n+1}$, whose points may be written as (t, y) with $t \in \mathbb{R}$ and $y \in \mathbb{R}^n$. Suppose $f : \Omega \to \mathbb{R}^n$ is a continuous function and is Lipschitz continuous with respect to y variables, that is, there is a positive constant L such that

$$\|f(t, y_1) - f(t, y_2)\| \leq L\|y_1 - y_2\|,$$

for all (t, y_1) and (t, y_2) in Ω and L does not depend on t. Here $\|\cdot\|$ denotes the norm introduced above. Suppose $(t_0, y_0) \in \Omega$. Choose $a, b > 0$ such that the *rectangle* $\mathcal{R}$ defined by

$$\mathcal{R} = \{(t, y) : |t - t_0| \leq a, \ \|y - y_0\| \leq b\}$$

is a subset of Ω. Let M be the maximum of $\|f(t, y)\|$ on $\mathcal{R}$ and $\alpha = \min\{a, b/M\}$. We can now state the following:

Theorem 2.34. The IVP for the n dimensional system

$$\dot{y} = f(t, y), \ y(t_0) = y_0, \tag{2.41}$$

has a unique solution in the interval $[t_0 - \alpha, t_0 + \alpha]$. □

Continuous Dependence of Solution on Initial Data and Dynamics: The initial data includes the given initial value y_0 and the dynamics f. In practical applications, it is important to know that how small errors in the initial data affect the solution. In other words, we would like to know that if the initial data is close to another initial data in appropriate norm, then the corresponding solutions are also close to each other. This is known as the continuous dependence of the solution on the initial condition and dynamics.

Theorem 2.35. Let $\mathcal{R}$ be as in Theorem 2.34. Suppose $f, \tilde{f} \in C(\mathcal{R})$ and be Lipschitz continuous with respect to y on $\mathcal{R}$ with Lipschitz constants $\alpha, \tilde{\alpha}$, respectively. Let y and $\tilde{y}$ be, respectively, the solutions of the IVP $\dot{y} = f(t, y), \ y(t_0) = y_0$ and $\dot{\tilde{y}} = \tilde{f}(t, \tilde{y}), \ \tilde{y}(\tilde{t}_0) = \tilde{y}_0$ in some closed intervals I_1, I_2 containing t_0 and $\tilde{t}_0$. For small $|t_0 - \tilde{t}_0|$, let I any finite interval containing t_0 and $\tilde{t}_0$, where both y and $\tilde{y}$ are defined. Then,

$$\max_{t \in I} \|y(t) - \tilde{y}(t)\| \leq \left(\|y_0 - \tilde{y}_0\| + |I| \max_{\mathcal{R}} \|f(t, y) - \tilde{f}(t, y)\| + M|t_0 - \tilde{t}_0| \right) e^{\alpha_0 |I|},$$

where $|I|$ is the length of the interval I, $M = \max\left(\max_{\mathcal{R}} \|f\|, \max_{\mathcal{R}} \|\tilde{f}\| \right)$ and $\alpha_0 = \min(\alpha, \tilde{\alpha})$. □

Proof We give a proof when $t_0 = \tilde{t}_0$. Observe that the solutions y and $\tilde{y}$ satisfy the following integral equations:

$$y(t) = y_0 + \int_{t_0}^{t} f(\tau, y(\tau)) \, dt, \quad \tilde{y}(t) = \tilde{y}_0 + \int_{t_0}^{t} \tilde{f}(\tau, \tilde{y}(\tau)) \, d\tau,$$

for all $t \in I$. Subtracting the second equation from the first, we get

$$y(t) - \tilde{y}(t) = y_0 - \tilde{y}_0 + \int_{t_0}^{t} (f(\tau, y(\tau)) - \tilde{f}(\tau, \tilde{y}(\tau))) \, d\tau.$$

When $t_0 \neq \tilde{t}_0$, an extra integral appears in the above equation, which can be estimated easily. Now add and subtract the term $f(\tau, \tilde{y}(\tau))$ in the above integral. We then obtain

$$\|y(t) - \tilde{y}(t)\| \leq \|y_0 - \tilde{y}_0\| + \int_{t_0}^{t} \|f(\tau, y(\tau)) - f(\tau, \tilde{y}(\tau))\| + \int_{t_0}^{t} \|f(\tau, \tilde{y}(\tau)) - \tilde{f}(\tau, \tilde{y}(\tau))\| \, d\tau.$$

Using the Lipschitz continuity of f with respect to y , the above inequality can be estimated as

$$\|y(t) - \tilde{y}(t) \leq \|y_0 - \tilde{y}_0\| + \alpha \int_{t_0}^{t} \|y(\tau) - \tilde{y}(\tau)\| \, dt + |I| \max_{\mathcal{R}} \|f(t, y) - \tilde{f}(t, y)\|.$$

Applying Gronwall's inequality, we get

$$\|y(t) - \tilde{y}(t)\| \leq \left(\|y_0 - \tilde{y}_0\| + |I| \max_{\mathcal{R}} \|f(t, y) - \tilde{f}(t, y)\| \right) \exp(\alpha |I|).$$

The same inequality is true when α is replaced by $\tilde{\alpha}$ as well if we add and subtract $\tilde{f}(\tau, y(\tau))$ instead of $f(\tau, \tilde{y}(\tau))$. This completes the proof. □

Continuation of Solution into Larger Intervals and Maximal Interval of Existence: Observe that the existence result, Theorem 2.34, is local in nature, that is, we could only claim the existence of a solution in a small interval containing the initial time t_0. However, when the differential equation is linear, that is $f(t, y) = A(t)y + b(t)$, where $A(t)$ and $b(t)$ are respectively, matrix-valued and vector-valued continuous functions defined on $[t_0 - a, t_0 + a]$, then f is defined on $\mathcal{R} = [t_0 - a, t_0 + a] \times \mathbb{R}$. In this case, the solution is defined on the entire interval $[t_0 - a, t_0 + a]$, that is, h can be taken as a itself. But, this is not true when f is not linear, that is, the solution may not exist in the interval where f is defined. At the same time, it is possible that $[t_0 - h, t_0 + h]$ may not be the largest possible interval of existence. This leads to the following question: Can we enlarge the domain of the solution y further? More generally, what is the largest possible interval of existence?

Example 2.36. Consider the IVP for the one-dimensional equation

$$\dot{y} = y^2, \; y(1) = -1.$$

Let $\mathcal{R} = \{(t, y) : |t - 1| \leq 1, \; |y + 1| \leq 1\}$. Here, $f(t, y) = y^2$ satisfies continuity and Lipschitz continuity assumptions on $\mathcal{R}$, $t_0 = 1, y_0 = -1$ and let $|t - 1| \leq h$, be the interval on which existence is guaranteed, which can be computed as $h = 1/4$. Hence by Theorem 2.34, the IVP has a solution on the interval $[3/4, 5/4]$. □

Now by the method of separation of variables, we can integrate the differential equation and use the initial condition to obtain the solution as $y(t) = -1/t$. Thus, the solution exists for

$0 < t < \infty$. In other words, we can continue the solution outside the interval $[3/4, 5/4]$. At the same time, since $|y(t)| \to \infty$ as $t \to 0+$, it ceases to exist to the left of 0, even though f is a differentiable function defined everywhere. This is the typical non-linear phenomena and f is not Lipschitz in the entire $\mathbb{R}$, but it is Lipschitz in any bounded interval.

Continuation of the Solution Outside the Interval $|t - t_0| \leq h$: The existence theorem (Theorem 2.34) guarantees that IVP (2.41) has a solution ϕ_0 on the interval $[t_0 - h, t_0 + h]$. Consider the right end point of $[t_0 - h, t_0 + h]$. Let $t_1 = t_0 + h$, $y_1 = \phi_0(t_1)$. The point (t_1, y_1) is inside $\mathcal{R}$, which is inside D. Now consider the IVP $\dot{y} = f(t, y)$, $y(t_1) = \phi_0(t_1) = y_1$. Now appealing to the existence theorem with this new initial condition, we obtain a solution ϕ_1 in the interval $t_1 - h_1 \leq t \leq t_1 + h_1$ for some $h_1 > 0$. Define

$$y(t) = \begin{cases} \phi_0(t), & t_0 - h \leq t \leq t_0 + h = t_1 \\ \phi_1(t), & t_1 \leq t \leq t_1 + h_1. \end{cases}$$

Then,

$$\phi_0(t) = y_0 + \int_{t_0}^{t} f(\tau, \phi_0(\tau))\, d\tau, \quad \text{for } t_0 - h \leq t \leq t_1$$

$$\phi_1(t) = \phi_0(t_1) + \int_{t_1}^{t} f(\tau, \phi_1(\tau))\, d\tau, \quad \text{for } t_1 \leq t \leq t_1 + h_1.$$

Thus, we have

$$y(t) = y_0 + \int_{t_0}^{t} f(\tau, y(\tau))\, d\tau$$

for $t \in [t_0 - h, t_1 + h_1] = [t_0 - h, t_0 + h + h_1]$. It is easy to see from these two expressions that y is differentiable at $t = t_1$ also, and verify that y indeed satisfies the DE in question. This solution $y(t)$ is called a *continuation* of the solution ϕ_0 to the interval $[t_0 - h, t_1 + h_1]$. Now repeating this process at the new end point $t_1 + h_1$, we obtain a solution on $[t_0 - h, t_1 + h_1 + h_2]$. In this manner, we may get longer intervals $[t_0 - h, t_n + h_n]$. Unfortunately, this still may not lead to the maximum interval of existence.

In Example 2.36, the function f is not Lipschitz in $\mathbb{R}$. When f is not globally Lipschitz, the bounds on the Picard's iterates may become larger and larger, thus reducing the interval of existence. This is really due to the *bad* non-linearity even though the function is very smooth as in Example 2.36. If f is Lipschitz globally, then we get the existence in the entire interval of definition as in the following theorem. The proof will follow along the same lines as in the Picard's existence theorem and we leave the details as an exercise to the reader. In fact, due to fact that f is defined on the entire real line in the second variable, we need not have to check the validity of the Picard's iterates as they are always defined.

Theorem 2.37. Let $f(t, y)$ be a bounded continuous vector-valued function defined in the unbounded domain $\mathcal{R} = \{(t, y) \ : \ a < t < b, \ y \in \mathbb{R}^n\}$. Let f be Lipschitz in y on $\mathcal{R}$. Then, a solution y of $\dot{y} = f(t, y)$, $y(t_0) = y_0$, $t_0 \in (a, b)$ is defined on the entire open interval $a < t < b$. In particular, if $a = -\infty$, and $b = +\infty$, then y is defined for all t in $\mathbb{R}$. □

Theorem 2.38. Suppose $f : D \to \mathbb{R}^n$ is bounded and y is the solution of the IVP (2.41) in some interval (a, b) containing t_0. If $b < \infty$, then, $y(b-) = \lim_{t \to b-} y(t)$ exists. If $(b, y(b-)) \in D$, then, the solution y may be continued to an interval $(a, \bar{b}]$ with $\bar{b} > b$. Similar statements hold at the left end point a. □

Proof Let M be the bound on $\|f\|$ on D and $t_0 < t_1 < t_2 < b$. Then,

$$\|y(t_2) - y(t_1)\| \leq \int_{t_1}^{t_2} \|f(s, y(s))\| ds \leq M|t_2 - t_1|.$$

Thus, if $t_1, t_2 \to b$ from the left, it follows that

$$\|y(t_2) - y(t_1)\| \to 0$$

which is the Cauchy criterion for the existence of the above said limit. Since $(b, y(b-)) \in D$, we can now consider the IVP with the initial condition at b as $y(b-)$ and the solution can be continued beyond b as asserted in the theorem. □

Maximal Interval of Existence: Consider the IVP (2.41). Assume the existence of a unique solution in a neighborhood t_0. Call such a neighborhood an interval of existence. Suppose I_1 and I_2 are intervals of existence containing t_0, then their union is an interval of existence (why?).

Definition 2.39. Let J be the union of all possible intervals of unique existence of (2.41). Then, J is an interval and, is called the *maximal interval of existence*. More precisely, let $\{I_\alpha\}$, be the collection of all intervals of unique existence, containing t_0. This collection is non-empty and $J = \bigcup_{\alpha} I_\alpha$. □

Indeed, we can define a unique solution y in J as follows: for any $t \in J$, t is in some interval I of existence and $y(t)$ is the value given by the solution in I. This is well defined by the uniqueness.

Proposition 2.40. The maximal interval J of existence is an open interval (α, β), where α can be $-\infty$ and β can be $+\infty$. □

Proof If not, suppose $J = (\alpha, \beta]$. In this case, $\beta < \infty$. Then, one can consider the IVP for the same ODE with initial condition at β, to get a solution in $[\beta, \beta + h]$ for some $h > 0$. This will produce a solution in $(\alpha, \beta + h]$ contradicting the maximality of J. Similar contradiction can be arrived if α is a point in J. □

Theorem 2.41. Let $J = (\alpha, \beta)$ be the maximal interval of existence and y be the solution to the IVP in J. Assume $\beta < \infty$. If K is any compact subset of $\mathbb{R}$ such that $[t_0, \beta] \times K \subset D$, then there exists a $t_1 \in (\alpha, \beta)$ such that $y(t_1)$ does not belong to K. Similar statements hold at the end point α. □

We infer the following from the conclusion of the theorem. Only one of the following statements is true:

- If $y(\beta-) = \lim_{t \to \beta-} y(t)$ exists, then $(\beta, y(\beta-)) \in \bar{D} \setminus D$, the boundary of D.
- The solution y becomes unbounded near β, that is, given any large positive number C, there exists $t_1 < \beta$ such that $\|y(t_1)\| \geq C$.

The proof of the theorem follows immediately from Theorem 2.38. The examples below illustrate both these situations.

Example 2.42. Consider the one-dimensional equation $\dot{y} = \dfrac{1}{ty}$.

Here, the function $f(t, y) = \dfrac{1}{ty}$ is defined in the entire $(t, y)-$ plane, except the t-axis and y-axis. We will consider the IVP in the first quadrant in the $(t, y)-$ plane: $y(t_0) = y_0$ where both t_0, y_0 are positive. The solution is given by $y = [2\log(t/t_0) + y_0^2]^{1/2}$. Therefore, the maximal interval of existence is (α, ∞), where $\alpha = t_0 e^{-y_0^2/2}$ and as $t \to \alpha+$, $y(t) \to 0$ with $(\alpha, 0)$ belonging to the boundary of the domain in question. □

Example 2.43. Consider the one-dimensional equation $\dot{y} = \dfrac{1}{t + y}$.

In this case, we take the domain as $\{(t, y) : t + y > 0\}$ and impose the initial condition as $y(t_0) = y_0$ with $t_0 + y_0 > 0$. By introducing a new variable $u(t) = t + y(t)$, we see that the solution is implicitly given by $\dfrac{e^{u(t)}}{1 + u(t)} = \dfrac{e^{y_0}}{1 + t_0 + y_0} e^t$. We notice that the maximal interval of existence in this case is given by (α, ∞), where $\alpha = t_0 + \log(1 + t_0 + y_0) - (t_0 + y_0) < t_0$. Again, it is not hard to see that as $t \to \alpha+$, $(t, y(t))$ approaches the boundary of the domain in question, that is, $t + y(t) \to 0$. □

Example 2.44. Consider the one-dimensional IVP: $\dot{y} = \dfrac{\pi}{2}(1 + y^2)$, $y(0) = 0$.

We now see that the solution $y(t) = \tan\left(\frac{\pi}{2} t\right)$ cannot be extended beyond the interval $(-1, 1)$. Note that if we take any rectangle $\{(t, y) : |t| \leq a, |y| \leq b\}$ around the origin $(0, 0)$, then as in the local existence theorem, we get the existence of a unique solution in an interval $[-h, h]$, where $h = \min\left(a, \frac{2}{\pi}\frac{b}{1+b^2}\right)$, which is always less than $\frac{1}{\pi}$. □

Autonomous Systems: We now discuss autonomous systems. Such systems occur in many situations including the characteristic equations arising in the study of first-order PDE. Consider an autonomous system of n first-order equations:

$$\dot{x} = f(x), \tag{2.42}$$

or, explicitly written

$$\dot{x}_j = f_j(x_1, x_2, \ldots, x_n), \quad j = 1, 2, \ldots, n.$$

Usual assumptions on f are made so that the system (2.42) has unique solution defined in its maximal interval of existence. As the system is autonomous, this maximal interval essentially depend on the initial condition; see Theorem 2.34. The solution $x(t)$ has a geometrical meaning of a curve in $\mathbb{R}^n$ and equation (2.42) gives its tangent vector at every t. For this reason, (2.42) is also referred to as a *vector field* and the corresponding solution as an *integral curve* of the vector field. If x is a solution of (2.42), we say that x passes through $x^0 \in \mathbb{R}^n$ if $x(t_0) = x^0$, for some $t_0 \in \mathbb{R}$.

Definition 2.45. Given a solution x of (2.42) passing through $x^0 \in \mathbb{R}^n$ with $x(t_0) = x^0$, for some $t_0 \in \mathbb{R}$, the *orbit through* x^0, is the set $\mathcal{O}(x^0)$ defined by

$$\mathcal{O}(x^0) = \{x(t) \in \mathbb{R}^n : t \in I(x^0)\},$$

where $I(x^0)$ denotes the maximal interval of existence of the solution x and the *positive orbit through* x^0, is the set $\mathcal{O}^+(x^0)$ defined by

$$\mathcal{O}^+(x^0) = \{x(t) \in \mathbb{R}^n : t \geq t_0,\ t \in I(x^0)\}.$$

□

Lemma 2.47 below shows that any solution passing through x^0 may be used to define $\mathcal{O}(x^0)$ or $\mathcal{O}^+(x^0)$ unambiguously. Generally speaking, the phase space (plane) analysis is about describing all the (positive) orbits of (2.42). The other terminologies used for orbit are *trajectory* and *path*. We will now discuss some important properties of solutions of autonomous systems. Below the statements regarding t refer to all t in the maximal interval of existence of the appropriate solution.

Lemma 2.46. If x is a solution of (2.42), define x_c by $x_c(t) = x(t + c)$ for any fixed c and for all t. Then x_c is also a solution of (2.42). □

Proof Direct differentiation. □

We remark that the above lemma is not true for a non-autonomous system.

Lemma 2.47. If x and y are solutions of (2.42) passing through $x^0 \in \mathbb{R}^n$ with $x(t_0) = y(t_1) = x^0$, for some $t_0, t_1 \in \mathbb{R}$, then,

$$y(t) = x(t + t_0 - t_1) \text{ and } x(t) = y(t + t_1 - t_0), \text{ for all } t.$$

□

Thus, $\mathcal{O}(x^0)$ or $\mathcal{O}^+(x^0)$ is the same set whether x or y is used in its definition.

Proof Define $z(t) = x(t + t_0 - t_1)$ for all t. By Lemma 2.46, z is a solution of (2.42) and $z(t_1) = x(t_0) = y(t_1)$. By uniqueness, $z \equiv y$. This completes one part of the proof and the other part is similar. □

Corollary 2.48. If $x^0, x^1 \in \mathbb{R}^n$ and $x^1 \in \mathcal{O}(x^0)$ (respectively $\mathcal{O}^+(x^0)$), then $\mathcal{O}(x^0) = \mathcal{O}(x^1)$ (respectively $\mathcal{O}^+(x^0) \supset \mathcal{O}^+(x^1)$). □

Lemma 2.49. If $x^0, x^1 \in \mathbb{R}^n$, then either $\mathcal{O}(x^0) = \mathcal{O}(x^1)$ or $\mathcal{O}(x^0) \cap \mathcal{O}(x^1) = \phi$, the empty set. Similar statements may be made regarding the positive orbits. □

Proof If $\tilde{x} \in \mathcal{O}(x^0) \cap \mathcal{O}(x^1)$, then by Corollary 2.48, it follows that $\mathcal{O}(x^0) = \mathcal{O}(\tilde{x}) = \mathcal{O}(x^1)$ and the proof is complete. □

Lemma 2.50. Suppose x is a solution of (2.42) and there exist t_0 and $T > 0$ such that $x(t_0 + T) = x(t_0)$. Then $x(t + T) = x(t)$ for all t. □

Proof Define x_T by $x_T(t) = x(t + T)$. Then x_T is a solution and by hypothesis, $x_T(t_0) = x(t_0)$. The proof is complete by uniqueness. □

Remark 2.51. The solution in Lemma 2.50 is termed as a *periodic solution*, with a period T. The smallest such a $T > 0$ is called the *period* of x. The orbit of a periodic solution is called a *periodic orbit* or *closed orbit*. If a periodic orbit is *isolated* in the sense that there is no other periodic orbit in its immediate neighborhood, then the periodic orbit is called a *limit cycle*. For example, the orbits of the one-dimensional equation $\ddot{x} + x = 0$ are all periodic orbits but, none of them is a limit cycle. Limit cycles can only occur in non-linear systems. The existence of periodic solutions to (2.42) is an important aspect of the qualitative theory and two important results, namely Poincarè–Bendixon Theorem and Leinard's Theorem give sufficient conditions for the existence of periodic solutions in $2D$ systems. See Nandakumaran et al. (2017) and references therein. □

2.4 FOURIER TRANSFORM, CONVOLUTION AND MOLLIFIERS

In this section, we briefly introduce some important topics, namely *Fourier transform (FT), convolution and mollifiers* and their properties. These concepts will be used in the main text, especially in chapters on Conservation Laws and Laplace equation. The reader wishing to have a more detailed exposition to these topics may consult Kesavan (1989), Folland (1992), and Rund (1973) among others.

Given a function $f : \mathbb{R}^n \to \mathbb{R}$ or $\mathbb{C}$, the formal definition of FT is given by

$$\mathcal{F}f(\xi) = \hat{f}(\xi) = \frac{1}{(2\pi)^{n/2}} \int_{\mathbb{R}^n} e^{-ix\cdot\xi} f(x)\, dx, \tag{2.43}$$

where $x, \xi \in \mathbb{R}^n, x \cdot \xi = \sum_{i=1}^{n} x_i \xi_i$. The choice of the constant $(2\pi)^{-n/2}$ is for the reason of symmetry. The same constant appears in the inverse FT, defined subsequently. If we omit the constant in (2.43), then the constant $(2\pi)^{-n}$ appears in the inverse transform.

It is formal in the sense that the integral on the right-hand side may not make sense for general f, even if f is smooth. But, if f is continuous with compact support, then $\hat{f}(\xi)$ is well defined. Our intention is to define the FT for general integrable functions, namely in $L^1(\mathbb{R}^n)$ and $L^2(\mathbb{R}^n)$. Indeed, if $f \in L^1(\mathbb{R}^n)$, then

$$|\hat{f}(\xi)| \leq (2\pi)^{-n/2} \|f\|_1$$

for $\xi \in \mathbb{R}^n$. In fact, we have the following lemma:

Lemma 2.52 (Riemann–Lebesgue Lemma). Assume $f \in L^1(\mathbb{R}^n)$, then $\hat{f}$ is a uniformly continuous function which vanishes at ∞; that is $\hat{f}(\xi) \to 0$ as $|\xi| \to \infty$. Furthermore, we have

$$\|\hat{f}\|_\infty \leq (2\pi)^{-n/2} \|f\|_1. \qquad \square$$

Description of $L^p(\Omega)$ Spaces: Let Ω be an open subset of $\mathbb{R}^n$. The space $L^p(\Omega)$, $1 \leq p < \infty$ is the space of all real or complex-valued (Lebesgue) measurable functions f defined on Ω such that

$$\|f\|_p = \left(\int_{\Omega} |f(x)|^p \, dx \right)^{1/p} < \infty.$$

For $p = \infty$, the space $L^\infty(\Omega)$ consists of all the measurable functions f which are *essentially bounded.* A function f is said to be essentially bounded on Ω if there is a finite $M > 0$ such that the set $\Omega_M = \{x \in \Omega : |f(x)| > M\}$ has measure zero. The *essential supremum* of f is then defined by

$$\|f\|_\infty = \inf\{M : \Omega_M \text{ has measure zero}\}.$$

Description of $\mathcal{D}(\Omega)$ Space: The space $\mathcal{D}(\Omega)$ consists of all $C^\infty(\Omega)$ functions having compact support. The topology is the following:

A sequence φ_k converges to φ in $\mathcal{D}(\Omega)$, if there is compact set $K \subset \Omega$ such that $\text{supp}\varphi_k, \text{supp}\varphi \subset K$ for all k and $D^\alpha \varphi_k \to D^\alpha \varphi$ uniformly on K, for every multi-index α. This makes $\mathcal{D}(\Omega)$ a locally convex complete topological vector space, hence has a rich dual space. This dual space $\mathcal{D}'(\Omega)$ is the space of *distributions*. The only inconvenience about $\mathcal{D}(\Omega)$ is that it is not metrizable. That is, there is no metric that can be defined on $\mathcal{D}(\Omega)$ that is compatible with the convergence criterion defined above.

It is known that $\mathcal{D}(\Omega)$ is dense in $L^p(\Omega)$ for all $1 \leq p < \infty$. See also the discussion on mollifiers. A nice property of FT is the recovery of the function from its FT through an inversion formula. This makes its application very useful, especially in obtaining solutions to differential equations. If we work in the space $L^1(\mathbb{R}^n)$ or $\mathcal{D}(\mathbb{R}^n)$, this inverse process is not possible. In fact, the FT of an L^1 function need not be an L^1 function. To see this, take the characteristic function $f = \chi_{(-1,1)}$ in $\mathbb{R}$, then its FT is given by $\hat{f}(\xi) = \frac{1}{\pi}\frac{\sin \xi}{\xi}$ which is not in $L^1(\mathbb{R}^n)$. The case of $\mathcal{D}(\mathbb{R}^n)$ is more serious in the sense that $\hat{f}$ is not in $\mathcal{D}(\mathbb{R}^n)$ for any $f \in \mathcal{D}(\mathbb{R}^n)$. This is part of the well-known *Paley–Wiener theorem*.

On the other hand, if we wish to work with the space L^2, the integral in (2.43) may not be finite. However, it is possible to interpret the integral in (2.43) in a different way. In fact, we can give interpretation in two different ways; either through the Schwartz class of functions $\mathcal{S}$ or through the space $L^1 \cap L^2$. Both the approaches need density arguments. There is an advantage working in the space $\mathcal{S}$ as the FT will be a continuous isomorphism of $\mathcal{S}$ onto $\mathcal{S}$. Also $\mathcal{S}$ happens to be the space of smooth functions that is dense in L^p for all $1 \leq p < \infty$. Thus, it may be possible to extend many arguments that hold true in $\mathcal{S}$ to L^p spaces by density arguments. The Schwartz space $\mathcal{S}(\mathbb{R}^n)$ is defined by

$$\mathcal{S} = \mathcal{S}(\mathbb{R}^n) = \left\{ f \in C^\infty(\mathbb{R}^n) : \sup_{x \in \mathbb{R}^n} \left| x^\alpha D^\beta f \right| < \infty \text{ for all multi-indices } \alpha, \beta \right\}.$$

A metric can be introduced in $\mathcal{S}$ via the following family of semi-norms:

$$\|f\|_{k,m} = \sup_{|\alpha| \leq k, |\beta| \leq m} \sup_{x \in \mathbb{R}^n} \left| x^\alpha D^\beta f \right|,$$

for $k, m = 0, 1, \ldots$. The dual $\mathcal{S}'$ of $\mathcal{S}$ is called the *space of tempered distributions*. Trivially $\mathcal{D}(\mathbb{R}^n) \subset \mathcal{S}$. Using the fact that $\int_{\mathbb{R}^n} (1 + |x|)^{-k}\, dx < \infty$ when $k > n/2$, it is easily verified that $f \in L^p(\mathbb{R}^n)$ for all $1 \leq p \leq \infty$. Moreover, if p is a polynomial and α is a multi-index, then the mappings $f \mapsto pf$ and $f \mapsto D^\alpha f$ are linear continuous from $\mathcal{S}$ into $\mathcal{S}$. Also, $\mathcal{S}$ is dense in $L^p(\mathbb{R}^n)$ for $1 \leq p < \infty$. That is, if $f \in L^p(\mathbb{R}^n)$, then there is a sequence f_k in $\mathcal{S}$ such that $\|f_k - f\|_p \to 0$ as $k \to \infty$.

Proposition 2.53. For any $p \in [1, \infty]$, the inclusion map

$$\mathcal{S}(\mathbb{R}^n) \hookrightarrow L^p(\mathbb{R}^n)$$

is a continuous linear mapping. Thus, $\mathcal{S}$ is continuously embedded into $L^p(\mathbb{R}^n)$ for any $p \geq 1$. □

For $f \in \mathcal{S}(\mathbb{R}^n)$, its inverse transformation $\check{f}$ is defined by

$$\check{f}(x) = \frac{1}{(2\pi)^{n/2}} \int_{\mathbb{R}^n} e^{i\xi \cdot x} \hat{f}(\xi)\, d\xi \tag{2.44}$$

for $x \in \mathbb{R}^n$. We have the following main result regarding the FT:

Theorem 2.54. The Fourier transform $\mathcal{F} : \mathcal{S}(\mathbb{R}^n) \to \mathcal{S}(\mathbb{R}^n)$ is a continuous linear isomorphism with $\mathcal{F}^{-1}(f) = \check{f}$, for $f \in \mathcal{S}$. Furthermore, for $f, g \in \mathcal{S}(\mathbb{R}^n)$, the following properties hold:

1. (*weak Parseval relation*):

$$\int_{\mathbb{R}^n} \hat{f}(\xi) g(\xi)\, d\xi = \int_{\mathbb{R}^n} f(x) \hat{g}(x)\, dx. \tag{2.45}$$

2. (*strong Parseval relation*):

$$\int_{\mathbb{R}^n} \hat{f}(\xi) \hat{\bar{g}}(\xi)\, d\xi = \int_{\mathbb{R}^n} f(x) \bar{g}(x)\, dx, \tag{2.46}$$

where $\bar{g}$ is the complex conjugate of g. The strong Parseval relation is equivalent to the L^2 isometry in $\mathcal{S}(\mathbb{R}^n)$, that is $\|f\|_2 = \|\hat{f}\|_2$.

3. (*Relation between differentiation and FT*):

$$\begin{aligned} (x^\beta D_x^\alpha f)^\wedge(\xi) &= i^{|\alpha|+|\beta|} D_\xi^\beta \left(\xi^\alpha \hat{f}(\xi)\right) \\ \xi^\beta D_\xi^\alpha \hat{f}(\xi) &= i^{|\alpha|+|\beta|} \mathcal{F}\left(D_x^\beta \left(x^\alpha f\right)\right)(\xi). \end{aligned} \tag{2.47}$$

Thus, FT takes the multiplication operator by polynomials to linear differential operators with constant coefficients and vice versa. □

We now state the, following, Plancherel theorem:

Theorem 2.55 (Plancherel)**.** There exists a unique isometry $P : L^2(\mathbb{R}^n) \to L^2(\mathbb{R}^n)$ that is onto such that $P(f) = \hat{f}$ if $f \in \mathcal{S}(\mathbb{R}^n)$. □

Interpretation of $\hat{f}$ if $f \in L^2(\mathbb{R}^n)$: For $f \in L^2(\mathbb{R}^n)$, choose a sequence $f_k \in \mathcal{S}(\mathbb{R}^n)$ such that $f_k \to f$ in $L^2(\mathbb{R}^n)$. Since $\|f_k\|_2 = \|\hat{f}_k\|_2$, it is easy to see that $\hat{f}_k$ converges in $L^2(\mathbb{R}^n)$. Denote the limit by Pf. That is $\|Pf - \hat{f}_k\|_2 \to 0$. This is taken as the definition of FT of an $L^2(\mathbb{R}^n)$ function, which does not depend on a particular choice of f_k. Keeping this in mind, the formula (2.43) is interpreted as *limit in mean* if $f \in L^2(\mathbb{R}^n)$.

2.4.1 Convolution

This is another important operation between two functions. The fantastic combination of FT and convolution is a powerful tool in many direct applications including the analysis of PDE. For two measurable functions $f, g : \mathbb{R}^n \to \mathbb{R}$, the convolution of f and g is denoted by $f * g$ that is formally defined as

$$f * g(x) = \int_{\mathbb{R}^n} f(x-y)g(y)\, dy. \tag{2.48}$$

Indeed, the above integral is finite for any $x \in \mathbb{R}^n$, if f, g are continuous and one of them has compact support. Moreover $\operatorname{supp}(f * g) \subset \operatorname{supp} f + \operatorname{supp} g$. In fact, if $f, g \in L^1(\mathbb{R}^n)$, then using the translation invariance property of the Lebesgue measure, it is easy to see that $f * g \in L^1(\mathbb{R}^n)$ and

$$\|f * g\|_1 \leq \|f\|_1 \|g\|_1. \tag{2.49}$$

Note that $f * g = g * f$. Thus $L^1(\mathbb{R}^n)$ becomes a commutative normed algebra with $*$ as the multiplication.

Theorem 2.56 (Young's Inequality). Assume $f \in L^p(\mathbb{R}^n)$, $g \in L^q(\mathbb{R}^n)$ with $1 \leq p, q \leq \infty$ and $\frac{1}{p} + \frac{1}{q} \geq 1$. Let r be such that $\frac{1}{p} + \frac{1}{q} = 1 + \frac{1}{r}$. Then, $f * g \in L^r(\mathbb{R}^n)$ and

$$\|f * g\|_r \leq \|f\|_p \|g\|_q. \tag{2.50}$$

□

The convolution is useful to smoothen the functions that are not smooth and it can be used to approximate $L^p(\Omega)$ functions by $\mathcal{D}(\Omega)$ functions. See the following proposition:

Proposition 2.57. For $f, g \in \mathcal{S}(\mathbb{R}^n)$, the convolution $f * g \in \mathcal{S}(\mathbb{R}^n)$ and

$$D^\alpha(f * g) = D^\alpha f * g = f * D^\alpha g$$

for any multi-index α. More generally, if $f \in C^k(\mathbb{R}^n)$, $g \in C(\mathbb{R}^n)$ and one of them has compact support, then $f * g \in C^k(\mathbb{R}^n)$ and

$$D^\alpha(f * g) = (D^\alpha f) * g \tag{2.51}$$

for all α with $|\alpha| \leq k$. □

2.4.2 Mollifiers

These are a special class of smooth functions with compact support, which is also known as Friedrichs mollifiers. These are used to convolve with functions that need not be smooth to produce smooth approximations. Consider the function $\rho : \mathbb{R}^n \to \mathbb{R}$ defined by

$$\rho(x) = \begin{cases} c \exp\left(\frac{-1}{1-|x|^2}\right) \text{ if } |x| < 1 \\ 0 \text{ if } |x| \geq 1. \end{cases} \tag{2.52}$$

Then, $\rho \in \mathcal{D}(\mathbb{R}^n)$ with $\operatorname{supp}\rho = \overline{B_1(0)}$. Here the constant c is chosen so that $\int_{\mathbb{R}^n} \rho(x)\, dx = 1$. Now, the mollifiers $\rho_\varepsilon : \mathbb{R}^n \to \mathbb{R}$ is defined by

$$\rho_\varepsilon(x) = \varepsilon^{-n} \rho\left(\frac{x}{\varepsilon}\right) \tag{2.53}$$

for $x \in \mathbb{R}^n, \varepsilon > 0$. Then, $\rho_\varepsilon \in \mathcal{D}(\mathbb{R}^n)$, supp $\rho_\varepsilon = \overline{B_\varepsilon(0)}$ and $\int_{\mathbb{R}^n} \rho_\varepsilon(x)\, dx = 1$, for all $\varepsilon > 0$. We also have the following main results:

Theorem 2.58. Let f be continuous with compact support K. Then, there is a compact set $K_1 \supset K$, a sequence $f_k \in \mathcal{D}(\mathbb{R}^n)$ with supp $f_k \subset K_1$ and $f_k \to f$ uniformly in K_1 and hence in $\mathbb{R}^n$.

The proof follows by taking $f_k = \rho_{1/k} * f$. Indeed, supp $f_k \subset$ supp $\rho_{1/k}$ + supp $f \subset \overline{B_1(0)} + K$. □

In the integrable case, we have

Theorem 2.59. For $1 \leq p < \infty$, the space $\mathcal{D}(\mathbb{R}^n)$ is dense in $L^p(\mathbb{R}^n)$. More generally, $\mathcal{D}(\Omega)$ is dense in $L^p(\Omega)$ for any open set $\Omega \subset \mathbb{R}^n$. □

The proof involves the use of cut-off functions to make it a compactly supported functions and then convolve with mollifiers to smoothen the functions. We end this sub-section with the following result:

Theorem 2.60 (Product Formula). Let $f, g \in L^1(\mathbb{R}^n)$. Then, $f * g \in L^1(\mathbb{R}^n)$ and

$$(f * g)^\wedge = \hat{f}\hat{g}. \tag{2.54}$$

Furthermore, by taking the inverse FT, we can recover $f * g$ as

$$f * g = \mathcal{F}^{-1}(\hat{f}\hat{g})$$

whenever both the sides make sense. □

CHAPTER 3

First-Order Partial Differential Equations: Method of Characteristics

3.1 INTRODUCTION

The first-order equations with real coefficients are particularly simple to handle. The *method of characteristics* reduces the given first-order partial differential equation (PDE) to a system of first-order ordinary differential equations (ODE) along some special curves called the *characteristics* of the given PDE. This will, in turn, help us to prove the existence of a solution to the *Cauchy problem* or *initial value problem* (IVP) associated with the PDE. Complications do arise in case of quasilinear or non-linear equations resulting only in *local existence*; the geometry of the characteristics also becomes more involved and non-uniqueness of (smooth) solutions may also result. To motivate the ideas we begin by a simple example.

Example 3.1. Consider the transport equation in two independent variables t and x, namely

$$u_t(x, t) + cu_x(x, t) = 0, \tag{3.1}$$

for $t > 0$, $x \in \mathbb{R}$, where $c > 0$ is a given constant. This is a linear, first-order PDE. Consider the curve $x = x(t)$ in the (x, t) plane given by the slope condition $\dot{x} = \frac{dx}{dt} = c, t \geq 0$. These are straight lines with slope $1/c$ and are represented by the equation $x - ct = x_0$, where x_0 is the point at which the curve meets the line $t = 0$ (see Figure 3.1(a)). These curves, straight lines in this case, are called the *characteristic curves* or simply the *characteristics* of (3.1). When c is a function of t and x, the characteristic curves need not be straight lines.

Now restrict the solution $u(x, t)$ to a characteristic $x(t) = ct + x_0$, that is, consider the function of one variable $U(t) = u(x(t), t)$. By the chain rule, it is easy to see that

$$\frac{d}{dt}U(t) = \frac{d}{dt}(u(x(t), t)) = u_x\dot{x}(t) + u_t \cdot 1 = u_x \cdot c + u_t = 0, \tag{3.2}$$

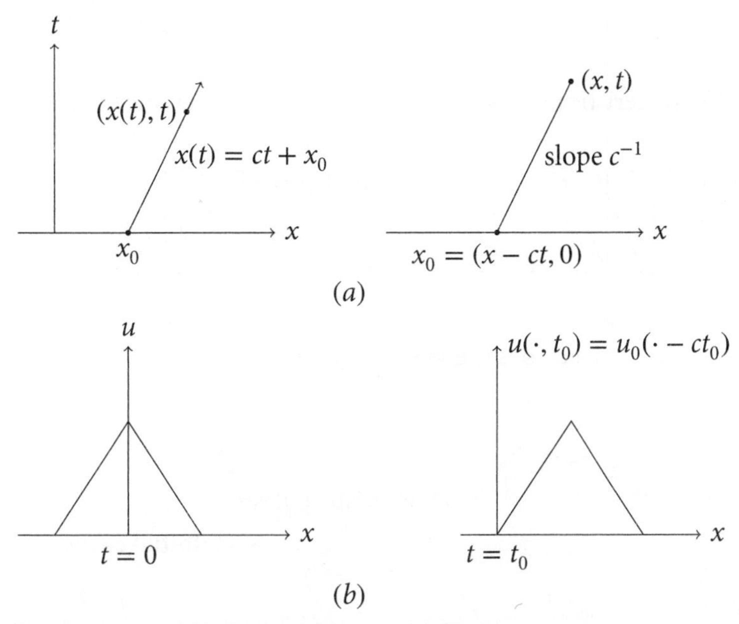

Figure 3.1 (a) Characteristic curves, (b) Solution curves

using (3.1). Therefore, $U \equiv$ constant and thus u is constant along the characteristic $x - ct = x_0$. This observation can be used to solve the IVP for the PDE (3.1) as follows:

Suppose the initial values of u are given on the line $t = 0$, that is, $u(x, 0) = u_0(x)$ is given and u_0 is a C^1 function. Now, for any point (x, t), $t > 0$ in the upper half plane, draw the characteristic passing through the point[1] (x, t). It is easy to see that this is given by the line with slope $1/c$ meeting the line $t = 0$ at the point $(x-ct, 0)$. As shown above, we have $u(x, t) = u(x - ct, 0) = u_0(x - ct)$.

It is easy to verify that this indeed is a solution of the PDE (3.1) satisfying the prescribed initial condition $u(x, 0) = u_0(x)$ on the line $t = 0$. □

We observe that the solution of the PDE was obtained by solving two ODEs. The initial curve, where initial value u_0 is assigned, namely the x-axis in this case, can be a smooth curve. However, it cannot be arbitrarily chosen. For example, the curve Γ shown in Figure 3.2(a) can be chosen as an initial curve, as it intersects all the characteristics, whereas the curve shown in Figure 3.2(b) cannot be considered as an initial curve as it coincides, locally, with a characteristic curve, thus becoming ambiguous with the fact that u is a constant along the characteristics. So, the initial curve will have to satisfy certain *transversality condition*, which we will discuss later.

[1]If c is a function of x and t, it may not be possible to draw a characteristic through every point (x, t), $t > 0$, thus obtaining the solution only in a subset of the upper half plane. If c is a function of u, then there may be two distinct characteristics passing through the same point.

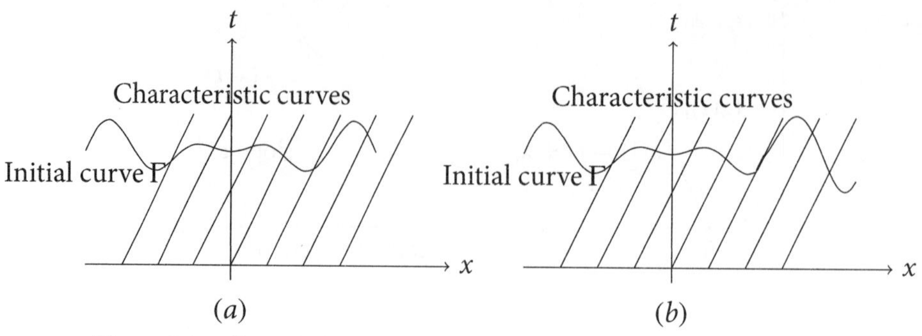

Figure 3.2 Characteristic and initial curves

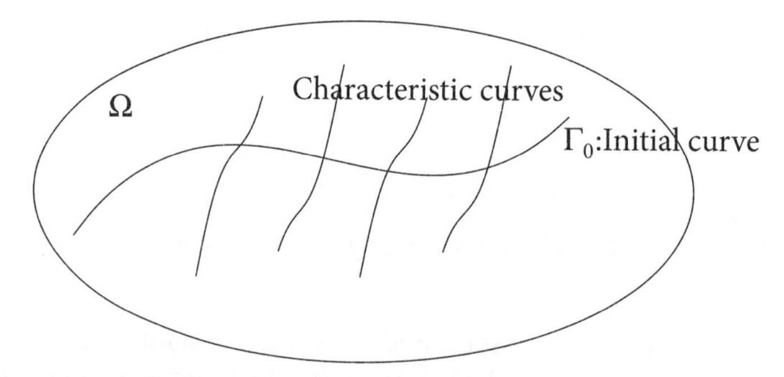

Figure 3.3 Characteristic and initial curves

3.2 LINEAR EQUATIONS

In Example 3.1, the way we constructed the characteristic curves, which were straight lines, may look bit ad-hoc or artificial. We will now explain the geometry behind this construction and see how it gets extended to general first-order equations. Thus, consider the general first-order linear PDE in two variables:

$$a(x,y)u_x + b(x,y)u_y = c(x,y)u + d(x,y), \tag{3.3}$$

where we have used x and y to denote the independent variables. Here $(x,y) \in \Omega$, a smooth bounded domain in $\mathbb{R}^2$; a, b, c, d are given smooth functions defined on Ω and $u = u(x,y)$ is the unknown function. Let $\Gamma_0 \subset \Omega$ be an initial curve, which is given in a parametric form (see Figure 3.3):

$$\Gamma_0 = \{(x_0(s), y_0(s)) \, : \, 0 \leq s \leq 1\},$$

where x_0 and y_0 are C^1 functions defined on $[0,1]$. Let $u_0 = u_0(s)$ be a given function defined on $[0,1]$, which will be served as an initial condition. The Cauchy problem or IVP for PDE (3.3) is as follows: Find $u = u(x, y)$ satisfying the PDE (3.3) together with the initial condition

$$u(x_0(s), y_0(s)) = u_0(s),\ s \in [0, 1]. \tag{3.4}$$

The problem of local solvability of IVP is to find u in a neighborhood of Γ_0 in Ω, satisfying (3.3) and (3.4).

We now give the geometric idea behind introducing the characteristics. First observe that for a fixed point $(x, y) \in \Omega$, the term on the left side of (3.3) is the directional derivative of u at (x, y) in the direction of the vector $(a(x, y), b(x, y))$. Thus, if we consider any curve $(x(t), y(t))$, parameterized by the t variable, in the (x, y) plane, such that the tangent at each point on this curve is the vector $(a(x, y), b(x, y))$ (see Figure 3.4), then the term on the left side of (3.3) will become the total derivative of u along this curve. Such curves are easy to construct by the requirement of the tangent vector at each of its points. These curves are the solutions of the following system of ODE

$$\frac{dx}{a(x, y)} = \frac{dy}{b(x, y)} \text{ or } \frac{dy}{dx} = \frac{b(x, y)}{a(x, y)} \tag{3.5}$$

or in the parametric form

$$\frac{dx}{dt}(t) = a(x(t), y(t)), \quad \frac{dy}{dt}(t) = b(x(t), y(t)), \tag{3.6}$$

with different initial conditions for $x(0)$ and $y(0)$. Along any such curve, u will satisfy the ODE

$$\frac{d}{dt}u(x(t), y(t)) = c(x(t), y(t))\, u(x(t), y(t)) + d(x(t), y(t)). \tag{3.7}$$

Note that t here is the parameter defining these curves. The one parameter family of curves defined by (3.6) are called the *characteristic curves* of the PDE (3.3).

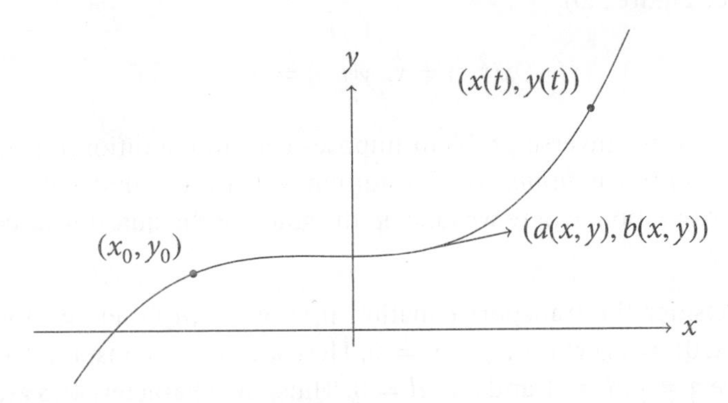

Figure 3.4 A characteristic curve

Let $s \in [0, 1]$ be fixed and consider the characteristic curve passing through the point $(x_0(s), y_0(s)) \in \Gamma_0$. This characteristic curve is obtained by solving the system of first-order ODE (3.6) with the initial values given by

$$x(0) = x_0(s), \ y(0) = y_0(s).$$

Under appropriate assumptions on a and b, for example, a, b are C^1 functions, the existence and uniqueness result from the ODE theory (see Chapter 2), there is a unique characteristic curve in a neighborhood of $t = 0$ (local existence). To emphasize the dependence of the solution on s, we denote it by $(x(t, s), y(t, s))$. More precisely, for fixed s the points $(x(t, s), y(t, s))$ moves along the characteristic curve for t small around $t = 0$. As we change $s \in [0, 1]$, we get different characteristic curves.

Now by restricting u along a characteristic curve (that is, fixing an s), we get an equation for u as

$$\frac{d}{dt}u(x(t,s), y(t,s)) = c(x(t,s), y(t,s))u(x(t,s), y(t,s)) + d(x(t,s), y(t,s)) \tag{3.8}$$

with the initial values

$$u(x(0,s), y(0,s)) = u_0(s). \tag{3.9}$$

Note that in the above ODE, s is merely a parameter, whereas t is the independent variable. Thus, we have obtained u along the fixed characteristic curve. By changing s, we obtain u along different characteristic curves. However, we still need to answer the question that whether this procedure indeed gives u in a neighborhood Ω_1 in Ω, containing the initial curve Γ_0, and then verify that u is a solution of the Cauchy problem. To achieve this, we need to assure that the family of characteristics $\{(x(t, s), y(t, s))\}$ obtained by taking all $s \in [0, 1]$, covers such an Ω_1. In other words, we need to solve the following inverse problem:

Given an arbitrary point $(x, y) \in \Omega_1$, find a characteristic curve passing through (x, y) and meeting the initial curve; more precisely, given $(x, y) \in \Omega_1$, find (t, s) and a solution of (3.6) such that (see Figure 3.5)

$$x(t,s) = x, \ y(t,s) = y. \tag{3.10}$$

A positive answer to this inverse problem imposes certain condition on the initial curve known as *transversality condition*. This condition will be discussed in a more general situation in the next section, where we take up the study of the quasilinear equations. Here are some examples.

Example 3.2. Consider the transport equation $u_y(x, y) + ku_x(x, y) = 0$ with the initial condition $u(x, 0) = u_0(x)$ on $\Gamma_0 : y = 0$. Here k is a real constant. Comparing with (3.3), we have $a = k$, $b = 1$ and $c = d = 0$. Thus, the characteristics are the solutions of $\dfrac{dy}{dx} = \dfrac{1}{k}$. This shows that the straight lines given by $x - ky = $ constant, are the characteristics. This has already been seen in Example 3.1. □

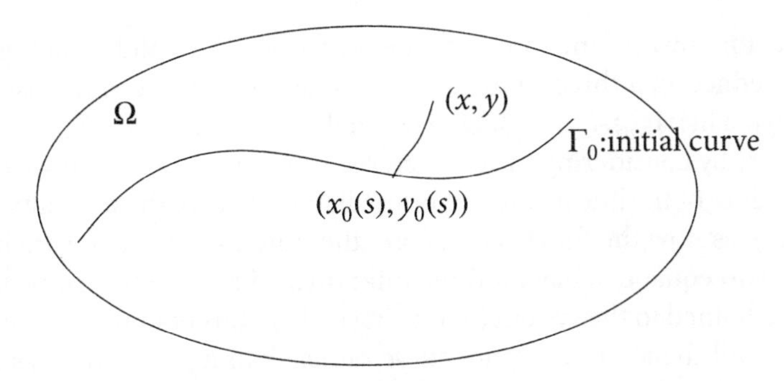

Figure 3.5 Characteristic through (x, y)

Example 3.3. Consider the PDE, $xu_x + yu_y = \alpha u$ and $u = \phi(x)$ on the initial curve $y = 1$. It is easy to see that $y = cx$, c constant, are the characteristic curves and along any of these curves, u satisfies

$$\frac{d}{dx}u(x, cx) = u_x(x, cx) + u_y(x, cx) \cdot c = u_x + \frac{y}{x}u_y = \frac{\alpha}{x}u(x, cx),$$

whose solution is given by $u(x, cx) = kx^\alpha$. As $k = k(c)$, depends on c it may differ from characteristic to characteristic. Thus, we have the general solution $u(x, y) = k\left(\frac{y}{x}\right)x^\alpha$, where k is an arbitrary function. Now applying the condition $u = \phi(x)$ at $y = 1$, we get

$$\phi(x) = k\left(\frac{1}{x}\right)x^\alpha \text{ or } k(x) = \phi\left(\frac{1}{x}\right)x^\alpha$$

and hence the required solution is

$$u(x, y) = \phi\left(\frac{x}{y}\right)y^\alpha.$$

There will be difficulties if, instead, we prescribe initial condition on the x-axis (why?). □

3.3 QUASILINEAR EQUATIONS

The general quasi-linear equation in two independent variables x and y, can be written as

$$a(x, y, u)u_x + b(x, y, u)u_y - c(x, y, u) = 0, \tag{3.11}$$

where u is the unknown function and the functions a, b and c are given smooth functions, now defined in a three-dimensional domain. Having seen the usefulness of the characteristics in the linear case, in reducing the PDE to a system of ODE, we wish to employ a similar procedure by considering the characteristic curves of (3.11). However, if we try to imitate the procedure of the linear case, we immediately observe the difficulty in solving the ODE system (3.6), as now the functions a, b are the functions of three variables – x, y and u, but with only two equations having three unknowns. Later we will show that one more equation can be adjoined to the system (3.6) using (3.11), thus obtaining a complete system. But, the solution will then define a *space curve* rather than a plane curve as in the case of (3.6). Let us begin a heuristic discussion with an example.

Example 3.4 (Burgers' Equation). Consider the quasilinear problem

$$u_t + uu_x = 0,\ x \in \mathbb{R},\ t > 0 \text{ and } u(x, 0) = u_0(x),\ x \in \mathbb{R}. \tag{3.12}$$

□

Though this equation looks simple, it poses non-trivial problems in the analysis and leads to a new phenomenon. Proceeding as in the linear case, we formally introduce the characteristic curve $C : x = x(t)$ through the equation $\frac{dx}{dt}(t) = u(x(t), t)$. Since u itself is unknown, the curve C cannot be determined as such. However, assuming such a C exists and restricting u to C and considering the function $U(t) = u(x(t), t)$, we immediately see that $\frac{dU}{dt} = 0$. Thus, the function U is a constant. Therefore, the solution u is a constant along C. The conclusion is that the characteristics of (3.12) are straight lines, perhaps with varying slopes (see Figure 3.6(a)). This is an important information we have obtained regarding the solution, though we are not able to solve the characteristic equation a priori. On the other hand, if two characteristics meet at a point P, each characteristic may carry different values from the initial values $u_0(x_1), u_0(x_2)$ as in Figure 3.6(b), thus creating discontinuity of the solution at P. Thus, in this situation the classical analysis fails. A discussion for equation (3.12) will be made below with specific initial values.

Now consider a point (x, t), $t > 0$ in the upper half plane. Assume that there is a unique characteristic passing through this point and meeting the initial axis $t = 0$ at x_0. From the

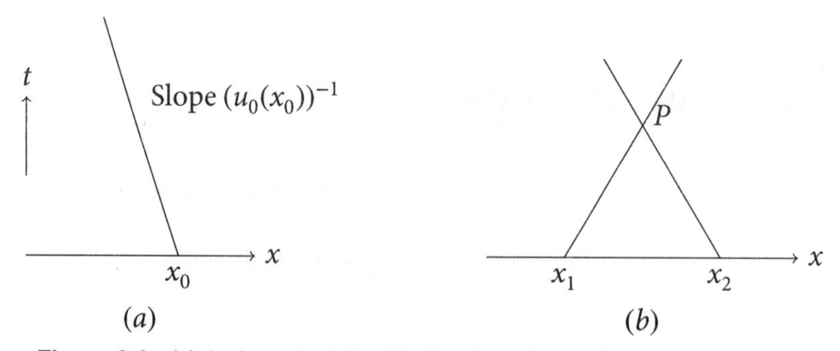

Figure 3.6 (a) A characteristic, (b) Meeting of two characteristics

discussion we just now had, we see that the following relations hold:

$$\begin{aligned} x &= tu_0(x_0) + x_0 \\ u(x,t) &= u_0(x_0), \end{aligned} \tag{3.13}$$

where u_0 is the initial function. Unlike the linear case, it is not evident how to obtain the initial point x_0 from the first equation in (3.13). By the implicit function theorem, there is a unique x_0, depending on x and t, satisfying the first equation in (3.13), provided that

$$1 + tu_0'(x_0) \neq 0. \tag{3.14}$$

Assuming the condition (3.14), it is now straightforward to verify that u as defined by the second equation in (3.13) is indeed a solution of the problem (3.12). If v is any other solution, then on the characteristic $x = tu_0(x_0) + x_0$, we have

$$v(x,t) = u_0(x_0) = u(x,t).$$

Thus, we have established the following theorem:

Theorem 3.5. Assume $u_0 \in C^1(\mathbb{R})$.

1. If $u_0'(\xi) \geq 0$ for all ξ, that is u_0 is a non-decreasing function, then the Cauchy problem (3.12) has a unique solution $u \in C^1(\mathbb{R} \times [0,\infty))$.
2. If either u_0 is non-increasing or u_0' changes sign and if

$$\sup_{u_0'(\xi)<0} |u_0'(\xi)| < \infty,$$

then the Cauchy problem (3.12) has a unique solution $u \in C^1(\mathbb{R} \times [0,T))$, where

$$T^{-1} = \sup_{u_0'(\xi)<0} |u_0'(\xi)|.$$

In either of the cases, the solution is given in the parametric form as

$$x = tu_0(x_0) + x_0, \; u(x,t) = u_0(x_0).$$

□

Some Comments on the Hypothesis in the above Theorem: First observe that the characteristic emanating from a point x_0 on the line $t = 0$ is the straight line having slope $1/u_0(x_0)$. Thus, if u_0 is non-decreasing, then the characteristics emanating from two distinct points on the line $t = 0$ will never meet for all $t > 0$ and thus have diverging slopes. An illustration is shown in Figure 3.7. For example, if we take $u_0(\xi) = \xi$ or $u_0(\xi) = \xi^3$ for $\xi \in \mathbb{R}$,

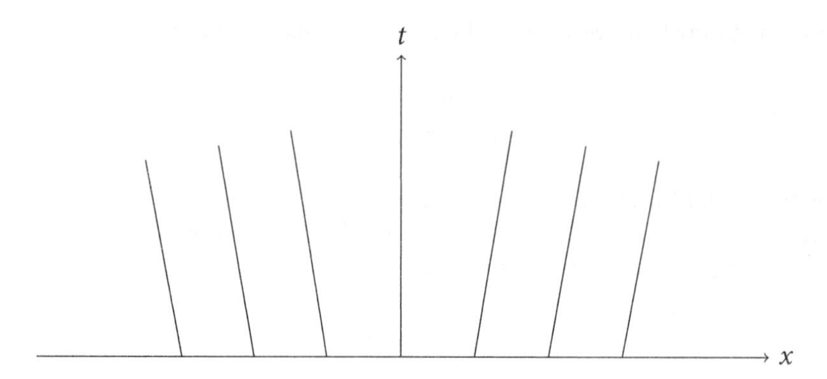

Figure 3.7 Diverging characteristics

it is not difficult to write down the solution of the Cauchy problem explicitly; the details are left as exercises.

On the other hand, if $u_0' < 0$ in an interval J, then the characteristics emanating from any two distinct points of J are going to intersect at a point (x, t) with $t > 0$. This is depicted in Figure 3.6(b). The second equation in (3.13) implies that the value $u(x, t)$ is ambiguous, that is, not uniquely defined. This is what that restricts the time of existence T in the above theorem, when we seek for C^1 solution. The reader should consider the examples of $u_0(\xi) = \xi^2$ and $u_0(\xi) = -\xi|\xi|$ and analyze the situation.

Specific Discontinuous Initial Values: As discussed above, if the initial function u_0 is non-decreasing and smooth, the characteristics emanating from distinct points on the x-axis do not intersect as depicted in Figure 3.7 and the IVP could be solved. In this scenario, for any point (x, t) in the upper half plane, there is a unique characteristic curve passing through (x, t) and meeting the x-axis at a point x_0. This line will have the slope $u_0(x_0)^{-1}$ and the solution at (x, t) is given by $u(x, t) = u_0(x_0)$. Let us take the specific example

$$u_0(x) = \begin{cases} 0 \text{ if } x \leq 0 \\ 1 \text{ if } x \geq 1 \end{cases}$$

and for $0 \leq x \leq 1$, u_0 is non-decreasing and smooth. Here the characteristics are as shown in Figure 3.8(a). But if u_0 is not smooth, there may be a region with no characteristics, as shown in the following example. Let

$$u_0(x) = \begin{cases} 0 \text{ if } x < 0 \\ 1 \text{ if } x \geq 0 \end{cases}$$

In this case, there is a region in the upper half plane without any characteristic (see Figure 3.8(b)). This region is known as *rarefaction*, and we need to define the solution here, perhaps in a non-classical way. The more serious situation is given by the following example. Let

$$u_0(x) = \begin{cases} 1 \text{ if } x < 0 \\ 0 \text{ if } x \geq 1 \end{cases}$$

and for $0 \leq x \leq 1$, u_0 is non-increasing and smooth. As the time evolves, the higher value of u_0 will result in higher speed (inverse of the slope) of the characteristic, eventually leading to the intersection of characteristics and we see the formation of a shock (see Figure 3.8(c)). More details are given in the chapter on conservation laws. It is to be noted that the concept of a classical solution fails, we may need to interpret the solution in a different way. This is the situation where the characteristics meet, resulting in the creation of discontinuities known as *shocks*. The evolution of $u(x, t)$ for various values of t are depicted in Figure 3.9. This can be interpreted as the evolution of waves coming from behind with high speeds, thus causing the discontinuity later.

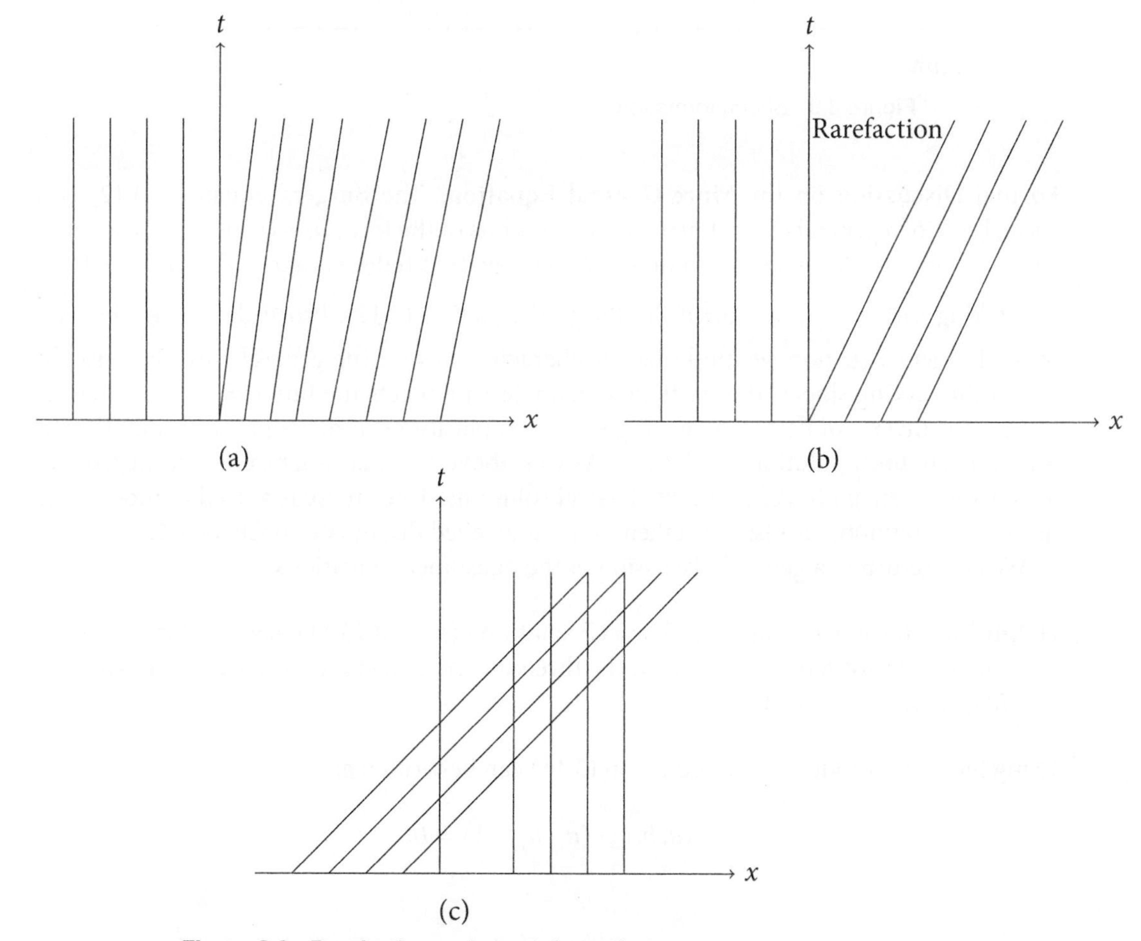

Figure 3.8 Rarefaction and shock formation

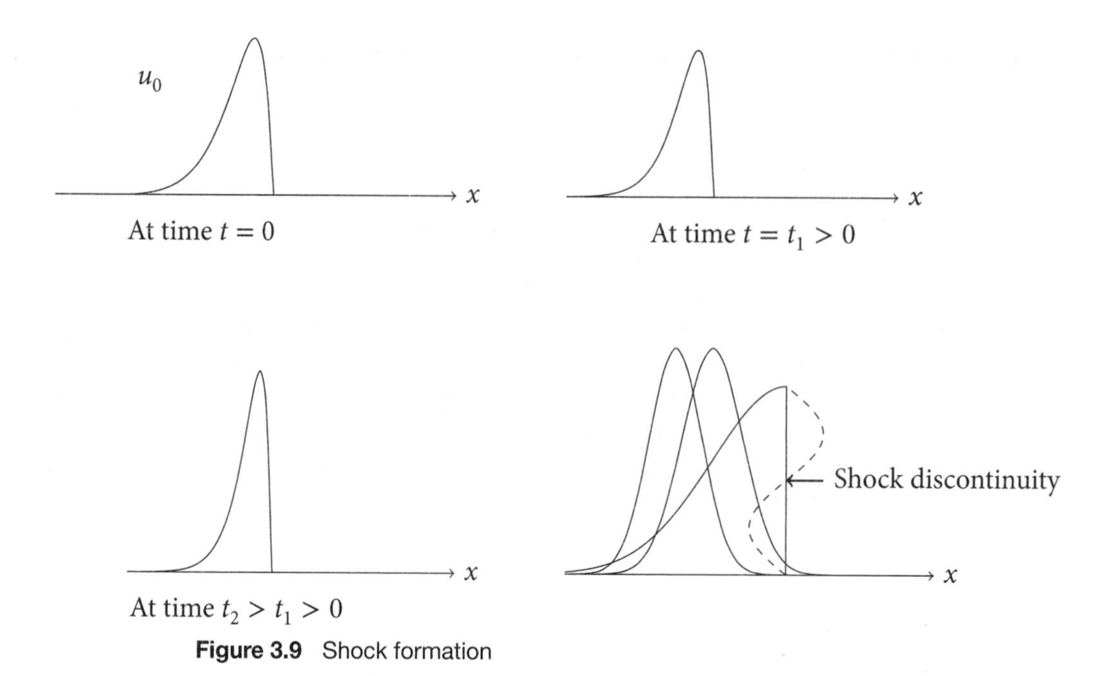

Figure 3.9 Shock formation

Formal Discussion on the More General Equation: The Burgers' equation (3.12) is a special case of a general class of first-order equations of the form $u_t + (f(u))_x = 0$, known as *conservation laws*. If f is a smooth function, the equation reduces to $u_t + f'(u)u_x = 0$. Note that taking $f(u) = \frac{u^2}{2}$, we obtain the Burgers' equation (3.12). Proceeding as in the case of the Burgers' equation, we find that the characteristics for the general f are also straight lines with varying slopes. If a particular characteristic meets the line $t = 0$ at x_0, then its slope is the inverse of $f'(u_0(x_0))$. Owing to their applications in many physical situations, it is required to find a solution for all $t > 0$. As seen above, this may not be possible, in general, if we wish to remain in the realm of classical solutions. Thus, there is a need to modify the notion of a solution. This issue is taken up for a detailed discussion in Chapter 5.

We now return to a general discussion of the quasilinear equations.

Definition 3.6 (Integral Surface). For a C^1 solution $u(x, y)$ of (3.11) in some domain of $\mathbb{R}^2$, the surface given by $z = u(x, y)$ in the three-dimensional (x, y, z) space, is known as an *integral surface* of (3.11). □

Using the scalar product in $\mathbb{R}^3$, equation (3.11) can be written as

$$(a, b, c) \cdot (u_x, u_y, -1) = 0. \tag{3.15}$$

Since the vector $(u_x, u_y, -1)$ is normal to the integral surface $z = u(x, y)$ at the point (x, y, z) on the surface, equation (3.11) or (3.15) can be interpreted as the condition that at each point on the integral surface, the vector (a, b, c) is tangent to the surface. Also, the PDE (3.11) defines a vector (direction) field (a, b, c) in $\mathbb{R}^3$, called the *characteristic directions*. Furthermore, a surface $z = u(x, y)$ is an integral surface of (3.11) if and only if at each point on the surface, the tangent plane contains the characteristic directions.

This motivates us to look at an integral surface as a family of space curves, whose tangent at any point on the curve coincides with the characteristic directions. These curves are known as *characteristic curves*. In other words, the tangential direction of the characteristic curve is given by the vector (a, b, c). Thus, introduce the family of space curves given by the system of ODE

$$\frac{dx}{a(x, y, z)} = \frac{dy}{b(x, y, z)} = \frac{dz}{c(x, y, z)}. \tag{3.16}$$

In the parametric form, the above equations can be written as

$$\left.\begin{aligned} \frac{dx}{dt}(t) &= a(x(t), y(t), z(t)) \\ \frac{dy}{dt}(t) &= b(x(t), y(t), z(t)) \\ \frac{dz}{dt}(t) &= c(x(t), y(t), z(t)). \end{aligned}\right\} \tag{3.17}$$

Note that in the linear case, a and b were independent of z and hence the solution $(x(t), y(t))$ of (3.6) defined plane curves in the x–y plane. In relation to (3.17), when a and b were independent of z, these plane curves are nothing but the projection of the space curves given by the solutions of (3.17). Using the existence and uniqueness result for ODE, under suitable conditions on a, b, c, we see that through each point (x_0, y_0, z_0), there passes a unique characteristic curve (integral curve)

$$x(t) = x(x_0, y_0, z_0, t), y = y(x_0, y_0, z_0, t), z = z(x_0, y_0, z_0, t),$$

defined for small t. Now, it is trivial to see that if a surface is generated by a family of characteristic curves, then, it is an integral surface as both have the same tangential directions. Conversely, if $z = u(x, y)$ is an integral surface S and $(x_0, y_0, z_0 = u(x_0, y_0))$ is a point on S, then, the integral curve through (x_0, y_0, z_0) will lie completely on S and thus, S is generated by a family of characteristic curves. To see this, consider the solution of

$$\frac{dx}{dt} = a(x, y, u(x, y)), \frac{dy}{dt} = b(x, y, u(x, y))$$

with $x = x_0, y = y_0$ at $t = 0$. Then, the corresponding curve

$$x = x(t), y = y(t), z = z(t) = u(x(t), y(t))$$

satisfies

$$\frac{dz}{dt} = u_x \frac{dx}{dt} + u_y \frac{dy}{dt} = au_x + bu_y = c.$$

Thus, the curve $(x(t), y(t), z(t)) = (x(t), y(t), u(x(t), y(t)))$ satisfies the system (3.17) and hence it is the characteristic through (x_0, y_0, z_0). Moreover, it lies on S by definition. Furthermore, if two integral surfaces intersect at a point, then the characteristic curve through the point would lie on both the surfaces and hence they intersect along the whole characteristic through this common point. With this detailed discussion, we can now formulate the IVP as follows:

IVP: As in the case of the linear equation, the IVP or the Cauchy problem consists of finding a solution u of (3.11), in a neighborhood of a given an initial curve Γ_0, in some two-dimensional domain, represented in the parametric form:

$$\Gamma_0 = \{(x_0(s), y_0(s)) : 0 \leq s \leq 1\},$$

where x_0 and y_0 are C^1 functions defined on $[0, 1]$ and satisfying the initial condition $u(x_0(s), y_0(s)) = u_0(s)$, $0 \leq s \leq 1$ on Γ_0, where u_0 is a given C^1 function[2] on $[0, 1]$. Since the given data is insufficient to solve (3.17), we *lift* the initial curve Γ_0 to a space curve $\overline{\Gamma}_0$ by adjoining the initial values as

$$\overline{\Gamma}_0 = \{(x_0(s), y_0(s), u_0(s)) : 0 \leq s \leq 1\}. \tag{3.18}$$

We call $\overline{\Gamma}_0$ the *initial space curve* (Figure 3.10)

Theorem 3.7 (Existence and Uniqueness). Consider the PDE (3.11) and assume the functions a, b and c are C^1 functions of the variables x, y, u, in a domain of $\mathbb{R}^3$. Suppose that along the initial curve Γ_0, the initial values $u = u_0(s)$ are prescribed, where x_0, y_0, u_0 are continuously differentiable functions in the interval $[0, 1]$. Further assume that the *transversality condition* holds:

$$a(x_0(s), y_0(s), u_0(s))\frac{dy_0}{ds}(s) - b(x_0(s), y_0(s), u_0(s))\frac{dx_0}{ds}(s) \neq 0, \tag{3.19}$$

[2]If we replace the interval $[0, 1]$ by any other interval in $\mathbb{R}$, we only need to use a different parameterization for Γ_0.

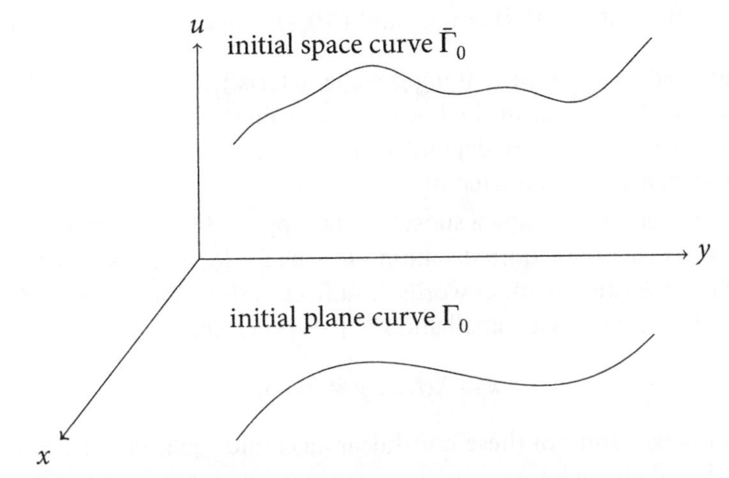

Figure 3.10 Initial plane and space curves

for all $0 \leq s \leq 1$. Then, there exists a unique solution $u(x, y)$ defined in some neighborhood of the initial curve Γ_0, which satisfies the PDE (3.11) and the initial condition

$$u(x_0(s), y_0(s)) = u_0(s),\ 0 \leq s \leq 1. \tag{3.20}$$

□

The theorem thus asserts that there is an integral surface through the space curve $\bar{\Gamma}_0$ in some neighborhood. It should also be noted that the transversality condition (3.19) only depends on the coefficients a, b, c and the Cauchy data (3.20).

Proof Consider the system of ODE

$$\frac{dx}{dt} = a(x, y, u),\ \frac{dy}{dt} = b(x, y, u),\ \frac{du}{dt} = c(x, y, u).$$

It follows then that through any fixed point $(x_0(s), y_0(s), u_0(s)) \equiv (x_0, y_0, z_0)$ for $s \in [0, 1]$, on the initial space curve, there is a unique family of characteristics

$$\begin{aligned} x &= x(x_0, y_0, u_0, t) \equiv X(t, s) \\ y &= y(x_0, y_0, u_0, t) \equiv Y(t, s) \\ u &= u(x_0, y_0, u_0, t) \equiv U(t, s) \end{aligned} \tag{3.21}$$

for small t in an interval around 0. Note that the variable s is a parameter on the initial curve and t, the parameter on the characteristics. Actually, the characteristics are the *orbits* of the above autonomous system of ODE. Therefore, the characteristics passing through distinct points on $\bar{\Gamma}_0$ never intersect. Note also that the functions X, Y and U are C^1 functions of s and t for $s \in [0, 1]$ and t small. These observations follow from ODE theory; see Chapter 2. Furthermore, note that

$$X(0,s) = x_0(s),\, Y(0,s) = y_0(s) \text{ and } U(0,s) = u(x_0(s), y_0(s)) = u_0(s)$$

and there is no need to solve for u along the characteristic curve separately, as u is also a part of the above ODE system. In the linear or semi-linear case (this is the case when the coefficients a and b in (3.11) do not depend on u), the characteristic equations are first solved and then u is solved along a characteristic.

Thus, the characteristics occupy a subset containing $\overline{\Gamma}_0$, in $\mathbb{R}^3$. In order to complete the proof, we now construct the required solution $u(x, y)$, for (x, y) in a small neighborhood of Γ_0, using the characteristics. In other words, it suffices to show that for given (x, y) in a small neighborhood of Γ_0, there exist t small and $s \in [0, 1]$ such that

$$x = X(t,s),\; y = Y(t,s).$$

Writing the solutions t and s of these non-linear algebraic equations as $t = t(x, y)$ and $s = s(x, y)$, we can then define u by

$$u(x,y) = U(t(x,y), s(x,y)).$$

Thus, there is a characteristic curve emanating from $(x_0(s), y_0(s), u_0(s))$ and reaching $(x, y, u(x, y))$ at time t. The procedure used to construct the characteristics proves that the function u defined as above, is the required solution.

Since the Jacobian of X, Y with respect s and t

$$\left.\frac{\partial(X,Y)}{\partial(s,t)}\right|_{t=0} = \left|\begin{matrix} \dfrac{\partial X}{\partial s} & \dfrac{\partial X}{\partial t} \\ \dfrac{\partial Y}{\partial s} & \dfrac{\partial Y}{\partial t} \end{matrix}\right|_{t=0} = b\,\frac{dx_0}{ds} - a\,\frac{dy_0}{ds}$$

does not vanish for all $s \in [0, 1]$, by the transversality condition, we can invoke the implicit function theorem to obtain t, s in terms of x, y, with t small and $s \in [0, 1]$. This completes the proof. □

The above proof actually gives a procedure to construct the solution explicitly, by solving a system of ODE and some non-linear algebraic equations. We illustrate this in the following example:

Example 3.8. Consider the PDE: $uu_x + u_y = 1$ with initial conditions $x = s, y = s$, $u = \frac{1}{2}s, 0 \le s \le 1$, that is, the initial value is given on a diagonal of the unit square in $\mathbb{R}^2$. Here $a = u, b = 1, c = 1$. Since

$$a(x_0(s), y_0(s), u_0(s))\frac{dy_0}{ds} - b(x_0(s), y_0(s), u_0(s))\frac{dx_0}{ds} = \frac{s}{2}\cdot 1 - 1\cdot 1 = \frac{s}{2} - 1$$

never vanishes for all $0 \leq s \leq 1$, the transversality condition (3.19) is satisfied. Solving the following ODE with the initial conditions:

$$\frac{dx}{dt} = u, \frac{dy}{dt} = 1, \frac{du}{dt} = 1, \; x(0,s) = s, y(0,s) = s, u(0,s) = \frac{s}{2},$$

we obtain the family of characteristic curves

$$x(t,s) = \frac{t^2}{2} + \frac{st}{2} + s, \; y(t,s) = t + s, \; u(t,s) = t + \frac{s}{2}$$

Now solving the first two algebraic equations for s and t in terms of x and y, we get $s = \dfrac{x - y^2/2}{1 - y/2}, t = \dfrac{y - x}{1 - y/2}$. Finally, the solution is given by

$$u(x,y) = \frac{2(y-x) + (x - y^2/2)}{2 - y},$$

for $0 \leq x, y \leq 1$. □

3.4 GENERAL FIRST-ORDER EQUATION IN TWO VARIABLES

We now consider the IVP for the general first order equation in two variables x, y:

$$F(x, y, u, u_x, u_y) = 0.$$

It is customary to use the notation $z = u(x,y), p = u_x = z_x$ and $q = u_y = z_y$. Thus, the general first order equation in two variables can be written as

$$F(x, y, z, p, q) = 0, \tag{3.22}$$

where we assume that the function F is a C^2 function in all its arguments in a domain in $\mathbb{R}^5$. The statement regarding the IVP remains the same as in the case of linear or quasilinear equations. The interesting and surprising fact is that, once again, we can reduce the study of IVP for (3.22) to that of a system of ODE. But the geometry is more involved now and requires much more intricate geometrical objects known as *strips* and *cones*.

As we are seeking real valued solutions (as a rule!), the equation

$$p^2 + q^2 + 1 = 0$$

does not have a solution. In the general case, there may be non-uniqueness as well. It is possible to get two integral surfaces passing through the same initial space curve. See Exercise 15. We are going to derive some sufficient conditions on F so that the IVP for (3.22) has a local solution. To proceed further, it is instructive to write the quasilinear equation (3.11) in the form (3.22) as

$$F(x, y, z, p, q) = a(x, y, z)p + b(x, y, z)q - c(x, y, z)$$

as we might be guided how to write the ODE system governing the characteristics in the present situation. First geometric interpretation of (3.22). Let $u = u(x, y)$ be a solution of (3.22) and (x_0, y_0, z_0) be a point in space. Consider an integral surface $z = u(x, y)$ through (x_0, y_0, z_0). The direction numbers $(p, q, -1)$ define the normal direction to the tangent plane at (x_0, y_0, z_0) of the integral surface. Then, equation (3.22) states that there is a relation

$$F(x_0, y_0, z_0, p, q) = 0 \tag{3.23}$$

between the direction numbers p and q. Thus, the differential equation (one relation with two numbers) will restrict its solutions to those surfaces having tangent planes belonging to a one-parameter family.

In general, this one-parameter family of planes will envelop a cone called the *Monge cone* (see Figure 3.11). Thus, the differential equation (3.22) describes a field of cones having the property that a surface will be an integral surface if and only if it is tangent to a cone at each point.

Remark 3.9. In the quasilinear case, the cone degenerates into a straight line whose direction is given by (a, b, c). □

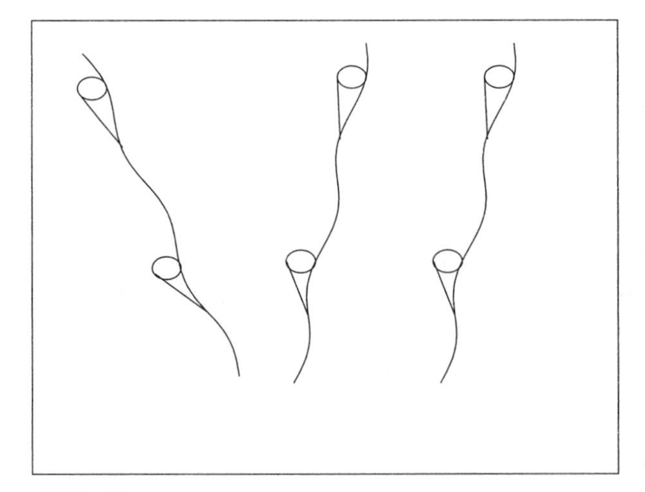

Figure 3.11 Monge cone

At each point, the surface will be tangent to a Monge cone. The line of contact of the surface and the cones define a field of directions on the surface called the *characteristic directions* and the integral curves of this field define a family of characteristic curves. The Monge cone at (x_0, y_0, z_0) is the envelope of the one-parameter family of planes (whose normal is $(p, q, -1)$) can be written as

$$z - z_0 = p(x - x_0) + q(y - y_0), \tag{3.24}$$

where p, q satisfy (3.23). By solving (3.23) for q in terms of p as $q = q(x_0, y_0, z_0, p)$, equation (3.24) can be written as

$$z - z_0 = p(x - x_0) + q(x_0, y_0, z_0, p)(y - y_0).$$

This is a one-parameter family of planes describing the Monge cone. By differentiating this equation with respect to p, we get

$$0 = (x - x_0) + (y - y_0)\frac{dq}{dp}.$$

On the other hand, a similar differentiation of (3.23) gives

$$0 = \frac{dF}{dp} = F_p + F_q\frac{dq}{dp}. \tag{3.25}$$

Eliminating $\frac{dq}{dp}$ from the above two equations, we obtain the following equations describing the Monge cone:

$$\left.\begin{aligned} &F(x_0, y_0, z_0, p, q) = 0 \\ &z - z_0 = p(x - x_0) + q(y - y_0) \\ &\frac{x - x_0}{F_p} = \frac{y - y_0}{F_q}. \end{aligned}\right\} \tag{3.26}$$

Given p and q, the last two equations in (3.26) give the line of contact between the tangent plane and the cone. These two equations can be written as

$$\frac{x - x_0}{F_p} = \frac{y - y_0}{F_q} = \frac{z - z_0}{pF_p + qF_q}. \tag{3.27}$$

Thus, on the given integral surface, at each point $p_0 = p(x_0, y_0), q_0 = q(x_0, y_0)$ are known, the tangent plane

$$z - z_0 = p_0(x - x_0) + q_0(y - y_0)$$

together with the third equation in (3.26) determines the line of contact with the Monge cone given by (3.27) or the characteristic direction. Thus, the characteristic curves are given as the solutions of the system of ODE

$$\frac{dx}{F_p} = \frac{dy}{F_q} = \frac{dz}{pF_p + qF_q}$$

or

$$\frac{dx}{dt} = F_p, \quad \frac{dy}{dt} = F_q, \quad \frac{dz}{dt} = pF_p + qF_q. \tag{3.28}$$

As there are five unknowns $x(t), y(t), z(t)$ and $p(t) = p(x(t), y(t)) = z_x(x(t), y(t))$, $q(t) = q(x(t), y(t)) = z_y(x(t), y(t))$, we need two more equations to make the system (3.28) a complete system. But along a characteristic curve on the given integral surface, we have

$$\left.\begin{aligned} \frac{dp}{dt} &= p_x\frac{dx}{dt} + p_y\frac{dy}{dt} = p_xF_p + p_yF_q \\ \frac{dq}{dt} &= q_xF_p + q_yF_q. \end{aligned}\right\} \tag{3.29}$$

As the functions p_x, p_y, q_x, q_y involve the second derivatives of the unknown function u, these are undesirable. In order to eliminate them from the above equations, we proceed as follows. By differentiating the given PDE (3.22) with respect to x and y, we obtain

$$F_x + F_zp + F_pp_x + F_qq_x = 0,$$
$$F_y + F_zq + F_pp_y + F_qq_y = 0.$$

Using these relations, equation (3.29) reduces to

$$\left.\begin{aligned} \frac{dp}{dt} &= -F_x - F_zp \\ \frac{dq}{dt} &= -F_y - F_zq, \end{aligned}\right\} \tag{3.30}$$

where we have used the relation $p_y = \dfrac{\partial^2 u}{\partial y \partial x} = q_x$. Thus, on the integral surface $z = u(x, y)$, we have a family of characteristic curves with co-ordinates $x(t), y(t), z(t)$ along with the numbers $p(t), q(t)$ and are given by the system (3.28), (3.30). Moreover along the characteristic curve, we have

$$\frac{dF}{dt} = F_x \frac{dx}{dt} + F_y \frac{dy}{dt} + F_z \frac{dz}{dt} + F_p \frac{dp}{dt} + F_q \frac{dq}{dt}$$

and we readily see that $\frac{dF}{dt} = 0$ using (3.28) and (3.30). Thus, F is constant along the characteristics. Thus, if $F = 0$ is satisfied at an initial point x_0, y_0, z_0, p_0, q_0 for $t = 0$, then (3.28), (3.30) will determine a unique solution $x(t), y(t), z(t), p(t), q(t)$ passing through this point and along which $F = 0$ will be satisfied for all t.

Hence, a solution can be interpreted using these five numbers and is called a *strip*. That is, the strip is a space curve $x = x(t), y = y(t), z = z(t)$ along with a family of tangent planes whose normal directions are $(p(t), q(t), -1)$. See Figure 3.12. For fixed t_0, the five numbers x_0, y_0, z_0, p_0, q_0 are said to define an *element* of the strip. That is, a point on the curve together with the tangent plane whose normal direction is $(p_0, q_0, -1)$. From (3.28), we get

$$\frac{dz}{dt}(t) = p(t)\frac{dx}{dt}(t) + q(t)\frac{dy}{dt}(t). \tag{3.31}$$

This is the condition that the planes are tangent to the curve and is called the *strip condition*. The strips that are solutions to (3.28), (3.30) are respectively called *characteristic strips* and the curves, *characteristic curves*.

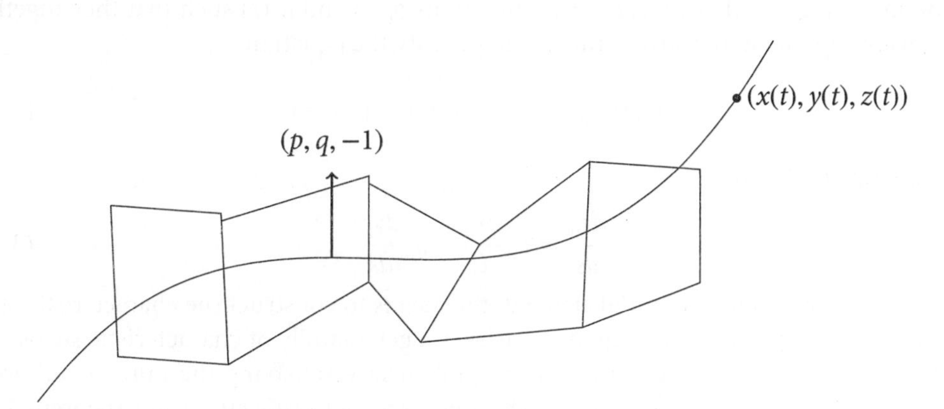

Figure 3.12 Characteristic strips

Furthermore, as in the case of quasilinear equations, if a characteristic strip has one element $(x_0, y_0, z_0, p_0, q_0)$ in common with an integral surface $z = u(x, y)$, then it lies completely on the surface. To see this, solve the ODE system

$$\frac{dx}{dt} = F_p(x, y, u(x, y), u_x(x, y), u_y(x, y))$$

$$\frac{dy}{dt} = F_q(x, y, u(x, y), u_x(x, y), u_y(x, y))$$

to obtain a curve $x = x(t), y = y(t)$ satisfying the initial conditions $x(0) = x_0, y(0) = y_0$. Then, by defining

$$z(t) = u(x(t), y(t)), \quad p(t) = u_x(x(t), y(t)), \quad q(t) = u_y(x(t), y(t)),$$

we see that

$$\frac{dz}{dt}(t) = p(t)F_p + q(t)F_q, \quad \frac{dp}{dt}(t) = -F_x - F_z u_x, \quad \frac{dq}{dt}(t) = -F_y - F_z u_y.$$

Therefore, $x(t), y(t), z(t), p(t), q(t)$ determine a characteristic strip and by definition, they lie on the surface. But, by uniqueness of the characteristic strip with the initial element x_0, y_0, z_0, p_0, q_0, this coincides with the given strip.

IVP: Suppose now an initial curve $\Gamma : x = x_0(s), y = y_0(s), z = z_0(s)$ be given. Furthermore, assume that we can assign functions $p_0(s)$ and $q_0(s)$ such that they together form an appropriate initial strip.[3] That is, they satisfy the equation

$$F(x_0(s), y_0(s), z_0(s), p_0(s), q_0(s)) = 0 \tag{3.32}$$

and the strip condition

$$\frac{dz_0}{ds} = p_0\frac{dx_0}{ds} + q_0\frac{dy_0}{ds}. \tag{3.33}$$

So, by fixing s and taking an initial element, the idea is to construct the characteristic strip starting from the given initial strip. As s varies, we get a family of characteristic strips (see Figure 3.13). This, in turn, will give the integral surface satisfying the initial conditions. Again, this requires the initial curve Γ_0 to be a non-characteristic curve; see Theorem 3.10, which we state now.

[3]The arbitrariness of picking the functions $p_0(s)$ and $q_0(s)$ may result in non-uniqueness. It is also a condition that is required on F in order to have a solution.

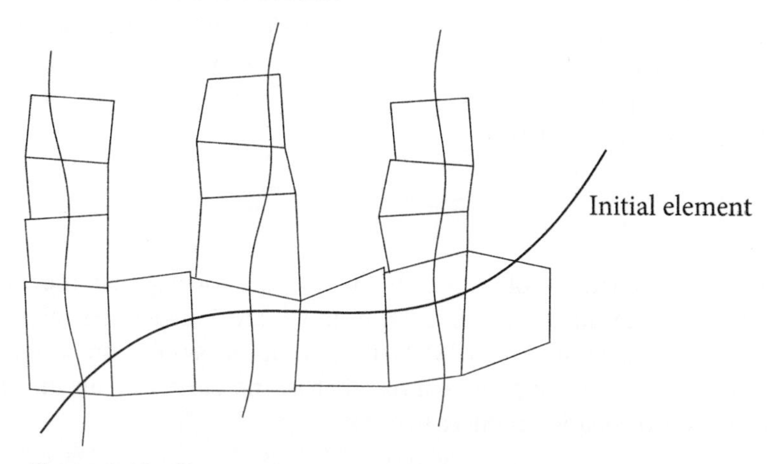

Figure 3.13 Characteristic and initial strips

Theorem 3.10. Let x_0, y_0, z_0, p_0, q_0 be as in (3.32) and (3.33). Assume x_0, y_0, z_0 have continuous derivatives and p_0, q_0 are continuously differentiable. Moreover, assume the following non-characteristic condition[4] along the initial curve:

$$\frac{dx_0}{ds}F_q(x_0, y_0, z_0, p_0, q_0) - \frac{dy_0}{ds}F_p(x_0, y_0, z_0, p_0, q_0) \neq 0.$$

Then, in some neighborhood of the initial curve there exists a solution $z = u(x, y)$ of (3.22) containing the initial strip. That is,

$$z(x_0(s), y_0(s)) = z_0(s), z_x(x_0(s), y_0(s)) = p_0(s), z_y(x_0(s), y_0(s)) = q_0(s).$$

□

Having given a detailed description, we omit the proof and ask the reader to fill in the details. In general, there is no uniqueness. See the following example:

Example 3.11. Consider the equation $p^2 + q^2 = 1$ with initial condition $u(x, y) = 0$ on the line $x + y = 1$. Then, there are two solutions given by $u(x, y) = \pm\frac{1}{\sqrt{2}}(x + y - 1)$. The details are left as an exercise. □

Remark 3.12. In the quasilinear case, we have

$$F(x, y, z, p, q) = a(x, y, z)p + b(x, y, z)q - c(x, y, z) = 0.$$

[4]This terminology coincides with earlier transversality condition in the quasilinear case.

Thus, $F_p = a, F_q = b$ and $pF_p + qF_q = c$ and hence the three equations in (3.28) are independent of p and q and they can be solved to determine the characteristic curves $(x(t), y(t), z(t))$. But, in the non-linear case we have to solve for $(x(t), y(t), z(t))$ together with the direction numbers $p(t)$ and $q(t)$. Moreover, in the quasilinear case, the Monge cone equations (3.27) reduce to

$$\frac{x - x_0}{a} = \frac{y - y_0}{b} = \frac{z - z_0}{c}.$$

These are the equations of a line in the space, showing that the Monge cone degenerates to a line. In the linear case, a and b are independent of z as well, so that the first two equations in (3.28) form a complete system for x and y, so that the characteristic curves are plane curves, that is, the curves lie on the (x, y) plane. Moreover, the third equation reduces to

$$\frac{du}{dt}(x(t), y(t)) = \frac{dz}{dt}(t) = c(x(t), y(t))$$

and is used to obtain u. □

3.5 FIRST-ORDER EQUATION IN SEVERAL VARIABLES

We can extend the theory of characteristics to the first-order equations in n variables, namely

$$F(x, z, p) = F(x_1, \dots x_n, z, p_1, \dots, p_n) = 0, \tag{3.34}$$

where $z = u(x_1, \dots, x_n)$ is the unknown and $p_i = \dfrac{\partial u}{\partial x_i}, 1 \leq i \leq n, p = (p_1, \dots, p_n)$. Here, the characteristic equations are given by

$$\frac{dx_i}{dt} = F_{p_i}, \frac{dz}{dt} = \sum_{i=1}^{n} p_i F_{p_i}, \frac{dp_i}{dt} = -F_{x_i} - p_i F_z, 1 \leq i \leq n$$

which is a system of $2n + 1$ equations in $2n + 1$ unknowns. We do not have the kind of geometry, that is available in the two variable case. Nevertheless, the theory can be developed in a similar fashion with new terminologies like *hyper-surface, characteristic* and *non-characteristic surfaces.* The Cauchy problem can be defined as follows: Given the PDE (3.34), a hyper-surface $S \subset \mathbb{R}^n$ and a real-valued function g defined on S, find u satisfying the PDE (3.34) in a neighborhood of S and satisfying $u = g$ on S. We now proceed step by step starting from the linear equations to quasilinear to fully non-linear equations to make the ideas clear.

A *hyper-surface* S is a subset of $\mathbb{R}^n$ given by the zero set of a C^1 function: $S = \{x \in \mathbb{R}^n : \psi(x) = 0\}$, where ψ is a C^1 function with $\nabla\psi(x) \neq 0$ for all $x \in S$. Thus the unit normal given by $\nu(x) = \dfrac{\nabla\psi(x)}{|\nabla\psi(x)|}$, $x \in S$ is well defined.

3.5.1 Linear First-Order Equation in Several Variables

A general first-order linear equation can be written as

$$Lu \equiv a \cdot \nabla u + a_0 u = f, \tag{3.35}$$

in an open subset Ω of $\mathbb{R}^n$, $a = a(x) = (a_1(x), \ldots, a_n(x))$ is a given vector field, $a \cdot \nabla u(x) = \sum_{i=1}^{n} a_i(x)\dfrac{\partial u}{\partial x_i}$ and $a_i, a_0, f \in C^1(\Omega)$, $1 \leq i \leq n$. The characteristic form of the operator L is defined by

$$\chi_L(x, \xi) = a(x) \cdot \xi,\ \xi \in \mathbb{R}^n, x \in \Omega$$

and the characteristic variety of L is defined as

$$char_x(L) = \{\xi \neq 0 : a(x) \cdot \xi = 0\}.$$

Thus, $char_x(L) \cup \{0\}$ is a hyper-surface (in fact, a hyper-plane), orthogonal to the vector field $a(x)$.

Definition 3.13. A hyper-surface S is called a *characteristic surface* for L at $x \in S$ if $\nu(x) \in char_x(L)$, where $\nu(x)$ is the normal to S at x. This is equivalent to the fact that the vector field a is tangent to S at x. A hyper-surface S is called *non-characteristic* for L if it is not characteristic for L at every point $x \in S$. □

Remark 3.14. The transversality condition given for the two variables in the previous section is same as the non-characteristic condition. □

With these terminologies, the analysis follows exactly as in the two-dimensional linear case. First introduce the characteristic curves $x(t) = (x_1(t), \ldots, x_n(t))$ in Ω given by the system of ODEs

$$\frac{dx}{dt} = a(x). \tag{3.36}$$

Along the characteristic curves, define a solution u of (3.35) as

$$\frac{du}{dt}(x(t)) = a(x(t)) \cdot \nabla u(x(t)) = f - a_0(x(t))u(x(t)). \tag{3.37}$$

Now solve (3.36) with the initial condition $x(0) = x_0$ for fixed $x_0 \in S$. A unique solution exists for small $|t|$. Then solve (3.37) with $u(0) = g(x_0)$. Since S is non-characteristic, $x(t) \notin S$, for $t \neq 0$. Furthermore, the inverse function theorem guarantees that for any x in a neighborhood of S, there is an $x_0 \in S$ such that the corresponding solution $x(t)$ starting at x_0 satisfies $x(t) = x$. Thus, all characteristic curves through S fill out a neighborhood of S and we have the following theorem:

Theorem 3.15 (Existence and Uniqueness). Let S be a non-characteristic hyper-surface for L of class C^1; a_i, a_0, f, g are real valued C^1 functions. Then, for a sufficiently small neighborhood Ω_0 of S in Ω, there is a unique solution $u \in C^1(\Omega_0)$ of (3.35) that satisfies $u = g$ on S. □

3.5.2 Quasilinear Equation in Several Variables

We next consider the quasilinear equation given by

$$Lu(x) \equiv a(x, u(x)) \cdot \nabla u(x) = a_0(x, u(x)) \tag{3.38}$$

for $x \in \Omega$, a domain in $\mathbb{R}^n$. Here $a(x, z) = (a_1(x, z), \ldots, a_n(x, z))$ and $a_0(x, z)$ not only depend on x, but also on the unknown $u(x)$. If u is a solution to (3.38), then $z = u$ is an integral surface that is sitting in $\mathbb{R}^{n+1}$. Thus, formally, introduce the curves $x(t)$ by

$$\frac{dx}{dt}(t) = a(x(t), u(x(t))). \tag{3.39}$$

Notice that the above ODE system is not complete as u is unknown.

IVP for (3.38)**:** Let $S \subset \Omega$ be a given hyper-surface and g be a function defined on S. Then, the IVP is to find a C^1 function defined in a neighborhood Ω_0 of S in Ω such that u satisfies (3.38) in Ω_0 with

$$u(x) = g(x), \text{ for all } x \in S. \tag{3.40}$$

We suppose that the hyper-surface is given in a parametric form $S = \{h(s) : s \in V\}$, where V is a connected open set in $\mathbb{R}^{n-1}$ and $h : V \subset \mathbb{R}^{n-1} \to \Omega$ is a (at least) C^1 function. Now, we lift the initial surface S to an initial surface (also called a *manifold* in $\mathbb{R}^{n+1}$) $\widetilde{S}$ by adjoining the initial values as

$$\widetilde{S} = \{(x, g(x)) : x \in S\} = \{(h(s), g(h(s))) : s \in V\}.$$

Note that $\widetilde{S}$ is an $n-1$ dimensional manifold in the space $\mathbb{R}^{n+1}$. The parameterization is possible because of the non-vanishing condition of the gradient describing the surface, at least locally.

Definition 3.16 (Non-characteristic)**.** The hyper-surface S as described above is called non-characteristic to the differential operator L if for any $s \in V$, we have

$$\det \begin{bmatrix} \frac{\partial h_1}{\partial s_1} \cdots & \cdots \frac{\partial h_1}{\partial s_{n-1}} & a_1(h(s), g(h(s))) \\ \cdots & \cdots & \cdots \quad \cdots \\ \cdots & \cdots & \cdots \quad \cdots \\ \frac{\partial h_n}{\partial s_1} \cdots & \cdots \frac{\partial h_n}{\partial s_{n-1}} & a_n(h(s), g(h(s))) \end{bmatrix} \neq 0. \tag{3.41}$$

□

Here $h = (h_1, \ldots, h_n)$ and $s = (s_1, \ldots, s_{n-1})$. Thus a solution to IVP (3.38), (3.40) is an integral surface $z = u(x)$ in $\mathbb{R}^{n+1}$ that passes through the lifted manifold $\widetilde{S}$. Now the system of ODE (3.39) can be completed by adjoining the equation for z. Since $z(t) = u(x(t))$, we get

$$\frac{dz}{dt}(t) = \sum_{i=1}^{n} \frac{\partial u}{\partial x_i} \frac{dx_i}{dt} = \sum_{i=1}^{n} a_i(x(t), u(x(t))) \frac{\partial u(x(t))}{\partial x_i} = a_0(x(t), u(x(t))). \tag{3.42}$$

The IVP (3.38), (3.40) can be solved as follows: For any point in $\widetilde{S}$ that is given by $(h(s), g(h(s)))$ for some $s \in V$, solve the complete ODE system (3.39), (3.42) with initial values $x(0) = h(s), z(0) = g(h(s))$. A unique solution $x = x(t; s), z(t) = z(t; s)$ for small t is guaranteed by the unique existence theorem for ODE systems (see Chapter 2). To complete the analysis, we need to invert an algebraic system as in two-dimensional case. That is, for $x \in \Omega_0$, a neighborhood close to S, we have to find a time t and $s \in V$ such that

$$x(t; s) = x. \tag{3.43}$$

The non-characteristic condition (3.41) together with inverse function theorem ensures the above claim. Let $t(x), s(x)$ solve (3.43), then $u(x) = z(t(x); s(x))$ solves the IVP (3.38), (3.40). Thus, we have the following theorem.

Theorem 3.17 (Existence and Uniqueness)**.** Consider the IVP (3.38), (3.40), where a, a_0, g are real-valued C^1 functions. Let S be a hyper-surface of class C^1 in $\mathbb{R}^n$ that satisfies the condition (3.41). Then, for a sufficiently small neighborhood Ω_0 of S, there is a unique solution $u \in C^1(\Omega_0)$ of (3.38), (3.40). □

3.5.3 General Non-linear Equation in Several Variables

Recall that the general first-order non-linear equation in n variables is given by

$$F(x, z, p) = F(x_1, \ldots, x_n, z, p_1, \ldots, p_n) = 0, \tag{3.44}$$

where $z = u$ is the unknown function and $p_i = \dfrac{\partial u}{\partial x_i}, 1 \leq i \leq n$. In the linear case, we have $F(x, z, p) = a(x) \cdot p + a_0(x)z - f(x)$ and in the quasilinear case $F(x, z, p) = a(x, z) \cdot p - a_0(x, z)$. Thus, in the either of the cases $a = \nabla_p F$. This together with the analysis of the general case in two variables, motivates us to define the characteristic curves $x(t)$ as the integral curves of the vector field $\nabla_p F$. That is, define $x = x(t)$ as

$$\frac{dx}{dt} = \nabla_p F. \tag{3.45}$$

Now, adjoin the equation for z as

$$\frac{dz}{dt} = \sum_{i=1}^{n} \frac{\partial u}{\partial x_i} \frac{dx_i}{dt} = p \cdot \nabla_p F. \tag{3.46}$$

Notice that in the linear case $\nabla_p F = a(x)$ does not depend on the unknown u and hence (3.45) is a complete system. In the quasilinear case $\nabla_p F = a(x, z)$ and thus (3.45) together with (3.46) is a complete system. In the general case $\nabla_p F$ and $p \cdot \nabla_p F$ may depend not only on the unknown z, but also on the n derivatives $p = \nabla u$. Hence, we need to derive n equations for $\dfrac{dp_i}{dt}$. Now, $p_i(t) = p_i(x(t))$ and compute

$$\frac{dp_i}{dt} = \sum_{i=1}^{n} \frac{\partial p_i}{\partial x_j} \frac{dx_j}{dt} = \sum_{i=1}^{n} \frac{\partial^2 u}{\partial x_j \partial x_i} \frac{\partial F}{\partial p_j} = \sum_{i=1}^{n} \frac{\partial p_j}{\partial x_i} \frac{\partial F}{\partial p_j}. \tag{3.47}$$

On the right hand side, we have the undesired second derivatives $\dfrac{\partial p_j}{\partial x_i}$. We need to eliminate them. Differentiating (3.44) with respect to x_i, we get

$$\frac{\partial F}{\partial x_i} + \frac{\partial F}{\partial z} \frac{\partial z}{\partial x_i} + \sum_{i=1}^{n} \frac{\partial F}{\partial p_j} \frac{\partial p_j}{\partial x_i} = 0.$$

Thus, we arrive at

$$\frac{dp_i}{dt} = -\frac{\partial F}{\partial x_i} - p_i \frac{\partial F}{\partial z} \tag{3.48}$$

for $1 \leq i \leq n$. Hence, we have a system of $2n+1$ equations given by (3.45), (3.46), (3.48) for $x(t), z(t), p(t)$. Thus, we are solving not only for the unknown, but for the derivatives $\frac{\partial u}{\partial x_i}$ as well, exactly what we have done in the two-dimensional case.

The setup is similar for IVP as in the quasilinear case like defining S where the initial values are defined and then lift it to the initial surface $\widetilde{S}$. Now important issue is the identification of n initial conditions for $p(0)$ for the system (3.48). This is done as follows: Recall $x(0) = x(0; s) = h(s), z(0) = z(0; s) = g(h(s))$. Further on S, we have

$$\frac{\partial z}{\partial s_i} = \sum_{i=1}^{n} \frac{\partial z}{\partial x_j} \frac{\partial (h_j(s))}{\partial s_i} = \sum_{i=1}^{n} p_j \frac{\partial h_j}{\partial s_i} \tag{3.49}$$

for $1 \leq i \leq n-1$ and

$$F(h(s), g(h(s)), p(h(s))) = 0. \tag{3.50}$$

The above system of n equations determines the initial conditions for p on S and should be thought as a requirement on F in order to have a solution. This may also lead to non-uniqueness of solutions as there may be more than one solution p to (3.50) as in PDE $p^2 + q^2 - 1 = 0$; lack of existence as in PDE $p^2 + q^2 + 1 = 0$

Unfortunately, these are non-linear equations that may or may not be solvable uniquely. In the quasilinear case, it was not necessary as (3.45), (3.46) was a complete system. Nevertheless, (3.49), (3.50) is indeed solvable uniquely in the quasilinear case as they are linear and S is non-characteristic. However, if (3.49), (3.50) is solvable for $p_1(0), \ldots p_n(0)$, we have $2n+1$ system of ODEs for $x(t), z(t), p(t)$ together with initial conditions $x(0) = x(0; s), z(0) = z(0; s), p(0) = p(0; s)$ for any $s \in V$, that is for any point on $\widetilde{S}$. Now the procedure is similar to the discussion as in the previous section. Of course S has to be non-characteristic to apply inverse function theorem in the sense that

$$\det \begin{bmatrix} \frac{\partial h_1}{\partial s_1} \cdots & \cdots \frac{\partial h_1}{\partial s_{n-1}} & \frac{\partial F}{\partial p_1}(h(s), g(h(s)), p(h(s))) \\ & \cdots \quad \cdots & \cdots \quad \cdots \\ & \cdots \quad \cdots & \cdots \quad \cdots \\ \frac{\partial h_n}{\partial s_1} \cdots & \cdots \frac{\partial h_n}{\partial s_{n-1}} & \frac{\partial F}{\partial p_n}(h(s), g(h(s)), p(h(s))) \end{bmatrix} \neq 0. \tag{3.51}$$

We can also state a similar theorem as in the quasilinear case, of course with the assumption of solvability of (3.49), (3.50). However, we confine ourselves to an example, which clearly explains a general procedure.

Example 3.18. Consider the Cauchy problem for the PDE

$$u_{x_1} - \left(u_{x_2}^2 + u_{x_3}^2\right)^{1/2} = 0.$$

Let x_{10} be a fixed real number and consider the initial condition

$$u(x_{10}, x_2, x_3) = u_0(x_2^2 + x_3^2),\ x_2, x_3 \in \mathbb{R},$$

where $u_0 \in C^1(0, \infty)$ with $u_0' > 0$. □

Here $n = 3$. Though we can work with the variables x_2 and x_3 to describe the initial manifold, we prefer to work with the notations introduced in the text. The initial manifold S in the present case is the affine plane $x_1 = x_{10}$ in the three-dimensional space that is described by the C^1 functions:

$$x_{10}(s_1, s_2) \equiv x_{10}, x_{20}(s_1, s_2) = s_1,\ x_{30}(s_1, s_2) = s_2$$

and s_1, s_2 vary over $\mathbb{R}$. The initial condition satisfied on S takes the form

$$u(x_{10}, s_1, s_2) = u_0(s_1^2 + s_2^2).$$

We first try to find the functions p_{i0}, $i = 1, 2, 3$ satisfying the conditions (3.36). We find that

$$\frac{\partial u_0}{\partial s_1} = p_{20},\ \frac{\partial u_0}{\partial s_2} = p_{30}$$

and

$$p_{10} = \left(p_{20}^2 + p_{30}^2\right)^{1/2}.$$

Therefore,

$$p_{20} = 2s_1 u_0',\ p_{30} = 2s_2 u_0' \text{ and } p_{10} = 2\left(s_1^2 + s_2^2\right)^{1/2} u_0'.$$

The characteristic equations (3.38) become

$$\begin{aligned}
&\tfrac{dx_1}{dt} = 1,\ \tfrac{dx_i}{dt} = -\tfrac{p_i}{p_1},\ i = 2, 3\\
&\tfrac{dp_i}{dt} = 0,\ i = 1, 2, 3 \text{ and } \tfrac{dz}{dt} = 0.
\end{aligned}$$

Solving this system of ODE with appropriate initial conditions, we obtain that

$$x_1 = t + x_{10}, x_2 = -\frac{s_1 t}{\sqrt{s_1^2 + s_2^2}} + s_1 \text{ and } x_3 = -\frac{s_2 t}{\sqrt{s_1^2 + s_2^2}} + s_2$$

with $z = u_0(s_1^2 + s_2^2)$. Our next task is to express t, s_1 and s_2 from the first three non-linear algebraic equations in terms of x_1, x_2, x_3. Trivially, $t = x_1 - x_{10}$. To solve for s_1, s_2, we use polar co-ordinates: $s_1 = r\cos\theta$, $s_2 = r\sin\theta$ with $r^2 = s_1^2 + s_2^2$. Therefore,

$$x_2 = (r - (x_1 - x_{10}))\cos\theta \text{ and } x_3 = (r - (x_1 - x_{10}))\sin\theta.$$

Thus,

$$\tan\theta = \tfrac{x_2}{x_1},\ r = \tfrac{x_2}{\cos\theta} + x_1 - x_{10} = \tfrac{x_3}{\sin\theta} + x_1 - x_{10},$$
$$s_1 = r\cos\theta = x_2 + (x_1 - x_{10})\cos\theta,$$
$$s_2 = r\sin\theta = x_3 + (x_1 - x_{10})\sin\theta.$$

Furthermore, it is not difficult to see that

$$\cos\theta = \frac{x_2}{\sqrt{x_2^2 + x_3^2}} \text{ and } \sin\theta = \frac{x_3}{\sqrt{x_2^2 + x_3^2}}.$$

Hence,

$$s_1 = x_2\left(1 + \frac{x_1 - x_{10}}{\sqrt{x_2^2 + x_3^2}}\right),\ s_2 = x_3\left(1 + \frac{x_1 - x_{10}}{\sqrt{x_2^2 + x_3^2}}\right).$$

Substituting these in the expression for z, we finally obtain the required solution in the explicit form:

$$u(x_1, x_2, x_3) = u_0(s_1^2 + s_2^2) = u_0(\xi) \text{ with } \xi = \left[x_1 - x_{10} + (x_2^2 + x_3^2)^{1/2}\right]^2.$$

Observe that this is a solution of the given PDE in $\mathbb{R}^3$ except for the line $x_2 = 0$ and $x_3 = 0$, that is, the x_1-axis.

3.6 HAMILTON–JACOBI EQUATION

In the previous sections of this chapter, we have seen a class of first-order equations that are quasilinear, known as conservation laws. We also saw that quite often, these equations do not admit smooth (classical) solutions. In fact, the physical solutions can develop discontinuities over time. In this section, we introduce the reader to another important class of equations, namely the Hamilton–Jacobi equation (HJE) that is a classical subject of more than 200 years old. These equations are relevant both in classical and non-classical way. Furthermore, they have very important applications. In fact, there is a more general equation in the modern era, namely, the *Hamilton–Jacobi–Bellman* (HJB) equation. These equations are non-linear and even simple equations do not admit classical solutions. Consider the following examples:

Example 3.19. Consider the ODE

$$|\dot{u}|^2 - 1 = 0 \text{ in } (-1, 1),$$

with the boundary conditions

$$u(-1) = u(1) = 0.$$

This equation has no classical solution $u \in C^1(-1,1) \cap C[-1,1]$. For, by continuity either $\dot{u}(t) = 1$ for all t or $\dot{u}(t) = -1$ for all t. In either case, both the boundary conditions cannot be satisfied.

However, the functions $u(x) = 1 - |x|$ and $u(x) = |x| - 1$ satisfy the equation except at the origin, along with the boundary conditions. Moreover, u is Lipschitz continuous as well. In fact, using the function $|x|$, it is possible to construct infinitely many such solutions satisfying the equation along with the boundary conditions, except at finitely many points. □

Example 3.20. Consider the one-dimensional Hamilton–Jacobi equation

$$u_t + u_x^2 = 0, \quad t > 0, x \in \mathbb{R}, u(x,0) = 0.$$

It is easy to see that the function

$$v(x,t) = \begin{cases} 0, & 0 < t < |x| \\ -t + |x|, & |x| \geq t \end{cases}$$

is continuous and satisfies the initial condition. It is also smooth and satisfies the equation off the lines $x = 0$ and $t = |x|$. Thus, v is completely distinct from the trivial classical solution $u \equiv 0$. □

We consider the first-order HJE given by

$$u_t + H(x, Du) = 0. \tag{3.52}$$

More generally, the term $H(x, Du)$ may be replaced by $H(t, x, u, Du)$, to include the t and u variables as well. The function H is referred to as a *Hamiltonian function* or simply a *Hamiltonian*, which is a real-valued smooth function. Our presentation here is deliberately vague, essentially because even the explanation of various terminologies involved is difficult and we will consider this equation like the conservation law in another chapter and present more details. However, because of the importance of the HJE in numerous applications, including optimal control problems, where it is referred to as HJB equation, we have included this brief introduction here to show the importance of the method of characteristics.

In the HJE, $u = u(t, x)$ is the unknown function of $n + 1$ variables, $t > 0$ and $x \in \mathbb{R}^n$, and $Du = (u_{x_1}, \dots u_{x_n})$ is the spatial gradient vector of u. Time-dependent HJE, that is $H = H(t, x, Du)$ has also been studied in the literature. This equation arises in the classical *calculus of variations*. In a smooth setup, the minimum value of an associated cost functional or energy functional, namely the *value function* is known to satisfy the HJ or HJB equation. We elaborate on this point a little later. Equation (3.52) is a first-order PDE in $n+1$ dimensions, and thus, its characteristic equations form a system of $2n + 3$ equations. But, due to the special structure of (3.52), the system of characteristic equations can be decoupled to get $2n$ equations known as *Hamilton's ODE System*

$$\left.\begin{aligned} \frac{dx}{dt} &= D_pH(x, p) \\ \frac{dp}{dt} &= -D_xH(x, p) \end{aligned}\right\} \tag{3.53}$$

together with an ODE for the unknown u. Furthermore, the variable t itself can be used as a parameter. For a particular H, the above system is indeed the Hamiltonian formalism corresponding to the Newton's law of motion in classical mechanics. We will see this soon. First, we will see an example from calculus of variations.

Example 3.21. Consider the rectangular region $Q = [0, 1] \times [-1, 1]$. For $t \in [0, 1]$, let $x : [t, 1] \to \mathbb{R}$ be a Lipschitz function. Thus, the points $(s, x(s))$ trace a curve as s varies over $[t, 1]$, starting at $(t, x(t))$. Let τ be the *exit time* of $x(s)$ from Q, that is,

$$\tau = \begin{cases} 1 \text{ if } x(s) \in (-1, 1) \text{ for all } s \in [t, 1] \\ T \text{ if } x(T) = \pm 1 \text{ for some } T \in [t, 1] \text{ and } x(s) \in (-1, 1) \text{ for all } s \in (t, T). \end{cases}$$

More precisely, $\tau = 1$ if the curve lies in Q for all $s \in [t, 1]$, otherwise it is the first time the curve reaches the boundary of Q (see Figure 3.14). □

Let X denote the class of Lipschitz functions described above and consider the minimization problem from calculus of variations:

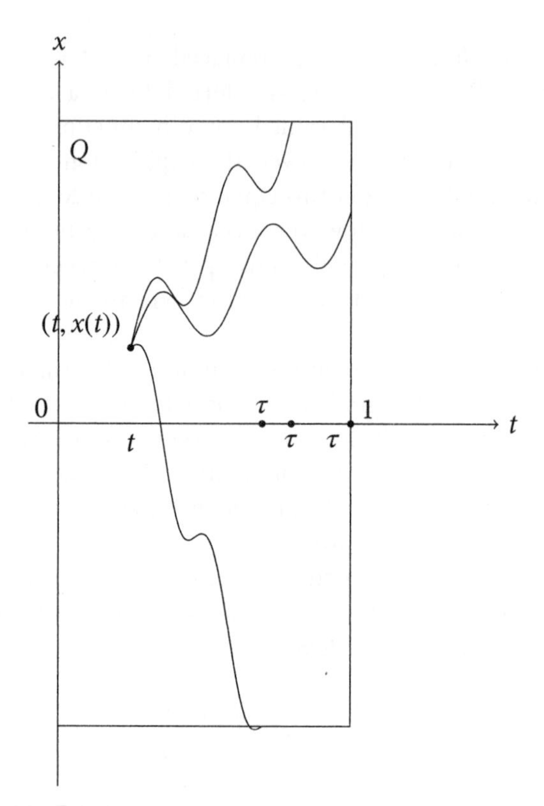

Figure 3.14 Exit times

$$\min \int_t^\tau \left(1 + \frac{1}{4}\dot{x}(s)^2\right) ds, \quad t \in [0, 1], \tag{3.54}$$

where the minimization sought over the space X and τ is the exit time of the function $x(s)$ described above. Define $L(v) \equiv L(t, x, v) = 1 + \frac{1}{4}v^2$, where v is any scalar.

The *value function* $V(t, x)$, for $t \in [0, 1]$ and $x \in [-1, 1]$, is defined as

$$V(t, x) = \min \left\{ \int_t^\tau L(t, x(s), \dot{x}(s))\, ds \right\},$$

where the minimum is taken over all $x(\cdot) \in X$ with the initial value $x(t) = x$. Given any $x(\cdot) \in X$, $x(t) = x$, consider the linear function (line segment) $\tilde{x}(\cdot)$ defined by

$$\tilde{x}(s) = x + v(s - t),$$

where $v = \dfrac{1}{\tau - t}\displaystyle\int_t^\tau \dot{x}(s)\,ds$ with τ as the exit time of $x(\cdot)$. It is then easy to verify that $\tilde{x}(\cdot) \in X$, $\tilde{x}(t) = x$ and τ is also the exit time of $\tilde{x}(\cdot)$. Also, $\dot{\tilde{x}}(s) = v$. Moreover, it is also easy to see that

$$\int_t^\tau L(\dot{\tilde{x}}(s))\,ds \leq \int_t^\tau L(\dot{x}(s))\,ds.$$

Thus, for the definition of the value function V, it suffices to consider only linear functions for minimization. Therefore, V is given by

$$V(t,x) = \min_v\{(\tau - t)L(v)\},$$

where v varies over all the real numbers. An elementary computation shows that

$$v^* = \begin{cases} 2 \text{ if } x \geq t \\ 0 \text{ if } |x| < t \\ -2 \text{ if } x \leq -t \end{cases},$$

gives rise to a minimizer for V, equivalently, the corresponding $\tilde{x}$ is a minimizer for (3.54) and is known as optimal solution. Furthermore, we have

$$V(t,x) = \begin{cases} 1 - |x| \text{ if } |x| \geq t \\ 1 - t \quad \text{if } |x| \leq t \end{cases}$$

Note that V is differentiable everywhere in the interior of Q, except on the lines $t = |x|$ and V satisfies the following HJE:

$$-V_t(t,x) + (V_x(t,x))^2 - 1 = 0, \tag{3.55}$$

for all $(t,x) \in (0,1) \times (-1,1)$ except when $t = |x|$ and also satisfies conditions

$$V(t,1) = V(t,-1) = 0,\ t \in [0,1] \tag{3.56}$$

and

$$V(1,x) = 0,\ x \in [-1,1]. \tag{3.57}$$

Thus, V is a Lipschitz solution of (3.55), (3.56), (3.57). Thus, if we look for a solution of the above HJE in the classical sense, we will not get the actual physical solution as it is not differentiable. Hence, we need to interpret the solution in a generalized sense. One possible way is to look for a Lipschitz continuous function as a solution, requiring that the equation need be satisfied only a.e. But, in this enlarged class of solutions, the uniqueness of the solution may be lost. To achieve uniqueness, we may have to impose additional condition(s). For example, there are infinitely many Lipschitz continuous functions, which are solutions to (3.55), (3.56), (3.57). To see this, define

$$W_k(t,x) = \min\{h_k(x), 1-t\},$$

where $h_k(x) = \frac{1}{2k+1} - \left|x - \frac{2i}{2k+1}\right|$, if $x \in \left[\frac{2i-1}{2k+1}, \frac{2i+1}{2k+1}\right]$. It is easy to see that W_k satisfies (3.55), (3.56), (3.57) at all the points of differentiability. Thus, we need a more robust notion of the solution and this is provided by the concept of *viscosity solution* whose discussion is beyond the scope of this book. Nevertheless, we will discuss more on the Lipschitz solution of HJE in Chapter 4.

We now briefly discuss the general problem of calculus of variations and the corresponding value function associated with the minimization problem gives rise to HJE.

Calculus of Variations: Given a *Lagrangian* $L = L(x,q)$, $x, q \in \mathbb{R}^n$, introduce the functional

$$J(w) := \int_0^T L(w(s), \dot{w}(s))\, ds, \tag{3.58}$$

where $w \in \mathcal{A}$, the space of admissible trajectories. For instance, we may take

$$\mathcal{A} = \left\{w \in C^2([0,T];\mathbb{R}^n) : w(0) = a,\ w(T) = b\right\}.$$

Here $T > 0$ is the terminal time, a and b are the given vectors in $\mathbb{R}^n$. The basic problem is to find a curve $x \in \mathcal{A}$, which solves the minimization problem

$$J(x) = \min_{w \in \mathcal{A}} J(w). \tag{3.59}$$

It is possible to extend the elementary result of the finite-dimensional optimization that extremal points occur at critical points, to the present minimization problem that is infinite dimensional, since we are working for trajectories belonging to the infinite dimensional space $\mathcal{A}$. If $x \in \mathcal{A}$ is a solution to the problem (3.59), known as *optimal trajectory*, then, x satisfies the *Euler–Lagrange(EL)* equations

$$\frac{d}{ds}\left(D_q L(x(s), \dot{x}(s))\right) + D_x L(x(s), \dot{x}(s)) = 0. \tag{3.60}$$

This is a system of n second-order equations. Thus, any minimizer $x \in \mathcal{A}$ of (3.59) solves EL equation, but the converse need not be true. A solution to the EL equations is called a *critical point* of the functional J.

Since EL is a system of n second-order equations, we can convert it into a system of $2n$ first-order system as follows: Introduce, $p(s) = D_q L(x(s), \dot{x}(s))$, called the *generalized momentum* corresponding to the position $x(\cdot)$ and velocity $\dot{x}(\cdot)$. These terminologies come from the classical mechanics. We need an important hypothesis to obtain the Hamiltonian formalism (ODE system), which we have obtained earlier with appropriate H.

Assumption: Suppose that, for given $x, p \in \mathbb{R}^n$, the equation $D_q L(x, q) = p$ can be uniquely and smoothly solvable for q as $q = q(x, p)$.

Now introduce the Hamiltonian

$$H(x, p) = p \cdot q(x, p) + L(x, q(x, p)). \tag{3.61}$$

Theorem 3.22. Let x solve the EL equations and p be the corresponding generalized momentum. Then, under the above assumption, the functions x and p satisfy the Hamilton's ODE system (3.53) with the Hamiltonian defined as in (3.61). Furthermore, the mapping $s \mapsto H(x(s), p(s))$ is constant. □

The main aim here was to show the application of characteristic curves in physical problems. In the next chapter, we present some more details on HJE including the Hopf–Lax formula. We remark that we can associate a value function and a Hamiltonian corresponding to the minimization problem and we expect the value function to satisfy a PDE associated to the Hamiltonian, which is highly non-linear. The examples suggest that we have to find a solution outside the realm of smooth solutions. The analysis for HJB is not a direct generalization of classical HJ equations and took almost two centuries to come up with a theory. Two theories emerged after 1950s; one due to Pontryagin in USSR and the other due to Bellman in the United States. The former one is based Hamiltonian ODE whereas the Bellman's theory is based on PDE. We now give the classical example from the Newtonian mechanics.

Example 3.23. Consider the motion of a particle of mass m under the influence of a force field f given by a potential V, that is $f = \nabla V$. Define the Lagrangian $L(x, q) = \frac{m}{2}|q|^2 - V(x)$, the difference in kinetic energy and potential energy. Then, the corresponding EL equations describe the Newton's second law of motion, namely $m\ddot{x}(s) = f(x(s)) = \nabla V(s)$. Here, the assumption that $p = D_q L(x, q) = mq$ holds trivially. It is easy to see that the Hamiltonian is the total energy $H(x, p) = \frac{1}{2m}|p|^2 + V(x)$. In Newton's theory, the position and the velocity formulation is given (i.e. Lagrangian formulation), whereas position and momentum (Hamiltonian formulation) are the unknowns given through the Hamiltonian system. It is possible to go from one formulation to the other in classical mechanics, whereas with general Lagrangian, this may not be possible due to the solvability issue as in the assumption. □

3.7 NOTES

The main aim of this chapter was to introduce the method characteristics which, in turn, imply the local solvability. The theory is very geometrical and general for first-order equations, which is demonstrated in this chapter. Such a general theory is not available for higher-order equations. The method can quite often be used to obtain explicit solutions. This statement has to be taken with a pinch of salt as we have seen very simple looking examples of first-order equations from conservation laws and HJE. We have also seen that we will not be able to reside in the comfortable zone of smooth functions, and move toward modern theories to understand such equations. Two of them are the theory of distributions and solutions in the sense of viscosity. These theories are beyond the scope of this first book on PDE. However, in the chapters that follow, we do indicate the notions of weak solutions to motivate the reader for future study.

For the present chapter, the reader can refer to Evans (1998), John (1978), Renardy and Rogers (2004), Prasad and Ravindran (1996), Smoller (1994), and Lax (1973) for further discussion and more details.

3.8 EXERCISES

1. Describe the characteristic curves of the PDE

$$(x+2)u_x + 2yu_y = 2u$$

in the (x, y)-plane and sketch few of them. Write the ODE for u along a characteristic curve with x as the parameter and then, solve the PDE with the initial condition $u(-1, y) = \sqrt{|y|}$.

2. Consider the PDE

$$xu_x + yu_y = 2u,$$

in the region $x > 0, y > 0$. Determine the characteristic curves. Solve the equation in the following domains, with the initial conditions given on the prescribed initial curve

   a. $u = 1$ on the hyperbola $xy = 1$; domain $xy > 1$.
   b. $u = 1$ on the circle $x^2 + y^2 = 1$; domain $x^2 + y^2 > 1$.

   Is it possible to solve the equation if the initial data is prescribed on the initial curve $y = e^x$? Justify.

3. Solve the IVP $u_t + u_x = 1$ with the initial condition $u = \phi(x)$ when $y = 2x$.
4. Find the general solution of $au_x + bu_y + cu = 0$, where a, b, c are constants by the method of characteristics.
5. Discuss the PDE $(x+1)^2 u_x + (y-1)^2 u_y = (x+y)u$ with the initial condition $u(x, 0) = -1 - x$ for $-1 < x < \infty$.

6. Write down the characteristic equations, verify the transversality condition and solve the following PDEs:

   a. $u_y = u_x^3,\ u(x, 0) = 2x^{3/2},\ x > 0.$
   b. $xu_x + yu_y = \frac{1}{2}\left(u_x^2 + u_y^2\right),\ u(x, 0) = \frac{1}{2}(1 - x^2).$

7. In the following, sketch the characteristic curves, the initial curve, verify the transversality condition and solve the Cauchy problem:

   a. $xu_x + yu_y = ku,\ x \in \mathbb{R},\ y \geq \alpha > 0;\quad u(x, \alpha) = F(x)$, where k, α are fixed constants and F is a given smooth function.
   b. $(x + 2)u_x + 2yu_y = \alpha u;\quad u(-1, y) = \sqrt{y}.$
   c. $yu_x - xu_y = 0;\quad u(x, 0) = x^2.$
   d. $x^2u_x - y^2u_y = 0;\quad u(1, y) = F(y).$

8. Find the characteristic curves of the following PDE:

   a. $(x^2 - y^2 + 1)u_x + 2xyu_y = 0.$
   b. $2xyu_x - (x^2 + y^2)u_y = 0.$

9. Solve the following IVP:

   a. $u_t + (x \cos t)u_x = 0,\ u(x, 0) = \frac{1}{1 + x^2} x \in \mathbb{R}, t > 0.$
   b. $u_t + x^2u_x = 0, u(x, 0) = \phi(x),\ x \in \mathbb{R}, t > 0.$
   c. $u_t + \frac{1}{1 + |x|}u_x = 0, u(x, 0) = \phi(x),\ x \in \mathbb{R}, t > 0.$
   d. $u_t + (x + t)u_x + t(x + 1)u = 0, u(x, 0) = \phi(x),\ x \in \mathbb{R}, t > 0.$

10. Solve the following quasilinear problems and verify the transversality condition in each case:

    a. $uu_x + u_y = 0,\quad u(x, 0) = x.$
    b. $uu_x + u_y = 1,\quad u(x, x) = x/2,\ x \in (0, 1].$

11. Sketch the characteristic curves of $uu_x + u_y = 0$ with the following initial conditions:[5]

    a. $u(x, 0) = \begin{cases} 0 & \text{if } x < 0 \\ 1 & \text{if } x \geq 0 \end{cases}$

    b. $u(x, 0) = \begin{cases} 1 & \text{if } x < 0 \\ 0 & \text{if } x \geq 0 \end{cases}$

    c. $u(x, 0) = \begin{cases} 0 & \text{if } x < 0 \\ 1 & \text{if } x \geq 1 \end{cases}$ and $u(x, 0)$ is smooth and increasing.

[5]The initial conditions given here are not continuous functions. Nevertheless, there is no difficulty in sketching the characteristics in the smooth regions. A little care is needed only at the point of discontinuity.

12. Find the integral surface of the equation $xu_x^2 + yu_y = u$ passing through the line $y = 1$, $x + z = 0$.

13. Solve the IVP for the Burgers' equation $u_t + uu_x = 0$, with the initial condition $u(x, 0) = x^3$, in the upper half plane $t > 0$. (Hint: Use (3.13). The mapping $x_0 \to tx_0^3 + x_0$, $t \geq 0$, is one-one from $\mathbb{R}$ onto $\mathbb{R}$. Thus, for each $x \in \mathbb{R}$, there is a unique x_0 such that $tx_0^3 + x_0 = x$. For $t > 0$, $x_0 = q_1^{1/3} + q_2^{1/3}$, where $q_1 = \dfrac{x\sqrt{t} + \sqrt{x^2t + 4/27}}{2t\sqrt{t}}$, $q_2 = \dfrac{x\sqrt{t} - \sqrt{x^2t + 4/27}}{2t\sqrt{t}}$. The solution is given by $u(x, t) = x_0^3 = \dfrac{x - x_0}{t}$, $x \in \mathbb{R}, t > 0$.)

14. Solve the IVP

$$u_t + u^2u_x = 0, u(x, 0) = x, \ x \in \mathbb{R}, t > 0.$$

15. Consider the equation $p^2 + q^2 = 1$ with initial condition $u(x, y) = 0$ on the line $x + y = 1$. Show that there are two solutions given by $u(x, y) = \pm\frac{1}{\sqrt{2}}(x + y - 1)$ using the method of characteristics.

16. Consider the PDE $x^2u_x + y^2u_y = (x + y)u$. Show that there are general solutions of the following forms:

(i) $F\left(\dfrac{x - y}{u}, \dfrac{xy}{u}\right) = 0$,

(ii) $u = xyf\left(\dfrac{x - y}{u}\right)$,

(iii) $u = xyg\left(\dfrac{x - y}{xy}\right)$,

where F, f, g are arbitrary functions.

17. Obtain the decoupled system (3.53).

18. Derive the Euler–Lagrange equations (3.60).

CHAPTER 4

Hamilton–Jacobi Equation

In Chapter 3, we have briefly introduced the Hamilton–Jacobi equation (HJE)

$$u_t(x,t) + H(x, u(x,t), Du(x,t)) = 0$$

as an example of a first-order equation to derive the characteristic curves, which form the well-known system of Hamilton's ordinary differential equations (ODE). We also have seen there an example of a minimization problem, where the minimum value (known as the *value function*) need not be differentiable at some points. However, at the points where it is differentiable, the value function does satisfy HJE. Hence, it is essential to look for functions *outside* the class of smooth functions, while attempting to solve the HJE and interpret the solution in a generalized sense. Uniqueness is also an added issue here.

The modern theory of distributions and Sobolev spaces provides a way to interpret the derivatives in a weak sense. The *distribution theory* is essentially a linear theory though it can also be applied to certain non-linear problems where the equation is in the conservation form like, in conservation laws, or in a variational form. There is also an attempt to extend the calculus of distributions to handle non-linear problems using the paradifferential calculus developed by Bony; see Hörmander (1988, 1997) and the references therein. In general, fully non-linear partial differential equations (PDE) quite often cannot be handled via distribution theory. See the sample problems in examples (3.19), (3.20). Crandall, Lions and Evans (see Crandall and Lions, 1981, 1983; Crandall et al., 1984, Lions, 1982) have initiated the theory of *viscosity solutions* in 1980s to study non-linear PDE in a general framework for dealing with non-smooth value functions arising in the dynamic optimization problems, like optimal control and differential games. Even a brief discussion of viscosity solutions is not included here because of technicalities involved.

In this chapter, we consider a very particular case of HJE, where an explicit formula for the solution will be derived. This is known as *Hopf–Lax formula* for the value function. Of course, the value function need not be smooth, but it satisfies certain special properties. Using these special properties, we define a weaker notion of a solution. We also present a uniqueness result under this new notion of a solution, to capture the physically relevant non-smooth solution.

The examples studied in Chapter 3, suggest that a possible way to define solution is as follows: Recall that a function $\phi : (X, d) \to (Y, \rho)$ is said to be *Lipschitz continuous* or simply *Lipschitz* if $\rho(\phi(x), \phi(y)) \le Cd(x, y)$ for all $x, y \in X$, for some constant $C > 0$; here (X, d) and (Y, ρ) are two metric spaces. If $X = \mathbb{R}^n$ or an open set in $\mathbb{R}^n$ and $Y = \mathbb{R}$, it is an important theorem due to Rademacher (see Evans, 1998), which asserts that a Lipschitz function is differentiable everywhere except on a set of measure zero. A set of measure zero, for example in $\mathbb{R}^2$, can be any countable set or union of finite number of lines. We may thus allow a Lipschitz function to be a solution of HJE at all the points where the function is differentiable. There could be many Lipschitz functions satisfying the HJE. The special properties enjoyed by the value function given by the Hopf–Lax formula, provide us with a uniqueness result. We will derive Hopf–Lax formula in the special case when $H = H(Du)$ and show the additional properties enjoyed by the value function, like convexity. To treat the case of general H, we need to enter the realm of viscosity solutions, but as mentioned earlier, we do not get into this topic.

4.1 HAMILTON–JACOBI EQUATION

The general HJE or more generally Hamilton–Jacobi–Bellman (HJB) equation from optimal control theory is a first-order non-linear equation of the form

$$u_t + H(x, u, Du) = 0, \tag{4.1}$$

where $x \in \mathbb{R}^n, t > 0$. Here $u : \mathbb{R}^n \times [0, \infty) \to \mathbb{R}$ is the unknown function and $Du = \nabla u = \left(\frac{\partial u}{\partial x_1}, \dots \frac{\partial u}{\partial x_n}\right)$ is its gradient. The non-linear function H is called the Hamiltonian, a terminology borrowed from Hamiltonian mechanics. The unknown u can be thought of as the minimum value of a cost functional or energy functional as the case may be. In optimization, u is also known as the *value function*. In mechanics, the Hamiltonian H is the total energy of the system. We restrict our study to a special case, where H depends only on Du, that is, $H : \mathbb{R}^n \to \mathbb{R}$, though some of the discussion is also applicable for $H = H(x, Du)$. We begin by considering the following initial value problem (IVP)

$$\begin{cases} u_t + H(Du) = 0 \text{ in } \mathbb{R}^n \times (0, \infty) \\ u(x, 0) = g(x) \text{ in } \mathbb{R}^n. \end{cases} \tag{4.2}$$

Here $g : \mathbb{R}^n \to \mathbb{R}$ is the given initial condition. Consider the more general PDE

$$F(x, t, u, u_t, Du) \equiv u_t(x, t) + H(x, Du(x, t)) = 0.$$

Since there are $n + 1$ variables, the characteristic equations consist of an ODE system of $2(n + 1) + 1 = 2n + 3$ equations. Due to the special form of the above system, it is straight forward to decouple the characteristic equations into the following Hamiltonian system of $2n$ equations:

$$\begin{cases} \dfrac{dx}{dt} = D_p H(x,p) \\ \dfrac{dp}{dt} = -D_x H(x,p), \end{cases} \tag{4.3}$$

where $p = D_x u$. The aim of this chapter is as follows.

- An explicit formula will be derived for u via the value function of an associated minimization problem of an integral functional as in (3.58). This is known as Hopf–Lax formula.
- The minimal solution u will be shown to satisfy special properties, namely u is *semi-concave* and Lipschtiz. Furthermore, u satisfies HJE a.e. Uniqueness will be established in the class of semi-concave Lipschitz functions.
- We will identify the relation between the characteristic curves and minimization problem and also obtain the necessary condition for optimal solution of the minimization problem.
- The integrand L in the integral functional is known as Lagrangian and we show the Lagrangian L and Hamiltonian H are connected via the Legendre transformation.

4.2 HOPF–LAX FORMULA

Consider the following minimization problem: Find $\overline{w} \in \mathcal{A}_t$ such that

$$J(\overline{w}) = \min_{\mathcal{A}_t} J(w), \tag{4.4}$$

where the functional $J(w)$ is defined by

$$J(w) = \int_0^t L(\dot{w}(s))ds + g(w(0)). \tag{4.5}$$

Here $\dot{w} = \dfrac{dw}{ds}$. The Lagrangian $L : \mathbb{R}^n \to \mathbb{R}$ is a given function and the admissible class $\mathcal{A}_t$ of functions, for fixed t and $x, y \in \mathbb{R}^n$, is defined by

$$\mathcal{A}_t = \left\{ w \in C^2([0,t]; \mathbb{R}^n) : w(0) = y, w(t) = x \right\}. \tag{4.6}$$

Thus, $\mathcal{A}_t$ consists of all *trajectories* connecting two points y and x in $\mathbb{R}^n$. In the minimization problem, we fix x, t and vary y. Finally, $g : \mathbb{R}^n \to \mathbb{R}$ is a given function that eventually will be the initial data for IVP. The minimizing term can be interpreted as follows: The integral term is the running cost and $g(w(0))$ is the initial cost.[1] The solution $\overline{w}$, when exists, is called the *optimal solution* and $J(\overline{w})$, the *minimum* or *optimal cost*. The minimum cost $J(\overline{w})$, depends on x and t, which we denote by $u(x, t)$, the value function. That is

$$u(x, t) := \inf_{\mathcal{A}_t} \left\{ \int_0^t L(\dot{w}(s))ds + g(w(0)) \right\}. \tag{4.7}$$

The above infimum is taken over a class of functions which, in general, is infinite-dimensional. This can be transformed to an infimum problem over the Euclidean space, under certain assumptions on L:

Assumption 4.1. Assume that the Lagrangian L satisfies the following:

- The mapping $q \mapsto L(q)$ is continuous and convex.
- The mapping L is *coercive* in the sense that $\lim_{|q|\to\infty} \dfrac{L(q)}{|q|} = \infty$. □

The second condition means L has super-linear growth, that is, L roughly behaves like $|q|^{1+\varepsilon}$ for some $\varepsilon > 0$ and for large $|q|$.

Theorem 4.2 (Hopf–Lax Formula). Assume that $L : \mathbb{R}^n \to \mathbb{R}$ satisfies the Assumption 4.1 and u be defined by (4.7). If g is continuous, then, u can be represented as

$$u(x, t) \equiv \inf_{y\in\mathbb{R}^n} \left\{ tL\left(\frac{x-y}{t}\right) + g(y) \right\}. \tag{4.8}$$

□

Proof For $y \in \mathbb{R}^n$, consider the linear trajectory $w(s) = y + \frac{s}{t}(x - y) \in \mathcal{A}_t$. Thus, by definition, we get

$$u(x, t) \le \int_0^t L(\dot{w}(s))ds + g(y) = tL\left(\frac{x-y}{t}\right) + g(y).$$

[1] If $g(w(0))$ is replaced by $g(w(t))$, it will be termed as terminal cost. These terminologies come from optimal control problems.

By taking infimum on right-hand side over $y \in \mathbb{R}^n$, we get the one-way inequality in (4.8). To prove the reverse inequality, let $w \in \mathcal{A}_t$. Applying Jensen's inequality for L, we arrive at

$$L\left(\frac{1}{t}\int_0^t \dot{w}(s)ds\right) \leq \frac{1}{t}\int_0^t L(\dot{w}(s))ds.$$

Since $y = w(0)$ and $x = w(t)$, we get

$$tL\left(\frac{x-y}{t}\right) + g(y) = tL\left(\frac{1}{t}\int_0^t \dot{w}(s)ds\right) + g(y) \leq \int_0^t L(\dot{w}(s))ds + g(w(0)).$$

First take the infimum over $w \in \mathcal{A}_t$, and then over $y \in \mathbb{R}^n$, to get the reverse inequality in (4.8). This completes the proof. □

We now derive a functional identity satisfied by u that is an important idea from *Dynamic Programming Principle (DPP)* in optimal control/optimization theory. If we want to compute $u(\cdot, t)$ at time t, first compute $u(\cdot, s)$ for any $s < t$ and then use $u(\cdot, s)$ as the initial condition in the interval $[s, t]$ and compute $u(\cdot, t)$. This is stated as follows:[2]

Theorem 4.3 (Functional Identity). The function u given by the Hopf–Lax formula (4.8) satisfies the functional identity

$$u(x, t) = \inf_{y \in \mathbb{R}^n} \left\{(t-s)L\left(\frac{x-y}{t-s}\right) + u(y, s)\right\} \tag{4.9}$$

for any $0 < s \leq t$. □

Proof The convexity of L is crucial in the proof. We fix $0 < s \leq t$ and $y \in \mathbb{R}^n$. From the Hopf–Lax formula, we can choose a $y^* \in \mathbb{R}^n$ such that

$$u(y, s) = sL\left(\frac{y-y^*}{s}\right) + g(y^*).$$

Again, by Hopf–Lax formula,

$$u(x, t) \leq tL\left(\frac{x-y^*}{t}\right) + g(y^*).$$

[2]Compare this with the (semi) group property enjoyed by an ODE system. More generally, this property is a part of the definition of an *abstract dynamical system*.

Writing $\frac{x-y^*}{t}$ as $\frac{x-y^*}{t} = \left(1-\frac{s}{t}\right)\left(\frac{x-y}{t-s}\right) + \frac{s}{t}\left(\frac{y-y^*}{s}\right)$ and applying the convexity of L, we obtain

$$u(x,t) \le (t-s)L\left(\frac{x-y}{t-s}\right) + u(y,s).$$

Taking the infimum over y on the right-hand side, we get the one-way inequality

$$u(x,t) \le \inf_{y\in\mathbb{R}^n}\left\{(t-s)L\left(\frac{x-y}{t-s}\right) + u(y,s)\right\}.$$

To get other way inequality, we choose x^* satisfying

$$u(x,t) = tL\left(\frac{x-x^*}{t}\right) + g(x^*).$$

If we choose $y = \frac{s}{t}x + \left(1-\frac{s}{t}\right)x^*$, we see that $\frac{x-y}{t-s} = \frac{x-x^*}{t} = \frac{y-x^*}{s}$ and thus

$$\begin{aligned}(t-s)L\left(\frac{x-y}{t-s}\right) + u(y,s) &\le (t-s)L\left(\frac{x-x^*}{t}\right) + sL\left(\frac{y-x^*}{s}\right) + g(x^*)\\ &= tL\left(\frac{x-x^*}{t}\right) + g(x^*) = u(x,t).\end{aligned}$$

Thus the infimum in (4.9) is also $\le u(x,t)$ and this completes the proof. □

Theorem 4.4 (Lipschitz Continuity). Assume g is Lipschitz continuous with Lipschitz constant k. Then, the function $u(\cdot,t)$ given by the Hopf–Lax formula is Lipschitz continuous in $\mathbb{R}^n$ with Lipschitz constant k, that is,

$$|u(x_1,t) - u(x_2,t)| \le k|x_1 - x_2| \tag{4.10}$$

for all $x_1, x_2 \in \mathbb{R}^n$. Further, u satisfies the initial condition $u(x,0) = g(x)$. □

Proof The Lipschitz continuity proof is not difficult. Given $x_1 \in \mathbb{R}^n$, let x_1^* be minimizing point in Hope–Lax formula. Then,

$$\begin{aligned}u(x_2,t) - u(x_1,t) &= \inf_{y\in\mathbb{R}^n}\left\{tL\left(\frac{x_2-y}{t}\right) + g(y)\right\} - tL\left(\frac{x_1-x_1^*}{t}\right) - g(x_1^*)\\ &\le g(x_2 - x_1 + x_1^*) - g(x_1^*) \le k|x_1 - x_2|\end{aligned}$$

by the choice of $y = x_2 - x_1 + x_1^*$. Reversing the role of x_1 and x_2, we get the Lipschitz inequality.

To get the value at $t = 0$, we proceed as follows: Using the Lipschitz property of g, we have

$$\begin{aligned} tL\left(\frac{x-y}{t}\right) + g(y) &\geq tL\left(\frac{x-y}{t}\right) - k|x-y| + g(x) \\ &= g(x) - t(k|z| - L(z)), \end{aligned}$$

where $z = \dfrac{x-y}{t}$. Taking infimum over $y \in \mathbb{R}^n$, which is equivalent to taking infimum over z, we see that

$$u(x,t) \geq g(x) + \inf_{z\in\mathbb{R}^n}[-t(k|z| - L(z))] \geq g(x) - t\sup_{z\in\mathbb{R}^n}(k|z| - L(z)) = g(x) - tL(k).$$

Using coercivity of L, we see that

$$\sup_{z\in\mathbb{R}^n}(k|z| - L(z)) \leq \sup_{|z|\leq R}(k|z| - L(z)) := C_1$$

for R large enough. This implies

$$u(x,t) - g(x) \geq -C_1 t.$$

Since $u(x,t) \leq tL(0) + g(x)$, we see that

$$|u(x,t) - g(x)| \leq Ct,$$

where $C = \max\{|L(0)|, C_1\}$. This proves that g is the limiting value of u as $t \to 0$. □

We now proceed to show that u satisfies HJE. For this purpose, we need to connect the Hamiltonian H in HJE and the Lagrangian L. This is achieved by the Legendre transformation. Before proceeding to define Legendre transformation, we derive necessary conditions for the optimality of the optimal solution $\overline{w}$. This is given by the Euler–Lagrange (E–L) equations.

4.3 EULER–LAGRANGE EQUATIONS

In this section, we consider more general Lagrangian $L : \mathbb{R}^n \times \mathbb{R}^n \to \mathbb{R}$ and recall the minimization problem (4.4) with the functional given by

$$J(w) = \int_0^t L(w(s), \dot{w}(s))ds. \tag{4.11}$$

We write $L = L(x,q)$ and use the notation $D_qL := \left(L_{q_1}, \ldots L_{q_n}\right)$ and $D_xL := \left(L_{x_1}, \ldots L_{x_n}\right)$.

The derivation of E–L equations is similar to the elementary result in one variable calculus, namely, if $f \in C^1(a, b)$ and $f(x_0) = \min_{x \in (a,b)} f(x)$, then $f'(x_0) = 0$. However, the analysis is now in a function space. Let $\overline{w}$ be a minimal solution for the functional in (4.11). Let $v : [0, t] \to \mathbb{R}^n$ be C^2 such that $v(0) = v(t) = 0$. Then for any $\tau \in \mathbb{R}$, the function $w = \overline{w} + \tau v$ is a test function in $\mathcal{A}_t$. Thus

$$J(\overline{w}) \leq J(\overline{w} + \tau v).$$

Now for v fixed, define $f : \mathbb{R} \to \mathbb{R}$ by $f(\tau) = J(\overline{w} + \tau v)$. Then, clearly $f(0) \leq f(\tau)$ for all τ and hence 0 is a minimum point for f. Thus, if f is differentiable, then $f'(0) = 0$. We now compute $f'(\tau)$, assuming L differentiable. We have

$$f(\tau) = \int_0^t L(\overline{w}(s) + \tau v(s), \dot{\overline{w}}(s) + \tau \dot{v}(s))ds.$$

Hence,

$$f'(\tau) = \int_0^t D_x L(\overline{w}(s) + \tau v(s), \dot{\overline{w}}(s) + \tau \dot{v}(s)) \cdot v(s)ds + \int_0^t D_q L(\overline{w}(s) + \tau v(s), \dot{\overline{w}}(s) + \tau \dot{v}(s)) \cdot \dot{v}(s)ds.$$

Consequently

$$f'(0) = \int_0^t D_x L(\overline{w}(s), \dot{\overline{w}}(s)) \cdot v(s)ds + \int_0^t D_q L(\overline{w}(s), \dot{\overline{w}}(s)) \cdot \dot{v}(s)ds.$$

Integrating the second term on the right-hand side by parts, we arrive at

$$f'(0) = \int_0^t \left[\left(-\frac{d}{ds} D_q L + D_x L \right) (\overline{w}(s), \dot{\overline{w}}(s)) \right] \cdot v(s)ds.$$

We write this as $f'(0) = J'(\overline{w})v$ and $J'(\overline{w})$ is termed as the Frechét derivative of J at $\overline{w}$. Since $f'(0) = 0$ and v is arbitrary, we get the Euler–Lagrange equations

$$J'(\overline{w}) = \left(-\frac{d}{ds} D_q L + D_x L \right) (\overline{w}(s), \dot{\overline{w}}(s)) = 0. \tag{4.12}$$

This is a system of n second-order ODE and we have the following theorem:

Theorem 4.5. Let $L : \mathbb{R}^n \times \mathbb{R}^n \to \mathbb{R}$ be a smooth Lagrangian and J be defined as in (4.11). If $\overline{w}$ is an optimal solution, that is,

$$J(\overline{w}) = \min_{w \in \mathcal{A}_t} J(w),$$

then, $\overline{w}$ satisfies the Euler–Lagrange system of ODEs (4.12). □

Remark 4.6. Any w that satisfies the E–L system (4.12) or equivalently $J'(w) = 0$ is called a *critical point* of J. The above theorem tells us that every minimizer of J is a critical point of J, but the converse is not true in general. In fact, the converse is also not true even in one-dimensional calculus (see also the Exercise 5.). □

Example 4.7. The Newtonian mechanics provides the important example. Consider a particle of mass m moving under the force filed F generated by a potential ϕ, that is $F(x) = D\phi(x)$. Define the Lagrangian $L(x, q) = \frac{m}{2}|q|^2 - \phi(x)$, which is the difference in kinetic and potential energy. It can be easily seen that the corresponding E–L system is given by

$$m\ddot{\overline{w}}(s) = F(\overline{w}(s)) = D\phi(\overline{w}(s)).$$

This is the Newton's second law of motion. Here $\overline{w}$ is the trajectory that minimizes the action potential $\int_0^t \left[\frac{m}{2}|\dot{w}(s)|^2 - \phi(w(s)\right] ds$ among all possible trajectories. □

Example 4.8 (Brachistochrone Problem). This well-known and famous problem, is due to Johann Bernoulli. A frictionless bead located in a vertical plane at a point $A(x_0, y_0)$ slides along a wire under the force of gravity alone whose other end is fixed in the vertical plane at $B(x_f, y_f)$. The problem is to find the shape of the curve (wire) so that the bead slides from A to B in shortest possible time interval (Greek: *brachistos* means *shortest* and *chronos* means *time*).

We will now convert this problem to a minimization problem using the conservation of energy. Let the positive *y-axis* point downwards and let A and B be placed at $(a, 0)$ and (x_f, y_f), respectively with $a < x_f$, $y_f > 0$ (see Figure 4.1). Since the total energy initially is zero, we have at any point of time

$$\frac{mv^2}{2} - mgy = 0,$$

where $y = y(x)$ is a curve such that $y(a) = 0, y(x_f) = y_f$ and $v = y'$.

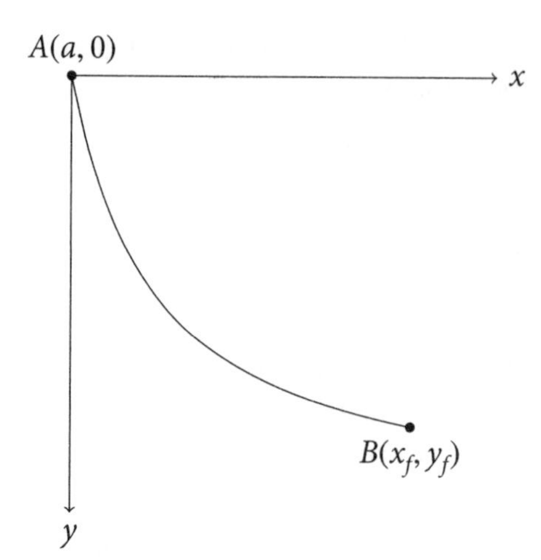

Figure 4.1 Brachistochrone

Normalizing with suitable units, we may assume $m = 1$, $g = 1/2$. Then, we get $v = \sqrt{y}$. Thus, our problem reduces to that of minimizing the functional

$$J(y) = \int_a^b \frac{1 + y'(s)^2}{\sqrt{y(s)}} ds,$$

where $b = y_f$. Johann Bernoulli in 1696 posed this problem to his contemporaries including Isaac Newton and correct solutions were obtained by Leibnitz, Newton, Jacob Bernoulli and others including Johann Bernoulli himself. The optimal curves are the *cycloids* (see Exercise 3.) given in parametric form by $x(t) = a + c(t - \sin t)$, $y(t) = c(1 - \cos t)$. This is the locus of a fixed point on a circle (e.g., the wheel of a cycle; hence the name cycloid) when it rolls without slipping on the horizontal axis. It is interesting to remark that the first solution by Johann Bernoulli was based on Snell's law of light refraction (see Simmons, 1991). □

Example 4.9 (Catenary). Consider a chain with a uniform mass density of given length hanging freely (under gravity) between two fixed points. What is the shape of the chain? This problem was posed by Galileo in 1630 and his anticipated answer that the shape is a parabola was proved to be wrong (see Figure 4.2). □

How do we describe this as an optimization problem and what is to be minimized? Let $y : [a, b] \to (0, \infty)$ with $y(a) = A, y(b) = B$ as the end points of the possible shape of the

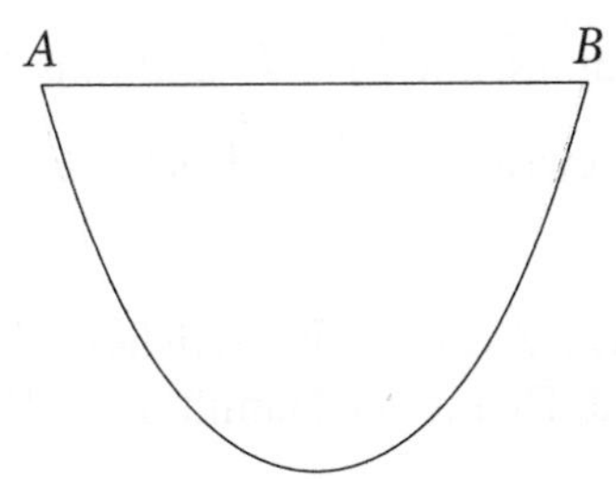

Figure 4.2 Catenary

chain. The only force acting on the chain is the potential energy and the chain will take the shape of minimal potential energy. The potential energy functional is given by

$$J(y) = \int_a^b y(x)\sqrt{1 + y'(x)^2}dx.$$

The catenary curve was obtained by Johann Bernoulli in 1670 and is given by $y(x) = c \cos h(x/c)$, $c > 0$ (see Exercise 4.) unless the chain touches the ground. The name catenary is derived from the Latin word *Catena* meaning *chain*.

Hamiltonian and Hamilton's ODE: Let $\overline{w}$ solve the E–L system (4.12). Introduce $p = p(s)$ by

$$p(s) = D_q L(\overline{w}(s), \dot{\overline{w}}(s)) \tag{4.13}$$

for $0 \leq s \leq t$. In Newtonian mechanics (see Example 4.7), we have $p(s) = m\dot{\overline{w}}(s)$, the momentum vector. In analogy, we call p defined by (4.13), the *generalized momentum* vector. Recall that in classical mechanics, the Lagrangian formulation is the dynamics described in terms of the position $\overline{w}$ and velocity $\dot{\overline{w}}$, whereas the Hamiltonian formalism is the description using position $\overline{w}$ and momentum $p = m\dot{\overline{w}}$. Since, we can obtain momentum p from the velocity $\dot{\overline{w}}$ and vice versa, we can easily go from one formalism to another.

To obtain a similar transformation with general Lagrangian L in calculus of variations, we have the difficult task (perhaps impossible) of obtaining $\dot{\overline{w}}$ in terms of p and $\overline{w}$ from (4.13). This leads to the solvability issue of the non-linear algebraic equation $p = D_q L(w, q)$. Thus, we make the following assumption:

Assumption 4.10. Given $p, x \in \mathbb{R}^n$, assume that the equation $p = D_q L(x, q)$ can be uniquely solved for $q = q(x, p)$ as a smooth function. □

Definition 4.11. Under the Assumption 4.10, define the Hamiltonian $H : \mathbb{R}^n \times \mathbb{R}^n \to \mathbb{R}$ by

$$H(x, p) = p \cdot q(x, p) - L(x, q(x, p)). \tag{4.14}$$

□

In the case of Newtonian mechanics, $q(x,p) = \frac{p}{m}$ and thus $H(x,p) = \frac{|p|^2}{m} + \phi(x)$, the total energy of the system. Now, we rewrite the second-order E–L system for $\overline{w}$ and $\dot{\overline{w}}$ as a first-order system of dimension $2n$.

Theorem 4.12 (Hamilton's ODE). Assume $\overline{w}$ satisfies the E–L system (4.12) and Assumption 4.10 is satisfied. Define the Hamiltonian H as in (4.14). Then, $\overline{w}, p$ satisfy the Hamiltonian system

$$\begin{cases} \dot{\overline{w}}(s) = D_p H(\overline{w}(s), p(s)) \\ \dot{p}(s) = -D_x H(\overline{w}(s), p(s)). \end{cases} \tag{4.15}$$

Moreover, the mapping $s \mapsto H(\overline{w}(s), p(s))$ is a constant. □

Proof From the Assumption 4.10 and equation (4.13), we get

$$q = q(\overline{w}(s), p(s)) = \dot{\overline{w}}(s).$$

Thus, we have

$$H(\overline{w}(s), p(s)) = p(s) \cdot \dot{\overline{w}}(s) - L(\overline{w}(s), \dot{\overline{w}}(s)).$$

Now

$$\dot{p}(s) = \frac{d}{ds} D_q L(\overline{w}(s), \dot{\overline{w}}(s)) = D_x L(\overline{w}(s), \dot{\overline{w}}(s)) = -D_x H(\overline{w}(s), \dot{p}(s)).$$

On the other hand, $\frac{\partial H}{\partial p_i} = \dot{\overline{w}}_i(s)$. It is also easy to see that $\frac{d}{ds} H(\overline{w}(s), p(s)) = 0$, which proves the theorem. □

We remark that the Hamiltonian is constant along the optimal trajectories that are the characteristic curves of the HJE. In the special case of classical mechanics, the characteristic curves are the constant energy curves.

4.4 LEGENDRE TRANSFORMATION

In this section, we give a connection between L and H via the so-called Legendre transformation. Here, we assume $L = L(p)$ is a function p alone. We further make Assumption 4.10. Recall the way H is defined from L using equation (4.14), where $q = q(p)$ solves $p = D_q L(q)$. Observe that the latter equation corresponds to a critical point of the functional $F(q) = p \cdot q - L(q)$, $q \in \mathbb{R}^n$. In other words, $q = q(p)$ is an extremal point of the functional F and $H(p) = F(q(p))$. This motivates the following definition:

Definition 4.13 (Legendre Transformation). The Legendre transformation of L, denoted by L^*, is defined by

$$L^*(p) = \sup_{q \in \mathbb{R}^n} \{p \cdot q - L(q)\} = \sup_{q \in \mathbb{R}^n} F(q), \ p \in \mathbb{R}^n. \tag{4.16}$$

□

Using the coercivity assumption, it is easy to see that (Exercise 7) the supremum is attained at some point $q^* = q^*(p)$, that is

$$L^*(p) = p \cdot q^* - L(q^*) = F(q^*).$$

Since q^* is a maximum of $F(q)$, we have $D_q F(q^*) = 0$ whenever F is differentiable. In other words, $D_q F(q^*) = p - D_q L(q^*) = 0$ is solvable for $q^* = q^*(p)$. Hence

$$H(p) = p \cdot q^* - L(q^*) = F(q^*) = L^*(p).$$

The converse is also true and we have the following theorem:

Theorem 4.14 (Convex Duality Between L and H). Suppose L satisfies Assumption 4.1 and Assumption 4.10. Define $H(p) = p \cdot q(p) - L(q(p))$, where $q = q(p)$ solves $p - D_q L(q) = 0$. Then, H is also convex and coercive. Further, $H = L^*$ and $H^* = L$. □

Proof We have

$$H(p) = L^*(p) = \sup_{q \in \mathbb{R}^n} \{p \cdot q - L(q)\}$$

and the convexity is trivial to check. To see coercivity, let $\lambda > 0$. For $p \neq 0$, consider $q = \dfrac{\lambda p}{|p|}$.

Then, we get

$$H(p) \geq p \cdot \frac{\lambda p}{|p|} - L\left(\frac{\lambda p}{|p|}\right) \geq \lambda |p| - M,$$

where M is the supremum of $L(q)$ over the ball of radius λ. Now dividing by $|p|$ and taking $|p| \to \infty$, we see that $\liminf_{|p| \to \infty} \dfrac{H(p)}{|p|} \geq \lambda$. Since λ is arbitrary, we see that $\lim_{|p| \to \infty} \dfrac{H(p)}{|p|} = \infty$.

It remains to prove that $L = H^*$. For any $p, q \in \mathbb{R}^n$, we have $H(p) \geq p \cdot q - L(q)$, equivalently $L(q) \geq p \cdot q - H(p)$. Taking supremum over p, we get $L(q) \geq H^*(q)$. Conversely

$$\begin{aligned} H^*(q) &= \sup_p \left\{p \cdot q - H(p)\right\} \\ &= \sup_p \left\{p \cdot q - \sup_r \left\{p \cdot r - L(r)\right\}\right\} \\ &= \sup_p \inf_r \left\{p \cdot (q - r) + L(r)\right\} \\ &\geq \inf_r \left\{p \cdot (q - r) + L(r)\right\} \end{aligned}$$

for all $p \in \mathbb{R}^n$. Using convexity of L, we see that there is an $s \in \mathbb{R}^n$ such that (by Hahn–Banach separation theorem)

$$L(r) - L(q) \geq s \cdot (r - q).$$

Indeed, if L is differentiable, then $s = D_q(q)$. Taking $p = s$ in the previous inequality, we get

$$H^*(q) \geq L(q).$$

The proof of the theorem is complete. □

Next, we will show that the value function u provided by the Hopf–Lax formula satisfies HJE wherever it is differentiable.

Theorem 4.15. Assume that the Lagrangian L satisfies Assumption 4.1 and u, g be as in Theorem 4.2. Then u is differentiable a.e. and solves the IVP

$$\begin{aligned} &u_t + H(Du) = 0 \text{ a.e. in } \mathbb{R}^n \times (0, \infty) \\ &u = g \text{ in } \mathbb{R}^n. \end{aligned} \tag{4.17}$$

□

Proof Assume u is differentiable at (x, t). Now, fix $q \in \mathbb{R}^n$ and let $h > 0$. Applying the functional identity (4.9) and Hopf–Lax formula (4.8), we get

$$\begin{aligned} u(x + hq, t + h) &\leq \inf_{y \in \mathbb{R}^n} \left\{ hL\left(\frac{x + hq - y}{h}\right) + u(y, t)\right\} \\ &\leq hL(q) + u(x, t) \end{aligned}$$

by taking $y = x$. Thus,

$$\frac{u(x + hq, t + h) - u(x, t)}{h} \leq L(q).$$

As $h \to 0+$, we get

$$u_t(x,t) + q \cdot Du(x,t) - L(q) \leq 0.$$

Now, by taking maximum over $q \in \mathbb{R}^n$, we have

$$u_t(x,t) + H(Du(x,t)) \leq 0.$$

To get the reverse inequality, choose $z \in \mathbb{R}^n$ that minimizes $u(x,t)$ in the Hopf–Lax formula; that is,

$$u(x,t) = tL\left(\frac{x-z}{t}\right) + g(z).$$

Again applying Hopf–Lax formula, for fixed $y \in \mathbb{R}^n$, $s > 0$, we get

$$u(x,t) - u(y,s) \geq \left[tL\left(\frac{x-z}{t}\right) + g(z)\right] - \left[sL\left(\frac{y-r}{s}\right) + g(r)\right]$$

for any $r \in \mathbb{R}^n$. Choose $r = z, s = t - h, h > 0$ small and finally choose $y \in \mathbb{R}^n$ such that $\dfrac{y-r}{s} = \dfrac{y-z}{s} = \dfrac{x-z}{t}$. This gives $y = \dfrac{y}{s}x + \left(1 - \dfrac{y}{s}\right)z$. Thus, we have

$$\frac{1}{h}\left[u(x,t) - u\left(\left(1 - \frac{h}{t}\right)x + \frac{z}{t}h\right)\right] \geq L\left(\frac{x-z}{t}\right).$$

As $h \to 0+$, we get

$$\frac{x-z}{t} \cdot Du(x,t) + u_t(x,t) \geq L\left(\frac{x-z}{t}\right).$$

Hence

$$\begin{aligned} u_t + H(Du) &= u_t + \max_q\{q \cdot Du - L(q)\} \\ &\geq u_t + \frac{x-z}{t} \cdot Du - L\left(\frac{x-z}{t}\right) \geq 0. \end{aligned}$$

This proves that u satisfies HJE in (4.17) wherever u is differentiable and the statement $u(x,0) = g(x)$ has been proved in Theorem 4.4. □

In view of the above theorem, it seems reasonable to define a solution in a generalized sense as a Lipschitz continuous function satisfying the initial condition and satisfying the HJE a.e. However, this turns out to be inadequate as such solutions in the generalized sense need not be unique and thus we may not be able to recover the correct physical solution. Recall Example 3.19 and Example 3.20, where we have infinitely many solutions to very simple HJEs. These examples suggest that we need some additional assumptions to capture the correct solutions. As we expect the Hopf–Lax formula is a good representation via the Lagrangian, we should analyze the formula more closely. In fact, u given by the Hopf–Lax formula inherits a kind of second derivative estimate. This notion turns out to be semi-concavity (see Bardi and Capuzzo Dolcetta, 1997; Evans, 1998).

Definition 4.16 (Semi-concavity). A function $f : \mathbb{R}^n \to \mathbb{R}$ is said to be semi-concave if $f(x) - C|x|^2$ is concave for some constant $C > 0$. □

Proposition 4.17. Assume $f : \mathbb{R}^n \to \mathbb{R}$ is continuous. Then, f is semi-concave if and only if f satisfies the one-sided regularity estimate:

$$f(x - h) - 2f(x) + f(x + h) \le C|h|^2 \tag{4.18}$$

for some constant $C > 0$ and for all $x, h \in \mathbb{R}^n$. □

Proof If f is semi-concave, that is, $g(x) = f(x) - C|x|^2$ is concave for some constant $C > 0$, then

$$g\left(\frac{z+y}{2}\right) \ge \frac{1}{2}(g(z) + g(y)), \tag{4.19}$$

that is

$$-2f\left(\frac{z+y}{2}\right) + f(z) + f(y) \le C\left(|z|^2 + |y|^2 - 2\left|\frac{z+y}{2}\right|^2\right).$$

If $x, h \in \mathbb{R}^n$, then by taking $y = x + h, z = x - h$, we get the inequality (4.18). Conversely, if f satisfies (4.18), it is easy to see that g satisfies (4.19) with g as before. By continuity of f, the concavity of g follows. □

For a C^2 function, the following is a sufficient condition for convexity.

Definition 4.18 (Uniform Convexity). Let $\Omega \subset \mathbb{R}^n$ be a open set and $H \in C^2(\Omega)$. Then, H is said to be *uniformly convex* in Ω if there is a constant $\theta > 0$ such that $(D^2H(x)\xi, \xi) \ge \theta|\xi|^2$ for all $x \in \Omega,\ \xi \in \mathbb{R}^n$. Here $D^2H(x)$ denotes the Hessian $\left[\dfrac{\partial^2 H}{\partial x_i \partial x_j}(x)\right]$ of H at x. □

The semi-concavity of u is proved in the following theorem:

Theorem 4.19. Assume that either $g : \mathbb{R}^n \to \mathbb{R}$ is semi-concave or $H : \mathbb{R}^n \to \mathbb{R}$ is uniformly convex. Then, the solution $u(\cdot, t)$ of HJE with initial data g given by Hopf–Lax formula is semi-concave for each $t > 0$. □

Proof **Case 1:** Assume g is semi-concave. Let x^* be the minimizer for $u(x,t)$. That is, $u(x,t) = tL\left(\dfrac{x - x^*}{t}\right) + g(x^*)$. Then by taking $y = h + x^*$ and $y = h - x^*$, respectively, in

Hopf–Lax formula, we get

$$u(x+h,t) = tL\left(\frac{x-x^*}{t}\right) + g(h+x^*)$$

and

$$u(x-h,t) = tL\left(\frac{x-x^*}{t}\right) + g(h-x^*).$$

Thus, we have

$$u(x-h,t) - 2u(x,t) + u(x+h,t) \le g(h-x^*) - 2g(x^*) + g(h-x^*) \le C|h|^2$$

The later inequality follows from the semi-concavity of g.

Case 2: Now, we consider the case when H is uniformly convex. Let $p_0 = \dfrac{p_1+p_2}{2}$. Apply Taylor's theorem to get

$$H(p_1) = H(p_0) + \frac{1}{2}DH(p_0)\cdot(p_1-p_2) + \frac{1}{8}\left(D^2H(\xi_1)(p_1-p_2), p_1-p_2\right)$$

and

$$H(p_2) = H(p_0) + \frac{1}{2}DH(p_0)\cdot(p_2-p_1) + \frac{1}{8}\left(D^2H(\xi_2)(p_1-p_2), p_1-p_2\right)$$

for some $\xi_1, \xi_2 \in \mathbb{R}^n$. Adding these equations and using the uniform convexity of H, we get the inequality[3]

$$H\left(\frac{p_1+p_2}{2}\right) \le \frac{1}{2}H(p_1) + \frac{1}{2}H(p_2) - \frac{\theta}{8}|p_1-p_2|^2.$$

The continuity of H then proves that H is convex. Since H and L are dual to each other, we prove a reverse inequality for L. For given q_1, q_2, from the definition of $L = H^*$, there are p_1, p_2 such that

$$\begin{aligned}\frac{1}{2}L(q_1) + \frac{1}{2}L(q_2) &= \frac{1}{2}(p_1\cdot q_1 + p_2\cdot q_2) - \frac{1}{2}(H(p_1)+H(p_2)) \\ &\le \frac{1}{2}(p_1\cdot q_1 + p_2\cdot q_2) - H\left(\frac{p_1+p_2}{2}\right) - \frac{\theta}{8}|p_1-p_2|^2\end{aligned}$$

[3] Note that this is a stronger estimate than just convexity.

$$
\begin{aligned}
&\leq \frac{1}{2}(p_1 \cdot q_1 + p_2 \cdot q_2) - (p_1 + p_2)\left(\frac{q_1+q_2}{2}\right) \\
&\qquad + L\left(\frac{q_1+q_2}{2}\right) - \frac{\theta}{8}|p_1 - p_2|^2 \\
&\leq L\left(\frac{q_1+q_2}{2}\right) + \frac{1}{4}(p_1 \cdot q_1 + p_2 \cdot q_2) - \frac{1}{4}(p_1 \cdot q_2 + p_2 \cdot q_1) \\
&\qquad - \frac{\theta}{8}|p_1 - p_2|^2 \\
&\leq L\left(\frac{q_1+q_2}{2}\right) + \frac{1}{8\theta}|q_1 - q_2|^2.
\end{aligned}
$$

The last inequality follows by expanding the expression:

$$\left|\frac{1}{\sqrt{8\theta}}(q_1 - q_2) - \frac{\theta}{\sqrt{8}}(p_1 - p_2)\right|^2 \geq 0.$$

Let x^* be as in Case 1. Then

$$
\begin{aligned}
u(x-h,t) - 2u(x,t) + u(x+h,t) &\leq \left(tL\left(\frac{x-h-x^*}{t}\right) + g(x^*)\right) \\
&\qquad -2\left(tL\left(\frac{x-x^*}{t}\right) + g(x^*)\right) \\
&\qquad + \left(tL\left(\frac{x+h-x^*}{t}\right) + g(x^*)\right) \\
&\leq \frac{1}{\theta t}|h|^2.
\end{aligned}
$$

The last inequality follows from the earlier inequality derived for L choosing q_1, q_2 suitably. □

Generalized Solution and Uniqueness

Definition 4.20 (Generalized Solution). We say that a Lipschitz function $u : \mathbb{R}^n \times [0, \infty) \to \mathbb{R}$ is a *generalized solution* of the IVP (4.2) if u satisfies

1. $u(x, 0) = g(x)$ for all $x \in \mathbb{R}^n$.
2. $u_t(x, t) + H(Du(x, t)) = 0$ a.e. x in $\mathbb{R}^n$, $t > 0$
3. $u(x-h, t) - 2u(x) + u(x+h) \leq C(t)|h|^2$, for all $x, h \in \mathbb{R}^n$ and $t > 0$, where $C(t) = C\left(1 + \frac{1}{t}\right)$ for some $C > 0$.

□

Theorem 4.21 (Existence and Uniqueness). Consider the IVP (4.2), where the Hamiltonian $H : \mathbb{R}^n \to \mathbb{R}$ is convex and coercive. Assume the initial data g is Lipschitz continuous. Further, assume that either g is semi-concave or H is uniformly convex. Then, the function u defined by the Hopf–Lax formula

$$u(x,t) = \inf_{y \in \mathbb{R}^n} \left\{ tL\left(\frac{x-y}{t}\right) + g(y) \right\}$$

is the unique generalized solution of the IVP (4.2). Here the Lagrangian L is given by

$$L(q) = H^*(q) = \sup_{p \in \mathbb{R}^n} \{p \cdot q - H(p)\}.$$

□

Proof In view of Theorem 4.15, only uniqueness need to be proved. Let u_1, u_2 be two generalized solutions and let $w = u_1 - u_2$. Then, w satisfies

$$w_t + a(x,t)Dw = 0 \text{ a.e.,} \tag{4.20}$$

where

$$a(x,t) = \int_0^1 DH(rDu_1 + (1-r)Du_2)dr.$$

Indeed w is differentiable a.e. and at a differentiable point (x,t), where u_1, u_2 are differentiable, we get from the HJEs for u_1, u_2:

$$w_t = -H(Du_1) + H(Du_2) = \int_0^1 \frac{d}{dr} H(rDu_1 + (1-r)Du_2)dr$$

which in turn gives (4.20). Now, we choose a smooth function $\phi : \mathbb{R} \to [0,\infty)$ such that $\phi = 0$ in a neighborhood of 0 and ϕ is positive outside this neighborhood. Then, $v = \phi(w) \geq 0$. Since $v_t = \phi'(w)w_t$ and $Dv = \phi'(w)Dw$, we see from (4.20) that v satisfies

$$v_t + a(x,t)Dv = 0, \text{ a.e.} \tag{4.21}$$

Since u_1, u_2 and hence a need not be C^2, we smoothen these functions via mollifiers (see Chapter 2). Thus, we define $u_i^\varepsilon = \rho_\varepsilon * u_i$, for $i = 1, 2$, where ρ_ε is the mollifier in x, t variable. We can also deduce that

$$u_i^\varepsilon \to u_i, \ Du_i^\varepsilon \to Du_i, \text{ a.e.} \tag{4.22}$$

Further, Du_i^ε's are bounded by the Lipschitz constants k_i of u_i, that is

$$|Du_i^\varepsilon| \leq k_i, \ i = 1, 2. \tag{4.23}$$

Moreover, the one-sided estimate (4.18) implies that

$$|D^2 u_i^\varepsilon| \leq C\left(1 + \frac{1}{t}\right). \tag{4.24}$$

Now define a^ε as

$$a^\varepsilon(x,t) = \int_0^1 DH(rDu_1^\varepsilon + (1-r)Du_2^\varepsilon)dr.$$

Fix a point (x_0, t_0) and $M = \max\{DH(p) : \ |p| \leq \max\{k_1, k_2\}\}$. Define

$$z(t) = \int_B v(x,t)dx,$$

where B is the ball $B_{M(t_0-t)}(x_0)$ for $t < t_0$. We derive an IVP for z. We have

$$\dot{z} = \int_B v_t dx - M \int_{\partial B} v dS.$$

From equation (4.21) for v, we have

$$v_t + a^\varepsilon(x,t) \cdot Dv = (a^\varepsilon - a) \cdot Dv, \ a.e.$$

Thus

$$v_t + \mathrm{div}(va^\varepsilon) = v\mathrm{div}a^\varepsilon + (a^\varepsilon - a) \cdot Dv.$$

Hence, replacing v_t from the above expression and integrating by parts, we arrive at

$$\dot{z} = -\int_{\partial B} v(a^\varepsilon \cdot \nu + M)dS + \int_B (\mathrm{div}a^\varepsilon \cdot v + (a^\varepsilon - a) \cdot Dv)dx.$$

The first term $-\int_{\partial B} v(a^\varepsilon.\nu + M)dS$ is negative by the choice of M and (4.23). By differentiating a^ε, we get the estimate

$$\mathrm{div}a^\varepsilon \leq C\left(1 + \frac{1}{t}\right).$$

Thus,

$$\dot{z} \leq C\left(1 + \frac{1}{t}\right) z(t) + \int_B (\mathrm{div}a^\varepsilon \cdot v + (a^\varepsilon - a) \cdot Dv)dx.$$

As $\varepsilon \to 0$, we finally get the differential inequality

$$\dot{z} \leq C\left(1 + \frac{1}{t}\right) z(t)$$

a.e., $0 < t < t_0$. Now let $0 < \varepsilon < t_0$. Since $u_1(x, 0) = u_2(x, 0)$, we have

$$|w(x,t)||u_1(x,t) - u_2(x,t)| \leq |u_1(x,t) - u_1(x,0)| + |u_2(x,t) - u_t(x,0)| \leq \varepsilon(k_1 + k_2)$$

for $|t| = \varepsilon$. Thus, if we choose ϕ such that ϕ vanishes in the interval $[-\varepsilon(k_1+k_2), \varepsilon(k_1+k_2)]$, we get $v(x, \varepsilon) = \phi(w(x, \varepsilon)) = 0$. Consequently, we get $z(\varepsilon) = 0$. Now applying Gronwall's inequality over the interval $[\varepsilon, t]$, we see that

$$z(t) \leq z(\varepsilon) \exp\left(C \int_{\varepsilon}^{t} \left(1 + \frac{1}{t}\right) dt \right) = 0.$$

This gives $v = 0$ and hence

$$|u_1 - u_2| = |w| \leq \varepsilon(k_1 + k_2)$$

by the choice of ϕ. As ε is arbitrary, we get $u_1 = u_2$. This completes the proof. □

4.5 NOTES

In this chapter, we have studied a very special case of HJE when the Hamiltonian is a function of Du only. Further, we have assumed that H is convex. In general H can be a function of x, u, Du and there may not be any convexity. The study on such general equations is beyond the scope of this first book on PDE. Most of the material covered in this chapter is available in Evans (1998) and further references are Lions (1982), Rund (1973), Benton (1977), Lax (1973), and Morawetz (1981).

The perception of PDE has completely changed after the introduction of *distribution theory* in the middle of the last century. It was quite apparent in the first half of the last century that a new theory was inevitable to understand PDE and hence physical problems. Many physical quantities appearing as unknowns in PDE need not be differentiable or smooth. Hence it is important to give an appropriate interpretation of solution of PDEs, where it is possible to accommodate *non-smooth functions as solutions*. The distribution theory allows us to differentiate functions in a weak sense, which are not differentiable in the usual sense and look for solutions in appropriate space after giving suitable interpretations that are physically meaningful. This together with the modern functional analysis and operator theory made the study of PDEs in a wider perspective and it became extremely rich in the last 50 years. Loosely speaking distribution theory is a linear theory and in general, it is not applicable to highly non-linear PDEs. Another concept of *weak solutions*, which was developed in the late 1970s in the context of HJB equations connected with optimal control problems.

The *theory of viscosity solutions* for HJEs was initiated in the early 1980s by Crandall and Lions (1981, 1983). The HJE or HJB equations arise in the study of optimal control problems. The dynamic programming became a standard topic in deterministic optimal control theory as early as 1960s. Dynamic programming is a functional relation of the value function associated with an optimal control problem from which it is possible to derive various other results. The infinitesimal version of the *DPP* was derived formally in the form of a (first- or second-order) non-linear PDE satisfied by the value function. Indeed, it is easy to prove that if the value function is smooth, then it is the unique solution of HJB equation (smooth unique existence). However, the lack of smoothness of the value function in optimal control and differential game problems of continuous time processes, creates considerable restriction to the applicability of the classical Hamilton–Jacobi theory in the setup of calculus of variations, which was recognized as early as 1930s by C. Carathèodory. What is required is an appropriate interpretation of a weak concept of the solution of HJB equations. Thus realizing the non-smooth values as the solution of (viscosity sense) HJB equation. This was reasonably well handled after the introduction of viscosity solutions.

Later the theory of viscosity solutions was extended to a unified treatment of first- and second-order degenerate elliptic equations and scope of applications increased considerably. The survey paper by Crandall et al. (1992) and the book by Bardi and Capuzzo Dolcetta (1997) is very good for the study of viscosity solution. The beginner can also look into the book by Daniel Liberzon (2012). It gives excellent introduction to optimization, calculus of variations and optimal control.

4.6 EXERCISES

1. Consider the minimization problem from calculus of variation:

$$\min \int_t^\tau \left(1 + \frac{1}{4}\dot{x}(s)^2\right) ds, \quad t \in [0, 1],$$

over a suitable class of functions, say Lipschitz functions. Here τ is the exit time of $(s, x(s))$ from the region $[0, 1] \times [-1, 1]$. Define $L(v) := L(t, x, v) = 1 + \frac{1}{4}v^2$, where v is any scalar. Let $x(\cdot)$ be the trajectory with the initial value $x(t) = x$, then the *value function* is defined as

$$u(t, x) = \min\left\{ \int_t^\tau L(t, x(s), \dot{x}(s))ds \;:\; x(\cdot) \text{ Lipschitz continuous} \right\}.$$

Show that u is given by

$$u(t,x) = \min_{v}\{(\tau - t)L(t,x,v)\},$$

where the minimization over the real numbers. Further show that

$$v^* = \begin{cases} 2 \text{ if } x \geq t \\ 0 \text{ if } |x| < t \\ -2 \text{ if } x \leq -t \end{cases},$$

is a minimizing solution and the corresponding value function is given by

$$u(t,x) = \begin{cases} 1 - |x| \text{ if } |x| \geq t \\ 1 - t \quad \text{if } |x| \leq t \end{cases}$$

Find the differentiable region and show that u satisfies the following HJE wherever it is differentiable:

$$-u_t(t,x) + (u_x(t,x))^2 - 1 = 0,$$

and satisfies conditions

$$u(t,1) = u(t,-1) = 0, t \in [0,1]; u(1,x) = 0, x \in [-1,1].$$

2. **a.** Consider the Lagrangian $L(q) = 1 + \frac{1}{4}|q|^2, q \in \mathbb{R}^n$ and derive the Hamiltonian via the Legendre transformation.

   b. Now define the minimal value

$$u(x,t) = \min\left[\int_0^t \left(1 + \frac{1}{4}\dot{w}(s)^2\right) ds + \frac{1}{2}w(0)^2\right],$$

   where the minimum is taken over all smooth trajectories w satisfying $w(t) = x$. Using Hopf–Lax formula find u explicitly, write down HJE, show that u satisfies HJE and find the initial condition.
3. Recall the brachistochrone problem in Example 4.8. Show that the minimizing solution is a cycloid, using E–L equations.
4. In Catenary Example 4.9, derive the optimal trajectory.
5. Consider $J(y) = \int_0^1 y\dot{y}^2 dx$ subject to $y(0) = y(1) = 0$. Show that the E–L equation is $\dot{y}^2 = \frac{d}{dx}(2y\dot{y})$ and $y = 0$ is, in fact, a solution, that is, $y = 0$ is a critical point. But conclude that $y = 0$ is neither a minimum nor a maximum.

6. The following example is known as the Dirichlet Principle:
   The optimization problem involves finding a surface that minimizes a given integral functional $J(y) = \int_\Omega \nabla y(x) \cdot \nabla y(x) dx$ over surfaces y satisfying $y(x) = \bar{y}(x)$ on $\partial\Omega$. Here Ω is a given open bounded set in $\mathbb{R}^2$, $\partial\Omega$ is the boundary of Ω and $\bar{y}$ are given. If y is an optimal solution, show that y satisfies the Laplace equation $\Delta y(x) = 0$ in Ω, $y = \bar{y}$ on $\partial\Omega$. This is also given as the equation satisfied by the electric potential in which a static two-dimensional electric field is distributed. The functional of the form $\frac{1}{2}\int \nabla y \cdot \nabla y - \int fv$ also represents the strain energy functional.
7. Assume that $L : \mathbb{R}^n \to \mathbb{R}$ is continuous and satisfies the coercivity condition in Assumption 4.1. Consider the maximization problem

$$H(p) := \sup_{q \in \mathbb{R}^n} \left\{ p \cdot q - L\left(q\right) \right\}.$$

   Given $p \in \mathbb{R}^n$, show that there exists a $q = q(p) \in \mathbb{R}^n$ such that

$$H(p) = p \cdot q(p) - L(q(p)).$$

8. Let E be a closed subset of $\mathbb{R}^n$. Applying Hopf–Lax formula, show that the solution of the HJE

$$\begin{aligned} &u_t + |Du|^2 = 0, \text{ in } \mathbb{R}^n \times (0, \infty) \\ &u(x, 0) = g(x) \text{ in } \mathbb{R}^n, \end{aligned}$$

   is given by $u(x,t) = \frac{d(x)}{4t}$. Here g is the indicator function that is zero in E and ∞ otherwise and $d(x) = d(x, E)$ is the distance function.

CHAPTER 5

Conservation Laws

5.1 INTRODUCTION

Many physical laws, for example, conservation of mass, momentum and energy, occur as *conservation laws*. Dynamics of compressible fluids, both in one and three dimensions, is a rich source of conservation laws and offers many problems that are challenging; see, for example, Courant and Friedrichs (1976), Morawetz (1981), Whitham (1974), and Majda (1985). This subject has been one of the most active research fields, both in the theoretical and computational aspects, for the past more than six decades. Because of its important applications in the aerospace engineering, the interest in this field is only growing. Yet, certain theoretical aspects in the study of systems of conservation laws (e.g., Euler's equations of gas dynamics) have not been resolved satisfactorily and remain a challenge. The interested reader will find some advanced topics in Smoller (1994), Majda (1985), and Lax (1973), and the references therein.

A conservation law asserts that the rate of change of the *total amount* of a substance (e.g., a fluid) in a domain Ω in space (an interval in the one-dimensional case) is equal to its *flux* across the boundary $\partial\Omega$ of the domain Ω. If $u = u(x, t)$ denotes the density of the substance at time t and f the flux, the conservation law is expressed as

$$\frac{d}{dt}\int_{\Omega} u\, dx = -\int_{\partial\Omega} f \cdot \nu\, dS, \tag{5.1}$$

where ν denotes the outward unit normal on $\partial\Omega$ and dS is the surface measure on $\partial\Omega$. The integral on the right-hand side is the total amount of the outflow of the substance across $\partial\Omega$, hence the negative sign. Assuming that u and f are smooth, we obtain using the divergence theorem that

$$\int_{\Omega} (u_t + \text{div} f)\, dx = 0.$$

Since Ω is an arbitrary domain, by shrinking it to a point, we obtain the following first-order PDE

$$u_t + \text{div} f = 0, \tag{5.2}$$

satisfied by the density u and the flux f. Equation (5.2) is the conservation law expressed in the *differentiated form* and equation (5.1) in the *integral form*.

Note that (5.1) is equivalent to

$$\int_{\Omega} (u(x,t_2) - u(x,t_1))\,dx = -\int_{t_1}^{t_2}\int_{\partial\Omega} f\cdot\nu\,dSdt, \tag{5.3}$$

for all t_1 and t_2 with $t_1 < t_2$.

In the present chapter, we consider only a single conservation law in one (space) dimension. In the differentiated form, this is a first-order quasilinear equation of the form

$$u_t + f_x = 0, \tag{5.4}$$

where $f = f(u)$ is assumed to be a non-linear function of u. Our main plan is to derive an explicit formula for the *generalized* or *weak solution* of (5.4) (defined in the next section) using the Hopf–Lax formula for the solution of Hamilton–Jacobi equation (HJE), derived in Chapter 4. This is the well-known Lax–Oleinik formula. As we have seen in Chapter 3, smooth solutions may not exist for all $t > 0$ for the Burgers' equation $u_t + uu_x = 0$, or more generally for $u_t + f(u)_x = 0$, and hence the need to look for generalized or weak solutions.

Equation (5.4) being a quasilinear first-order equation, it can be solved using the method of characteristics, described in Chapter 3. The following observations easily follow as in the case of Burgers' equation, discussed earlier in the same chapter. The characteristic equations are given by

$$\frac{dx}{dt} = c(u),\ \frac{du}{dt} = 0,$$

where[1] $c(u) = f'(u)$. If we prescribe an initial condition $u(x,0) = u_0(x)$, $x \in \mathbb{R}$ with u_0 being a C^1 function, then the solution of (5.4) is given in the following parametric form:

$$x = c(u_0(x_0))t + x_0,\ u(x,t) = u_0(x_0), \tag{5.5}$$

where x_0 is the point on the line $t = 0$ at which the characteristic through (x,t), $t > 0$ meets the line $t = 0$. Note that the characteristics are straight lines, with varying slopes. We recall that by implicit function theorem, a sufficient condition for solving for x_0 in terms of x and t is

$$1 + u_0'ct \neq 0.$$

[1] As a convention, we use $'$ to denote the differentiation of any function of a *single variable*. If f is a linear function of u, then c will be a constant function.

Thus, we are able to obtain the solution u, at least for $t > 0$ small enough. Observe that we, then have

$$u_t = -\frac{u_0'c}{1 + u_0'ct}, \text{ and } u_x = \frac{u_0'}{1 + u_0'ct}. \tag{5.6}$$

This immediately shows that u defined by (5.5) is indeed a solution of (5.4), at least for small $t > 0$, satisfying the initial condition $u(x, 0) = u_0(x)$. Now let us assume that equation (5.4) is *genuinely non-linear*, that is, $c(u)$ is not zero for all u. We may assume that[2]

$$c(u) > 0, \text{ for all } u. \tag{5.7}$$

Then, if $u_0' \geq 0$, we have a solution defined for all $t > 0$. On the other hand, if $u_0' < 0$, then the solution cannot be defined beyond a certain T, by the method of characteristics. If we choose the initial values u_0 from the important class of functions having compact supports, they inevitably satisfy the condition $u_0' < 0$ in an interval. Thus, in this case, if we wish to define a solution for all $t > 0$, we need to move outside the class of smooth functions. But then, such a function cannot satisfy equation (5.4) point-wise. This issue will be taken up in the next section. Before proceeding further, we discuss an interesting example. We shall now formulate the *traffic flow problem*, leading to a conservation law, first proposed by Lighthill and Whitham (see Lighthill and Whitham, 1955; Morawetz, 1981; Whitham, 1974).

Example 5.1 (Traffic Flow Problem). Imagine a long road, identified as the x-axis, crowded by vehicles. Let $u(x, t)$ denote the density, namely the number of vehicles passing through the position x at time t along the road. Let $v(x, t)$ be the average local velocity of the vehicles. Assuming that in any section $[x_1, x_2]$ of the road, the total number of vehicles is preserved (conservation), we get the equation (see (5.3)) □

$$\int_{x_1}^{x_2} u(x, t_2)dx - \int_{x_1}^{x_2} u(x, t_1)dx = \int_{t_1}^{t_2} u(x_1, t)v(x_1, t)dt - \int_{t_1}^{t_2} u(x_2, t)v(x_2, t)dt.$$

Assume u, v are smooth. Then, dividing the above equation by $t_2 - t_1$ and letting $t_1, t_2 \to t$, we get the integral equation

$$\int_{x_1}^{x_2} u_t(x, t)dx = (vu)(x_1, t) - (vu)(x_2, t).$$

[2]If c has isolated zeros or vanishes on an interval, then additional difficulties arise in the analysis.

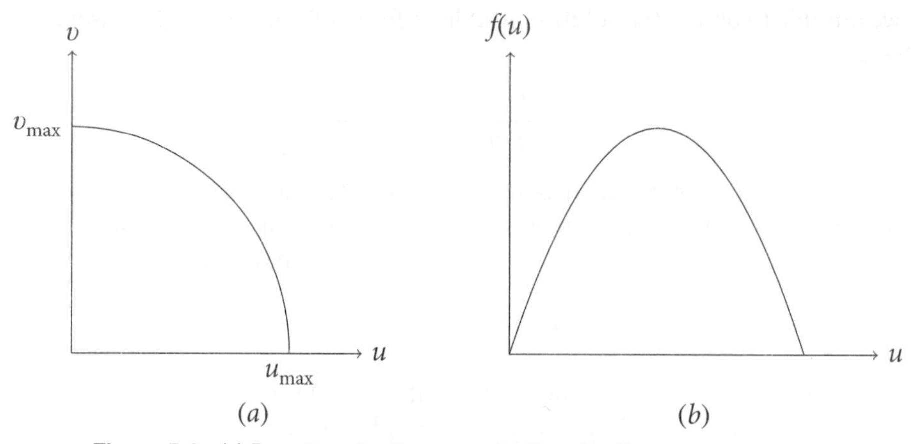

Figure 5.1 (a) Density–velocity curve, (b) Density–flux curve

Comparing this with equation (5.1), we see that the expression vu represents the flux. This is the integral form of the conservation law equation. Now, divide the above equation by $x_2 - x_1$ and let $x_1, x_2 \to x$, to get

$$u_t + (vu)_x = 0. \tag{5.8}$$

If we assume v is a function of u, say, $v = v(u)$, which is a reasonable assumption as the drivers are supposed to increase or decrease the vehicle speed according as the density decreases or increases, respectively, then equation (5.8) can be written in the form of equation (5.2), with $f(u) = uv(u)$. Observe that the maximum of $v(u)$ (maximum means maximum speed allowed) is attained when the density $u = 0$. And, when the density u is maximum, the drivers need to stop their vehicles (travelling with almost zero speed). Consequently v is zero. Thus, $f(u) = 0$ when both $u = 0$ or u is maximum (since, in this case $v = 0$.). The density–velocity and density–flux curves schematically depicted in Figure 5.1.

5.2 GENERALIZED SOLUTION AND RANKINE–HUGONIOT (R–H) CONDITION

For the conservation law (5.4), the method of characteristics provides a smooth solution, under the transversality condition on the initial curve, at least for small $t > 0$. It is evident from the discussion in the previous section and Chapter 3 that there may not exist a smooth solution for all time $t > 0$. This forces us to look for a different concept of the solution, admitting non-differentiable functions as solutions. Obviously, such a solution cannot satisfy (5.4) point-wise. To get a better feeling of this new concept of solution, let us begin first with a *weak* differentiation concept.

Let $h \in C^1[a,b]$ and $\psi \in C^1[a,b]$ such that $\psi(a) = \psi(b) = 0$ be an arbitrary function. All such ψ are referred to as *test functions*. Then, an integration by parts yields

$$\int_a^b h'(t)\psi(t)dt = -\int_a^b h(t)\psi'(t)dt.$$

There are no boundary terms as $\psi(a) = \psi(b) = 0$. In other words, the function h' can be interpreted by its action on all test functions ψ, namely by the integral $\int_a^b h'(t)\psi(t)dt$ that is the same as $-\int_a^b h(t)\psi'(t)dt$. The advantage of the representation $\int_a^b h(t)\psi'(t)dt$ is that we do not require h to be differentiable. Thus, we interpret the function h' by its actions $-\int_a^b h(t)\psi'(t)dt$ for all test functions ψ. Of course, when h is differentiable in the usual sense, it is possible to recover h' by this interpretation, that is, the two notions are the same if h is differentiable. We use this idea to obtain a weak formulation of the conservation law (5.4). Consider the initial value problem (IVP) for the conservation law

$$\begin{cases} u_t + (f(u))_x = 0 \text{ in } \mathbb{R} \times (0,\infty) \\ u(x,0) = u_0(x),\ x \in \mathbb{R}. \end{cases} \tag{5.9}$$

Here $f : \mathbb{R} \to \mathbb{R}$ is a smooth function and u_0 is a given C^1 function. Let $v : \mathbb{R}\times[0,\infty) \to \mathbb{R}$ be a smooth function with compact support in $\mathbb{R} \times [0,\infty)$. Such functions are referred to as *test functions*. It is to be remarked that the support of the test functions is in the closed set $\mathbb{R} \times [0,\infty)$ and *not* in the open set $\mathbb{R} \times (0,\infty)$. This is important to retrieve the initial condition, when the solution is smooth.

Let u be a C^1 solution of (5.9). Multiplying equation (5.9) by v and integrating by parts, we arrive at the integral equation

$$\int_0^\infty \int_{-\infty}^\infty \left(uv_t + f(u)v_x\right)\, dxdt + \int_{-\infty}^\infty g(x)v(x,0)\, dx = 0. \tag{5.10}$$

Note that the integral equation (5.10) does not require u or f to be differentiable. In fact, we can even admit discontinuous u, only requiring that all the integrals in (5.10) are well-defined. This leads to the following definition:

Definition 5.2 (Generalized or Weak Solution). A function $u \in L^\infty(\mathbb{R} \times (0,\infty))$ is a *generalized solution* or *weak solution* of (5.9) if the integral equation (5.10) is satisfied, for all the test functions v. □

Thus, if u is a classical solution (that is u is smooth and satisfies (5.9) point-wise), then u is a generalized solution. Conversely, if u is a generalized solution and u is a C^1 function, it is not difficult to show that u indeed satisfies (5.9) (see Exercise 5). However, there are generalized solutions that are not differentiable, as we will see the examples below. These solutions cannot therefore satisfy (5.9) at all points.

Before proceeding toward existence and uniqueness of generalized solutions, we discuss a simple class of generalized solutions: these solutions are differentiable everywhere except on a curve in the (x, t) plane.

Rankine–Hugoniot (R–H) Condition: Let $V \subset \mathbb{R} \times (0, \infty)$ be an open region and let γ be a smooth curve in V as shown in Figure 5.2, dividing the region V into two open sets V_ℓ and V_r. Assume that u is a generalized solution of (5.9). Assume further that u is C^1 in both the regions V_ℓ and V_r and has jump discontinuity across γ. It follows, by taking v with support in V_ℓ and V_r, respectively that

$$u_t + (f(u))_x = 0 \text{ both in } V_\ell \text{ and } V_r. \tag{5.11}$$

Let v be a test function having support in V. Using (5.10), we have

$$\begin{aligned} 0 &= \iint_V \left(uv_t + f(u)v_x\right)\, dxdt \\ &= \iint_{V_\ell} \left(uv_t + f(u)v_x\right)\, dxdt + \iint_{V_r} \left(uv_t + f(u)v_x\right)\, dxdt. \end{aligned}$$

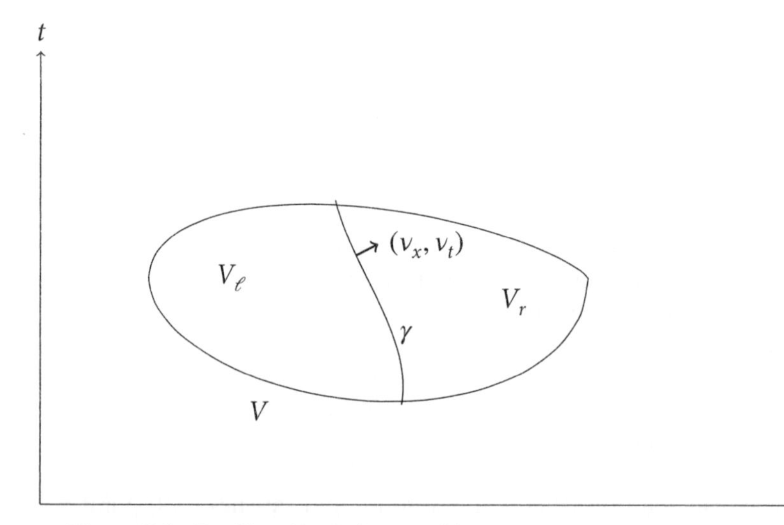

Figure 5.2 Rankine–Hugoniot condition

Since u is smooth in V_ℓ and V_r, we can integrate by parts in the two integrals on the right-hand side. Thus,

$$\iint_{V_\ell} \left(uv_t + f(u)v_x\right)\, dxdt = -\iint_{V_\ell} \left(u_t v + (f(u))_x v\right)\, dxdt + \int_\gamma (u_\ell \nu_t + f(u_\ell)\nu_x) v\, dS.$$

Here u_ℓ and u_r denote the limits of the function u on γ, when approached through the points from V_ℓ and V_r, respectively; $\nu = (\nu_x, \nu_t)$ is the unit normal on γ, which is outward to the domain V_ℓ, and hence inward to the domain V_r. The second integral on the right-hand side is the line integral along the curve γ. Using the equation on the domain V_ℓ in (5.11), we therefore have

$$\iint_{V_\ell} \left(uv_t + f(u)v_x\right)\, dxdt = \int_\gamma (u_\ell \nu_t + f(u_\ell)\nu_x) v\, dS.$$

Similarly, we have

$$\iint_{V_r} \left(uv_t + f(u)v_x\right)\, dxdt = -\int_\gamma (u_r \nu_t + f(u_r)\nu_x) v\, dS.$$

Adding these two equations, we arrive at

$$\int_\gamma \left([u]_\gamma \nu_t + [f(u)]_\gamma \nu_x\right) v\, dS = 0, \tag{5.12}$$

where $[u]_\gamma = u_\ell - u_r$ and $[f(u)]_\gamma = f(u_\ell) - f(u_r)$ denote the *jumps* of u and $f(u)$ across γ, respectively. Since (5.12) holds for all the test functions v, we conclude that

$$[u]_\gamma \nu_t + [f(u)]_\gamma \nu_x = 0$$

along the curve γ. If the curve γ has a parametric representation $x = s(t)$ where $s : [0, \infty) \to \mathbb{R}$ is a smooth function, then

$$\nu_x = \frac{1}{\sqrt{1+\dot{s}^2}} \text{ and } \nu_t = -\frac{\dot{s}}{\sqrt{1+\dot{s}^2}}.$$

Consequently, $\dfrac{\nu_t}{\nu_x} = -\dot{s}$ and the jump condition becomes

$$[f(u)]_\gamma = \sigma[u]_\gamma. \tag{5.13}$$

This is referred to as the Rankine–Hugoniot (R–H) condition[3] and $\sigma = \dot{s}$ is referred to as the speed with which the discontinuity (of u) propagates. Thus, for a piece-wise smooth solution of (5.4), possibly having jump discontinuities along a curve in the (x, t) plane, R–H condition is a necessary condition.

Remark 5.3. For the conservation law

$$u_t + f(u)_x = 0,$$

the R–H condition or the *jump condition* is given by

$$-\sigma[u] + [f(u)] = 0,$$

where σ is the speed of the shock discontinuity and $[\cdot]$ denotes the jump across the shock discontinuity. There is a nice correspondence between the PDE and the R–H condition:

$$\frac{\partial}{\partial t} \leftrightarrow -\sigma[\cdot], \ \frac{\partial}{\partial x} \leftrightarrow [\cdot].$$

The R–H condition can also be written as

$$\sigma = \frac{f(u_r) - f(u_l)}{u_r - u_l}.$$

It is important to note that the *direct* association of the R–H condition with the PDE in conservation form is *not* unique. For example, consider the Burgers' equation

$$u_t + uu_x = 0.$$

This can be written in conservation form as

$$u_t + \left(\frac{1}{2}u^2\right)_x = 0$$

and also as

$$\left(\frac{1}{2}u^2\right)_t + \left(\frac{1}{3}u^3\right)_x = 0.$$

Obviously, the R–H conditions of these two PDE are different. Thus, we have to choose the appropriate R–H condition only from the physical considerations of the problem and the *original* integrated form of the conservation law. □

[3]This terminology is derived from gas dynamics.

Since we do expect the discontinuous solution to (5.4), possibly at a later time $T > 0$, we may as well start with a discontinuous initial condition and analyze a generalized solution. For this purpose, we consider the example of Burgers' equation ($f(u) = u^2/2$) with different discontinuous initial conditions.

Example 5.4 (Rarefaction). Consider the initial condition (see Figure 3.8(b))

$$u_0(x) = \begin{cases} 0 \text{ if } x < 0 \\ 1 \text{ if } x > 0. \end{cases}$$

Using the method of characteristics, we obtain that

$$u(x,t) = \begin{cases} 0 \text{ when } x < 0,\ t > 0 \\ 1 \text{ when } x > t > 0. \end{cases}$$

But, the region $0 < x < t$ in the (x,t) plane is not covered by any characteristic and thus u is not defined there. Now, define the function

$$u_1(x,t) = \begin{cases} 0 \text{ if } x < \frac{t}{2},\ t > 0 \\ 1 \text{ if } x > \frac{t}{2},\ t > 0 \end{cases}$$

and

$$u_2(x,t) = \begin{cases} 0 \text{ if } x < 0,\ t > 0 \\ \frac{x}{t} \text{ if } 0 < x < t,\ t > 0 \\ 1 \text{ if } x > t > 0. \end{cases}$$

It is easy to see that u_1 is a generalized solution; the line $x = \frac{t}{2}$ is its discontinuity curve and u_1 satisfies R–H condition across this discontinuity curve. The function u_2 is a continuous generalized solution. But u_2 is not differentiable across the lines $x = 0$ and $x = t$ (see Figure 5.3). □

Later we will see that the solution u_1 is not a physically relevant solution as the characteristics emanating from either side of the discontinuity curve move away from the discontinuity curve. In other words, the discontinuity is not formed by the intersection of two different characteristics. The second solution u_2 is called a *rarefaction wave*. Later, this will be shown to be the physically relevant solution.

The above example also exhibits non-uniqueness of a generalized solution. To pick a physically relevant solution and perhaps to prove uniqueness, we thus need to impose

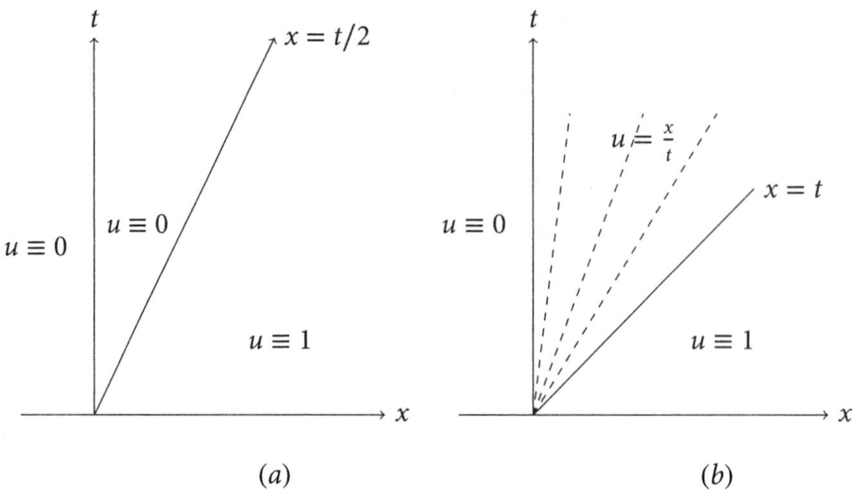

Figure 5.3 (a) Line of discontinuity, (b) Rarefaction wave

additional condition(s). This turns out to be the *entropy condition*, a terminology borrowed from thermodynamics.

Example 5.5 (Shock Discontinuity). The initial condition here is the non-decreasing function u_0 given by □

$$u_0(x) = \begin{cases} 1 \text{ if } x \leq 0 \\ 1 - x \text{ if } 0 \leq x \leq 1 \\ 0 \text{ if } x \geq 1. \end{cases}$$

See Figure 5.4. As shown in Chapter 3, the characteristics emanating from any two distinct points in the interval where u_0 is decreasing bound to intersect in the region $t > 0$, thus making the definition of the solution u after some time ambiguous. Drawing various characteristics, it is easy to show that the characteristics do not intersect till $t = 1$. Thus, using the method of characteristics, we can construct the solution u up to $t = 1$; it is given by

$$u(x,t) = \begin{cases} 1 \text{ if } x < t \\ \dfrac{1-x}{1-t} \text{ if } t < x < 1 \\ 0 \text{ if } x > 1, \end{cases}$$

for $0 \leq t < 1$. This procedure breaks down for $t \geq 1$. We now construct a curve of discontinuity $s(t)$ for $t \geq 1$ satisfying $s(1) = 1$. Define $u = 1$ on the left side of $s(t)$ and

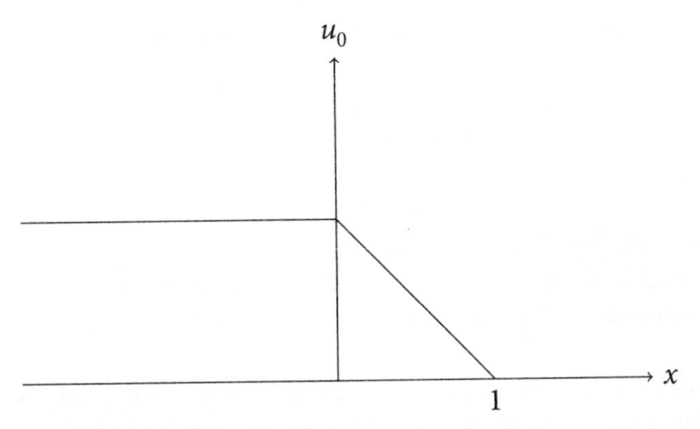

Figure 5.4 Shock discontinuity

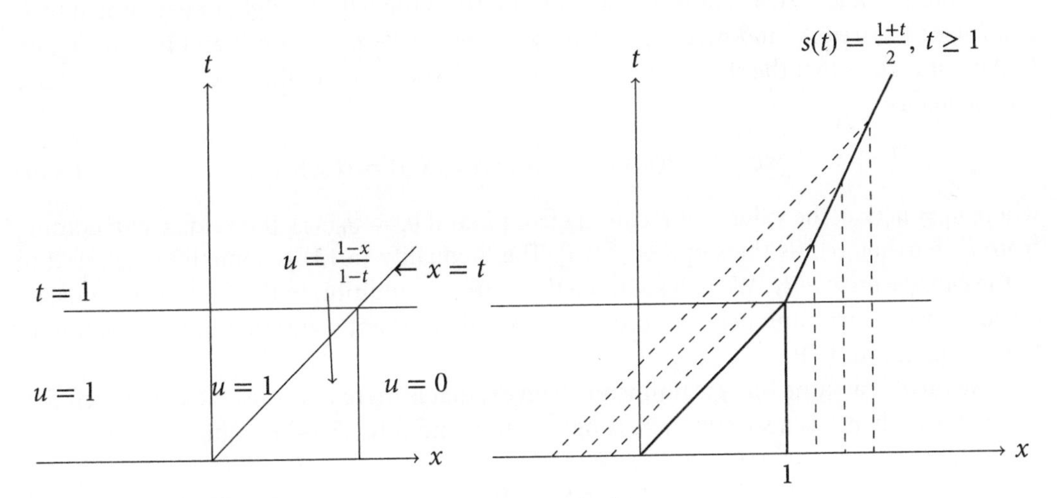

Figure 5.5 Solution and line of discontinuity

$u = 0$ on the right side of $s(t)$. For u to be a generalized solution, the R–H condition need be satisfied and this immediately shows that $s(t) = \dfrac{1+t}{2}$. Thus, we define u, for $t \geq 1$ as

$$u(x, t) = \begin{cases} 1 \text{ if } x < s(t) \\ 0 \text{ if } x > s(t). \end{cases}$$

It can be easily verified that u defined by the above two expressions for $t \in [0, 1)$ and $t \geq 1$, is a generalized solution, satisfying the R–H condition (see Figure 5.5). The curve $s(t)$ is known as *shock discontinuity*.

A similar analysis can be done if we take the initial condition u_0 as

$$u_0(x) = \begin{cases} 1 \text{ if } x < 0 \\ 0 \text{ if } x > 0. \end{cases}$$

In this case, the line of discontinuity is the line $x = t$; $u = 1$ on its left and $u = 0$ on its right. We remark that both the cases of rarefaction wave and shock wave can be observed in a *shock tube* experiment, where u represents the pressure of the gas in the shock tube.

Entropy Condition. A heuristic argument: Consider a smooth flux f and initial condition u_0. Then, from any point $(x_0, 0)$ on the initial line $t = 0$, the characteristic curve (of (5.4)) is a straight line with slope $(c(u_0(x_0)))^{-1}$ with $c = f'$. Now consider a curve γ of discontinuity of the solution u and two characteristics coming from the left and right of γ, denoted by l_1 and l_2 (see Figure 5.6) and meeting at P on γ. Indeed, $u \equiv u_0(x_1)$ on l_1 and $u \equiv u_0(x_2)$ on l_2. Also, it is clear that the slopes of l_1, γ and l_2 should be in increasing order. Thus, we have the inequality

$$c(u_l) = c(u_0(x_1)) > \sigma > c(u_0(x_2)) = c(u_r), \tag{5.14}$$

where $u_l = u_0(x_1)$, the value u at P coming from l_1 and $u_r = u_0(x_2)$, the value u at P coming from l_2. Further, σ^{-1} is the slope of γ at P. The inequality (5.14) is sometimes referred to as the *entropy inequality*. We will soon see the entropy condition in the form of a regularity estimate, which will uniquely capture the physically relevant solution and thus leading to the uniqueness of IVP.

If we further assume that f is uniformly convex, that is there is a constant $\kappa > 0$ such that $f'' \geq \kappa > 0$, then, c is a strictly increasing function and thus (5.14) is equivalent to

$$u_l > u_r \tag{5.15}$$

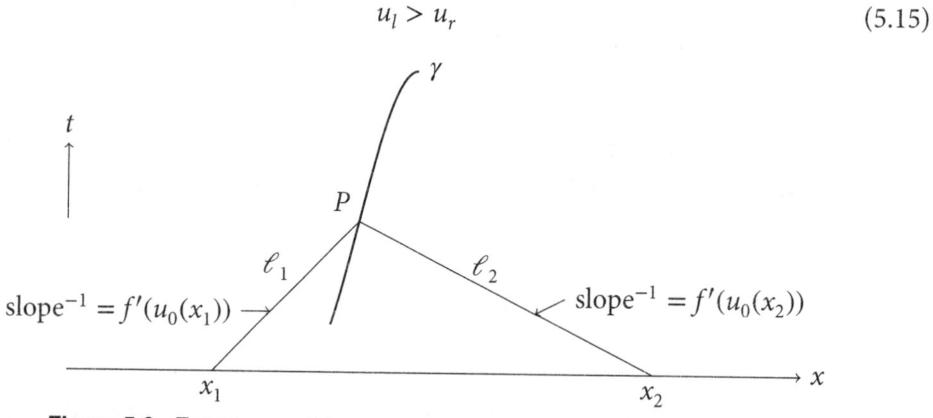

Figure 5.6 Entropy condition

along the curve of discontinuity. this is also a necessary condition that we will use in another form to define a weak solution and prove uniqueness.

We now obtain a formula, known as Lax–Oleinik formula, for a generalized solution of the IVP (5.10). This is obtained using Hopf–Lax formula, derived in Chapter 4.

5.3 LAX–OLEINIK FORMULA

Consider the IVP (5.10), where we assume that the flux function f is uniformly convex, $f(0) = 0$ and u_0 is a C^1 function. First we will connect the conservation law and the following HJE with f as the Hamiltonian:

$$\begin{cases} w_t + f(w_x) = 0 \text{ in } \mathbb{R} \times (0, \infty) \\ w(x, 0) = w_0(x) \text{ for } x \in \mathbb{R}, \end{cases} \tag{5.16}$$

with a given smooth initial function w_0. To get an idea about the connection, assume that f is a C^2 function and w is the unique smooth C^2 solution of the HJE (5.16). Differentiating the HJE with respect to x, we get

$$\begin{cases} (w_x)_t + (f(w_x))_x = 0 \text{ in } \mathbb{R} \times (0, \infty) \\ w_x(x, 0) = w_{0x}(x) \text{ on } \mathbb{R}. \end{cases} \tag{5.17}$$

Therefore, if we take $w_0(x) = \int_0^x u_0(y)dy$ and if w is the corresponding solution of (5.16), then, $u(x, t) = w_x(x, t)$ would be a solution of the IVP (5.10).

We know from the results of Chapter 4 that the solution w has a representation given by the Hopf–Lax formula (4.9):

$$w(x, t) = \inf_{y \in \mathbb{R}} \left\{ tL\left(\frac{x - y}{t}\right) + w_0(y) \right\}, \tag{5.18}$$

where L is the Legendre transformation of f, given by

$$L(y) = f^*(y) = \sup_{x \in \mathbb{R}} \{xy - f(x)\}. \tag{5.19}$$

As a consequence, we formally have

$$u(x, t) = \frac{\partial}{\partial x} \left[\inf_{y \in \mathbb{R}} \left\{ tL\left(\frac{x - y}{t}\right) + w_0(y) \right\} \right]. \tag{5.20}$$

The difficulty arises as the solution of the HJE need not be differentiable everywhere. Thus, we expect the expression on the right side of (5.20) to make sense only a.e. (x) and hope that the function u thus defined is going to be a possible candidate for a generalized solution of the IVP (5.10).

Let $y = y(x, t)$ be the minimizer of the right-hand side in (5.18). That is

$$w(x,t) = tL\left(\frac{x - y(x,t)}{t}\right) + w_0(y(x,t)).$$

Claim: For fixed $t > 0$, the mapping $x \mapsto y(x, t)$ is a non-decreasing function.

By a theorem of Lebesgue, it then follows that the function $y(\cdot, t)$ is differentiable a.e. (x).

To see the claim, let $x_1 < x_2$ and $y_1 = y(x_1, t), y_2 = y(x_2, t)$, respectively, be the minimizers corresponding to x_1 and x_2. We need to prove $y_1 \leq y_2$. In fact, we prove the following inequality

$$tL\left(\frac{x_2 - y_1}{t}\right) + w_0(y_1) < tL\left(\frac{x_2 - y}{t}\right) + w_0(y) \tag{5.21}$$

valid for all $y < y_1$, which in turn proves the claim. Using the convexity of L and by writing

$$x_2 - y_1 = \alpha(x_1 - y_1) + (1-\alpha)(x_2 - y) \text{ and } x_1 - y = (1-\alpha)(x_1 - y_1) + \alpha(x_2 - y),$$

with $\alpha = \dfrac{y_1 - y}{x_2 - x_1 + y_1 - y} \in (0, 1)$, we get

$$L\left(\frac{x_2 - y_1}{t}\right) \leq \alpha L\left(\frac{x_1 - y_1}{t}\right) + (1-\alpha)L\left(\frac{x_2 - y}{t}\right).$$

and

$$L\left(\frac{x_1 - y}{t}\right) \leq (1-\alpha)L\left(\frac{x_1 - y_1}{t}\right) + \alpha L\left(\frac{x_2 - y}{t}\right).$$

Adding these two inequalities, we have

$$L\left(\frac{x_2 - y_1}{t}\right) + L\left(\frac{x_1 - y}{t}\right) \leq L\left(\frac{x_1 - y_1}{t}\right) + L\left(\frac{x_2 - y}{t}\right).$$

A simple manipulation then gives

$$\left[tL\left(\frac{x_2 - y_1}{t}\right) + w_0(y_1)\right] + \left[tL\left(\frac{x_1 - y}{t}\right) + w_0(y)\right]$$
$$\leq \left[tL\left(\frac{x_1 - y_1}{t}\right) + w_0(y_1)\right] + \left[tL\left(\frac{x_2 - y}{t}\right) + w_0(y)\right].$$

Since y_1 is the minimizer for $tL\left(\frac{x_1 - y}{t}\right) + w_0(y)$, we get the required inequality (5.21). Since $y_2 = y(x_2, t)$ is the minimizer, the inequality (5.21) cannot be satisfied if we take $y = y_2$. Thus, $y_2 < y_1$ is not possible and consequently, $y_1 \leq y_2$.

Further, for $t > 0$ fixed, the mappings

$$x \mapsto L\left(\frac{x - y(x,t)}{t}\right) \text{ and } x \mapsto w_0(y(x,t))$$

are also differentiable a.e. (x). Hence, a.e. (x) we get

$$\begin{aligned} u(x,t) &= \frac{\partial}{\partial x}\left[tL\left(\frac{x-y(x,t)}{t}\right) + w_0(y(x,t))\right] \\ &= L'\left(\frac{x-y(x,t)}{t}\right)\left(1 - \frac{\partial y}{\partial x}(x,t)\right) + \frac{\partial}{\partial x}w_0(y(x,t)) \\ &= L'\left(\frac{x-y(x,t)}{t}\right) + \left[-L'\left(\frac{x-y(x,t)}{t}\right) + w_0'(y(x,t))\right]\frac{\partial y}{\partial x}(x,t). \end{aligned}$$

Next, for x, t fixed, since $y(x,t)$ is minimizer for $tL\left(\frac{x-y}{t}\right) + w_0(y)$, we have

$$\frac{\partial}{\partial y}\left[tL\left(\frac{x-y}{t}\right) + w_0(y)\right] = 0$$

at $y = y(x,t)$. That is,

$$-L'\left(\frac{x-y(x,t)}{t}\right) + w_0'(y(x,t)) = 0.$$

Thus, we have

$$u(x,t) = L'\left(\frac{x-y(x,t)}{t}\right).$$

Now, for a given $y \in \mathbb{R}$, let $x^* \in \mathbb{R}$ be such that the supremum is achieved in (5.19), that is, $L(y) = f^*(y) = x^*y - f(x^*)$. It is evident that $y = f'(x^*)$, since x^* is a maximizer. Since f is assumed to be uniformly convex, it follows that f' is strictly increasing and onto. Thus, if we put $G = \left(f'\right)^{-1}$ (the functional inverse), it follows that $x^* = \left(f'\right)^{-1}(y) = G(y)$. Hence,

$$L(y) = yG(y) - f(G(y)).$$

Thus,

$$L'(y) = G(y) + yG'(y) - f'(G(y))G'(y) = G(y)$$

as $f'(G(y)) = y$. Thus, we have

$$u(x,t) = G\left(\frac{x - y(x,t)}{t}\right). \tag{5.22}$$

This is known as the *Lax–Oleinik formula* for the solution u of the conservation law. The foregoing discussion is consolidated in the form of the following theorem:

Theorem 5.6 (Lax–Olenik formula). Assume the flux function f is a C^2, uniformly convex function and $u_0 \in L^\infty(\mathbb{R})$. Define $w_0(x) = \int_0^x u_0(x)\,dx$. For $x \in \mathbb{R}, t > 0$ fixed, let $y(x,t)$ be the minimizer in (5.18) that exists a.e. $(x, t > 0)$. Then, the following hold:

1. The mappings

$$x \mapsto y(x,t),\ x \mapsto w_0(y(x,t)) \text{ and } x \mapsto tL\left(\frac{x - y(x,t)}{t}\right)$$

are all differentiable a.e. $(x, t > 0)$.

2. The function u defined by (5.20) can be written as

$$u(x,t) = G\left(\frac{x - y(x,t)}{t}\right) = (f')^{-1}\left(\frac{x - y(x,t)}{t}\right). \tag{5.23}$$

□

We next verify that the function u given in the above theorem is indeed a generalized solution of the IVP (5.10).

Theorem 5.7 (Existence). Assume the hypotheses of Theorem 5.6. Then, the function u given by the Lax–Oleinik formula (5.23) is a generalized solution of the IVP (5.10). □

Proof We need to show that u satisfies the integral equation (5.10) for all the test functions v described there. Let w be the solution of the HJE (5.16). Then, from HJE theory (see Theorem 4.4), we know that w is a Lipschitz function and hence differentiable a.e. $(x, t > 0)$.

Let v be a test function. Multiplying equation (5.16) by v_x, we get

$$\int_0^\infty \int_{-\infty}^\infty \left(w_t v_x + f(w_x) v_x\right) dx\,dt = 0. \tag{5.24}$$

Integrating by parts with respect to the t variable, we get

$$\int_0^\infty \int_{-\infty}^\infty w_t v_x \, dx \, dt = -\int_0^\infty \int_{-\infty}^\infty w v_{tx} \, dx \, dt - \int_{-\infty}^\infty w_0(x) v_x(x,0) \, dx.$$

Note that the mapping $x \mapsto w(x,t)$ is Lipschitz continuous for each $t > 0$ and hence absolutely continuous. Thus, w_x exist a.e., $w(x,0) = w_0(x)$ and $w_x(x,0) = u_0(x)$. This justifies the above-performed integration by parts. Next, performing an integration by parts of the second and third terms on the right-hand side of the above equation with respect to x and using (5.24), we see that w satisfies

$$\int_0^\infty \int_{-\infty}^\infty \left(w_x v_t + f(w_x) v_x \right) dx \, dt - \int_{-\infty}^\infty u_0(x) v(x,0) \, dx = 0.$$

Since $u = w_x$, we conclude that u satisfies the integral equation (5.10). This completes the proof of the theorem. □

Example 5.8 (Rarefaction). Consider the Burgers' equation $u_t + u u_x = 0$. Here $f(u) = u^2/2$. It is easy to calculate that $L(y) = f^*(y) = y^2/2$. Thus, $u(x,t) = \dfrac{x - y(x,t)}{t}$, where $y(x,t)$ is the minimizer of

$$tL\left(\frac{x-y}{t}\right) + w_0(y) = \frac{(x-y)^2}{t} + w_0(y).$$

If we take the rarefaction case,

$$u_0(x) = \begin{cases} 0 \text{ if } x < 0 \\ 1 \text{ if } x > 0, \end{cases}$$

then

$$w_0(x) = \begin{cases} 0 \text{ if } x \leq 0 \\ y \text{ if } x > 0. \end{cases}$$

Indeed, there is a lack of differentiability of w_0 at $y = 0$. Nevertheless, we can compute the minimizer $y(x,t)$ as

$$y(x,t) = \begin{cases} x \text{ if } x < 0, \ t > 0 \\ 0 \text{ if } t > x \geq 0 \\ x - t \text{ if } x \geq t > 0. \end{cases}$$

Thus, we get the rarefaction solution as in Example 5.4. □

5.4 GENERALIZED SOLUTION AND UNIQUENESS

In the previous section, we have established the existence of a generalized solution for IVP (5.10) via the explicit Lax–Oleinik formula for the solution. But, we have seen examples that the generalized solutions are not unique, we may end up having non-physical solutions. However, we may expect the Lax–Oleinik formula provides the correct solution and hence we must deduce certain necessary conditions from such a solution to rule out other non-physical solutions.

In Hamilton–Jacobi theory, the Hopf–Lax formula provided a one-sided second derivative estimate for the solution. But, since the solution u for the conservation law is obtained as the derivative of the solution for HJE, we expect a one-sided first derivative estimate for u. This estimate is called the *entropy condition* and is given in the following proposition:

Proposition 5.9 (One-Sided Derivative Estimate-Entropy Condition). Under the hypotheses of Theorem 5.6, the solution u given by the Lax–Oleinik formula satisfies the inequality: there is a constant $C > 0$ such that

$$u(x_2, t) - u(x_1, t) \leq \frac{C}{t}(x_2 - x_1), \tag{5.25}$$

for all $x_1 < x_2$ and $t > 0$. □

Proof The proof is almost trivial. We use the monotonicity of $y(x, t)$ and $G(x, t)$ to get

$$G\left(\frac{x_2 - y(x_2, t)}{t}\right) - G\left(\frac{x_1 - y(x_2, t)}{t}\right) \leq \frac{k}{t}(x_2 - x_1),$$

where k is the Lipschitz constant of G; it is easily checked that $k \leq C^{-1}$, where C satisfies the estimate $f'' \geq C$. Thus,

$$\begin{aligned} G\left(\frac{x_2 - y(x_2, t)}{t}\right) &\leq G\left(\frac{x_1 - y(x_2, t)}{t}\right) + \frac{k}{t}(x_2 - x_1) \\ &\leq G\left(\frac{x_1 - y(x_1, t)}{t}\right) + \frac{k}{t}(x_2 - x_1), \end{aligned}$$

Hence,

$$G\left(\frac{x_2 - y(x_2, t)}{t}\right) - G\left(\frac{x_1 - y(x_1, t)}{t}\right) \leq \frac{k}{t}(x_2 - x_1).$$

This proves the proposition, using the Lax–Oleinik formula. □

Definition 5.10 (Entropy Solution). A function $u \in L^\infty(\mathbb{R}\times[0,\infty))$ is said to be an *entropy solution* of the IVP (5.9) with the initial condition $u(x,0) = u_0(x)$, where u_0 is an integrable function, if it satisfies the integral equation (5.10) for all the test functions v and satisfies the entropy condition (5.25). □

In what follows we make the convention that a test function is a C^1 function having compact support either in $\mathbb{R}\times[0,\infty)$ or $\mathbb{R}\times(0,\infty)$.

Theorem 5.11 (Existence and Uniqueness). Assume the hypotheses as in Theorem 5.6. Then the function u given by the Lax–Oleinik formula (5.23) is the unique entropy solution of the IVP (5.9). □

Proof We only need to prove the uniqueness. This lengthy proof will be divided into several steps.

Step 1: Let u_1, u_2 be two generalized solutions satisfying the entropy condition and satisfying the same initial condition. Put $w = u_1 - u_2$. We have to show that $w = 0$ a.e. in $t > 0$, or equivalently, to show that $\int_0^\infty \int_{-\infty}^\infty w\phi\, dxdt = 0$ for all the test functions ϕ; here the support of ϕ is a compact subset of $\mathbb{R}\times(0,\infty)$.

For brevity, we write

$$\iint\limits_{t\geq\alpha} dxdt = \int_\alpha^\infty \int_{-\infty}^\infty dxdt \text{ and } \iint\limits_{t\leq\alpha} dxdt = \int_0^\alpha \int_{-\infty}^\infty dxdt,$$

for $\alpha \geq 0$ and for $\alpha > 0$, respectively. Since both u_1 and u_2 satisfy the integral relation (5.10), upon subtraction we get

$$\iint\limits_{t\geq 0} wv_t + (f(u_1) - f(u_2))v_x dxdt = 0$$

for all test functions v having compact support in $\mathbb{R}\times[0,\infty)$. Now,

$$f(u_1(x,t)) - f(u_2(x,t)) = \int_0^1 f'(\sigma u_1(x,t) + (1-\sigma)u_2(x,t))w(x,t)\, d\sigma.$$

Define $\theta(x,t) = \int_0^1 f'(\sigma u_1(x,t) + (1-\sigma)u_2(x,t))\, d\sigma$, so that we get

$$\iint\limits_{t\geq 0} w(v_t + \theta v_x)\, dxdt = 0 \tag{5.26}$$

for all test functions v. Thus, for a given test function ϕ, if we could solve the equation $v_t + \theta v_x = \phi$, to get a test function v, then we are done and the uniqueness is proved.

Step 2: The difficulty is that the function θ is not even continuous. To overcome this difficulty, we need to begin with smooth approximations to u_1 and u_2, which in turn will give a smooth approximation to θ. This is accomplished by making use of the standard mollifiers η_ε in $\mathbb{R}^2$ (in x, t variables). The arguments are quite technical.

Define $u_i^\varepsilon = \eta_\varepsilon * u_i$ for $i = 1, 2$. The well-known properties of the mollifiers give the following:

$$u_i^\varepsilon \to u_i \text{ a.e. as } \varepsilon \to 0 \text{ and } \|u_i^\varepsilon\|_{L^\infty} \leq \|u_i\|_{L^\infty}. \tag{5.27}$$

Now, consider the smooth approximation to θ defined by

$$\theta_\varepsilon(x,t) = \int_0^1 f'(\sigma u_1^\varepsilon(x,t) + (1-\sigma)u_2^\varepsilon(x,t))d\sigma.$$

Suppose, for a given test function ϕ, we are able to solve the first-order equation $v_t^\varepsilon + \theta_\varepsilon v_x^\varepsilon = \phi$, to get a test function v^ε, then

$$\begin{aligned}
\iint_{t\geq 0} w\phi\, dxdt &= \iint_{t\geq 0} wv_t^\varepsilon\, dxdt + \iint_{t\geq 0} w\theta_\varepsilon v_x^\varepsilon\, dxdt \\
&= -\iint_{t\geq 0} (f(u_1) - f(u_2))v_x^\varepsilon\, dxdt + \iint_{t\geq 0} w\theta_\varepsilon v_x^\varepsilon\, dxdt \\
&= \iint_{t\geq 0} w(\theta_\varepsilon - \theta)v_x^\varepsilon\, dxdt.
\end{aligned}$$

The second equality holds as u_1 and u_2 satisfy the integral relation (5.10).

If $\theta_\varepsilon \to \theta$ in $L^1_{\text{loc}}(\mathbb{R}^2)$ as $\varepsilon \to 0$ and v_x^ε is bounded, independent of ε, then we can pass to the limit as $\varepsilon \to 0$ on the right-hand side and obtain the required relation that

$$\iint_{t\geq 0} w\phi\, dxdt = 0$$

for all the test functions ϕ.

Step 3: First, we prove $\theta_\varepsilon \to \theta$ in $L^1_{\text{loc}}(\mathbb{R}^2)$. This is not very difficult.
Suppose $|u_i|_{L^\infty} \le M$ for $i = 1, 2$. It follows from the definition of the mollifiers that

$$|u_i^\varepsilon(x,t)| = \left| \iint_{\mathbb{R}^2} \eta_\varepsilon(x-y, t-s) u_i(y,s)\, dyds \right|$$
$$\le M \iint_{\mathbb{R}^2} \eta_\varepsilon(y,s)\, dyds \;=\; M,\; i = 1, 2.$$

Next, we have

$$\theta_\varepsilon(x,t) = \int_0^1 [f'(\sigma u_1^\varepsilon(x,t) + (1-\sigma)u_2^\varepsilon(x,t)) - f'(0)]\, d\sigma + f'(0)$$
$$= \int_0^1 f''(\xi)(\sigma u_1^\varepsilon(x,t) + (1-\sigma)u_2^\varepsilon(x,t))\, d\sigma + f'(0),$$

for $\xi = \tau(\sigma u_1^\varepsilon(x,t) + (1-\sigma)u_2^\varepsilon(x,t))$ for some $\tau \in [0,1]$. Therefore, we have the estimate

$$|\theta_\varepsilon(x,t)| \le M_1 \equiv cM + |f'(0)|, \tag{5.28}$$

where $c = \sup\{f''(u) : |u| \le M\}$. In particular, θ_ε is bounded independent of ε. Similarly, θ is also bounded. Now consider

$$\begin{aligned}
&|\theta_\varepsilon(x,t) - \theta(x,t)| = \\
&\left| \int_0^1 [f'(\sigma u_1^\varepsilon(x,t) + (1-\sigma)u_2^\varepsilon(x,t)) - f'(\sigma u_1(x,t) + (1-\sigma)u_2(x,t))]\, d\sigma \right| \\
&= \left| \int_0^1 f''(\xi)[\sigma(u_1^\varepsilon - u_1) + (1-\sigma)(u_2^\varepsilon - u_2)]\, d\sigma \right| \\
&\le \int_0^1 |f''(\xi)| [\sigma |u_1^\varepsilon - u_1| + (1-\sigma)|u_2^\varepsilon - u_2|]\, d\sigma \\
&\le \tfrac{c}{2} \left(|u_1^\varepsilon(x,t) - u_1(x,t)| + |u_2^\varepsilon(x,t) - u_2(x,t)| \right).
\end{aligned}$$

It therefore follows that

$$\iint_K |\theta_\varepsilon - \theta| \to 0 \text{ as } \varepsilon \to 0 \tag{5.29}$$

for any compact subset K of $\mathbb{R}^2$.

We next analyze v^ε and its derivative. Recall that the requirements on v^ε are that it is a test function having support in $\mathbb{R} \times [0, \infty)$ and satisfies the first-order linear PDE

$$v_t^\varepsilon + \theta_\varepsilon v_x^\varepsilon = \phi, \tag{5.30}$$

for a given test function ϕ having compact support in $\mathbb{R} \times (0, \infty)$. This requires more delicate arguments, since we have to prove the uniform boundedness of the derivative of v^ε. First, we prove that indeed v^ε is a test function having support in $\mathbb{R} \times [0, \infty)$.

Step 4: Let S be the support of ϕ and assume that

$$S \subset (a, b) \times (\delta, T)$$

for some finite $a < b$ and $0 < \delta < T$. If we impose the *terminal* condition $v^\varepsilon(x, T) = 0$ for $x \in \mathbb{R}$ on the line $t = T$, then by the method of characteristics, we find that the unique solution to (5.30) is given by

$$v^\varepsilon(x, t) = -\int_t^T \phi(x_\varepsilon(s), s)\, ds, \tag{5.31}$$

where x_ε is the characteristic curve passing through (x, t), given by

$$\dot{x}_\varepsilon(s) \equiv \frac{dx_\varepsilon}{ds}(s) = \theta^\varepsilon(x_\varepsilon(s), s), \; x_\varepsilon(t) = x.$$

To emphasize the dependence on x, t and for later analysis, we denote this characteristic by $x_\varepsilon(s; x, t)$. In order to show that v^ε has compact support, let R be the domain bounded by the lines $t = 0$, $t = T$ and the lines ℓ_1, ℓ_2, as shown in Figure 5.7. Here, ℓ_1 is the line given by $x = A + M_1 t$ and ℓ_2 is the line given by $x = B - M_1 t$ in the upper half plane, where A and B are chosen on the line $t = 0$ such that $A + M_1 T < a$ and $B - M_1 T > b$. Thus, S lies in the interior of R. Recall that M_1 is the bound on θ^ε and is independent of ε. We show that the support of v^ε lies in R, which is also independent of ε.

Since $\phi(x, t) = 0$ for $t \geq T$, it is immediate that $v^\varepsilon(x, t)$ also vanishes for $t \geq T$. Suppose $(x, t) \notin R$ and $t < T$. For definiteness, assume $x > B$. Then, the characteristic curve given by $x_\varepsilon(s; x, t)$ lies to the right of the line ℓ_2 and crosses the line $t = T$ at a point outside R. Thus, ϕ vanishes along this characteristic and thereby proving, using the formula (5.31), that $v^\varepsilon(x, t) = 0$.

We now move to the delicate analysis of boundedness of v_x^ε. We divide the region of integration $t \geq 0$ into two parts: $t \geq \alpha$, $\alpha > 0$ and $0 \leq t < \alpha$. We deal with the first part using entropy inequality and we use a bound on the total variation of v^ε for the second part.

Step 5: We consider the case $t \geq \alpha$, $\alpha > 0$. If u denotes either of u_i, $i = 1, 2$, recall the entropy condition:

$$u(x + a, t) - u(x, a) \leq \frac{Ca}{t},$$

for all x and $a > 0$, $t > 0$, for some constant $C > 0$. Let $\alpha > 0$ be arbitrary. We claim that for fixed $t \geq \alpha$, the function $x \mapsto u(x, t) - \frac{Cx}{\alpha}$ is non-increasing. For, if $a > 0$ and $t \geq \alpha$,

$$\begin{aligned} u(x + a, t) - \frac{C(x + a)}{\alpha} - \left(u(x, t) - \frac{Cx}{\alpha}\right) &= u(x + a, t) - u(x, t) - \frac{Ca}{\alpha} \\ &\leq Ca\left(\frac{1}{t} - \frac{1}{\alpha}\right) \leq 0, \end{aligned}$$

using the entropy condition. Since the mollifiers η_ε are non-negative functions, the function (of x)

$$\eta_\varepsilon * \left(u(x, t) - \frac{Cx}{\alpha}\right) = u^\varepsilon - \frac{C}{\alpha}(\eta_\varepsilon * x)$$

is also non-increasing. Since the latter function is differentiable, we see that

$$\frac{\partial u^\varepsilon}{\partial x} \leq \frac{C}{\alpha}$$

for $t \geq \alpha$. Replacing u by u_1 and u_2, we obtain the bounds on $\frac{\partial u_i^\varepsilon}{\partial x}$, $i = 1, 2$. Next, using the definition of θ_ε, we obtain that

$$\begin{aligned} \frac{\partial \theta_\varepsilon}{\partial x} &= \int_0^1 f''(\sigma u_1^\varepsilon(x, t) + (1 - \sigma)u_2^\varepsilon(x, t)) \left[\sigma \frac{\partial u_1^\varepsilon}{\partial x} + (1 - \sigma)\frac{\partial u_2^\varepsilon}{\partial x}\right] d\sigma \\ &\leq \frac{C}{\alpha}c, \text{ using } f'' > 0 \\ &\leq K_\alpha \end{aligned}$$

for all $t \geq \alpha$ and K_α is independent of ε.

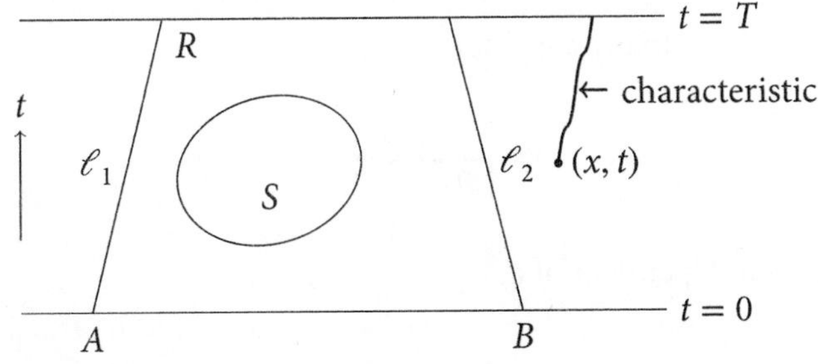

Figure 5.7 Support of v^ε

To estimate $\frac{\partial v^\varepsilon}{\partial x}$, we need to analyze $\frac{\partial x_\varepsilon}{\partial x}$ for the characteristic curve $x_\varepsilon(s; x, t)$; see (5.31). Now let

$$a_\varepsilon(s) = \frac{\partial x_\varepsilon}{\partial x}(s; \bar{x}, \bar{t})$$

where $(\bar{x}, \bar{t})$ is a point in the upper half plane. Note that

$$\frac{dx_\varepsilon}{ds} = \theta_\varepsilon(x_\varepsilon(s), s)) \text{ and } x_\varepsilon(\bar{t}; \bar{x}, \bar{t}) = \bar{x}.$$

This implies that

$$a_\varepsilon(\bar{t}) = 1. \tag{5.32}$$

Thus,

$$\frac{\partial a_\varepsilon}{\partial s} = \frac{\partial}{\partial s}\frac{\partial x_\varepsilon}{\partial x} = \frac{\partial}{\partial x}\frac{\partial x_\varepsilon}{\partial s} = \frac{\partial}{\partial x}\theta_\varepsilon(x_\varepsilon(s; \bar{x}, \bar{t}), s)) = \frac{\partial \theta_\varepsilon}{\partial x}\frac{\partial x_\varepsilon}{\partial x} = \frac{\partial \theta_\varepsilon}{\partial x}a_\varepsilon.$$

Using (5.32), we obtain the formula

$$a_\varepsilon(s) = \exp\left(\int_{\bar{t}}^{s} \frac{\partial \theta_\varepsilon}{\partial x}(x_\varepsilon(\tau), \tau)\, d\tau\right). \tag{5.33}$$

If we take $\alpha \le \bar{t} \le s \le T$, then using the bound on $\frac{\partial \theta_\varepsilon}{\partial x}$, we have

$$|a_\varepsilon(s)| = a_\varepsilon(s) \le \exp(K_\alpha(T - \alpha)).$$

Using the formula (5.31), we have

$$\frac{\partial v^\varepsilon}{\partial x} = \int_T^t \frac{\partial \phi}{\partial x_\varepsilon}\frac{\partial x_\varepsilon}{\partial x} = \int_T^t \frac{\partial \phi}{\partial x_\varepsilon}a_\varepsilon.$$

Using the bound on a_ε, we finally conclude that

$$\left|\frac{\partial v^\varepsilon}{\partial x}\right| \le \bar{K}_\alpha \tag{5.34}$$

for $t \ge \alpha$, where $\bar{K}_\alpha$ is independent of ε.

Step 6: Now, we take up the case of t near 0. We adopt a different strategy, namely, we estimate the total variation of v^ε near the line $t = 0$. Define

$$V^t(v^\varepsilon) = \int_{-\infty}^{\infty} \left| \frac{\partial v^\varepsilon}{\partial x} \right| dx \tag{5.35}$$

which is the total variation of v^ε as a function of x, for each fixed $t > 0$. Using the estimate (5.34) and the fact that $\text{supp}(v^\varepsilon) \subset R$ (which is independent of ε), we see that

$$V^t(v^\varepsilon) \leq C_\alpha, \text{ for } t \geq \alpha, \tag{5.36}$$

where the constant C_α is independent of ε. We need one more estimate, namely: There is a positive integer N such that for all $n > N$, we have

$$V^t(v^\varepsilon) \leq C_{1/n}, \text{ for all } t \text{ satisfying } 0 < t \leq \frac{1}{n} < \frac{1}{N}. \tag{5.37}$$

To see this, choose N such that $\frac{1}{N} < \delta$ so that $\phi(x, t) = 0$ if $t < \frac{1}{N}$. This is possible since, by assumption, $\text{supp}(\phi) \subset (a, b) \times (\delta, T)$. Therefore

$$\frac{\partial v^\varepsilon}{\partial t} + \theta_\varepsilon \frac{\partial v^\varepsilon}{\partial x} = 0, \text{ if } t < \frac{1}{N}. \tag{5.38}$$

Let $n > N$, so that (5.38) holds for $t \leq \frac{1}{n}$. Let $\sigma^t : \mathbb{R} \to \mathbb{R}$ be the bijection map defined for $t < \frac{1}{n}$ by the solution of the characteristic equation, that is, $\sigma^t(x) = x_\varepsilon\left(\frac{1}{n}; x, t\right)$; see Figure 5.8.

Now let $0 < t \leq \frac{1}{n} < \frac{1}{N}$. For any finite sequence $x_1 < x_2 < \cdots < x_p$, we have

$$\sum_{k=1}^{p-1} \left| v^\varepsilon(x_{k+1}, t) - v^\varepsilon(x_k, t) \right| = \sum_{k=1}^{p-1} \left| v^\varepsilon\left(\sigma^t(x_{k+1}), \frac{1}{n}\right) - v^\varepsilon\left(\sigma^t(x_k), \frac{1}{n}\right) \right|$$

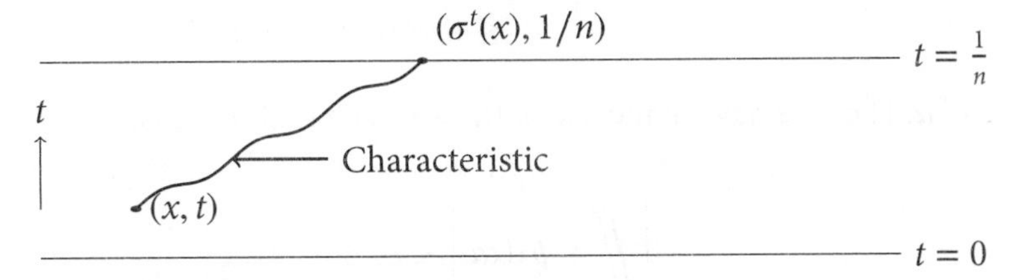

Figure 5.8 Bijective map σ^t

using the constancy of v^ε along a characteristic (see (5.38)). Using (5.38) and (5.36), we therefore conclude that

$$\sum_{k=1}^{p-1} \left| v^\varepsilon(x_{k+1}, t) - v^\varepsilon(x_k, t) \right| \le V^{1/n}(v^\varepsilon) \le C_{1/n},$$

proving the estimate (5.37).

Step 7: We can now complete the proof of the theorem, namely the uniqueness of an entropy solution. Let $\tilde{\varepsilon} > 0$ be arbitrarily small. With N and $n > N$ as chosen above, choose $\alpha > 0$ such that $\alpha < 1/n < 1/N$ and $4MM_1C_{1/n}\alpha < \tilde{\varepsilon}/2$. For this α, choose ε_0 small, so that

$$\iint_{t\ge\alpha} |w||\theta_\varepsilon - \theta||v_x^\varepsilon|\, dxdt < \tilde{\varepsilon}/2 \tag{5.39}$$

for $\varepsilon < \varepsilon_0$. This is possible since $|w| \le 2M$, $\theta_\varepsilon \to \theta$ in L^1_{loc} and v^ε have compact support in R with $|v_x^\varepsilon| \le C_\alpha$, independent of ε. Next, we write

$$\left| \iint_{t\ge 0} w\phi\, dxdt \right| \le \iint_{t\ge\alpha} |w||\theta_\varepsilon - \theta||v_x^\varepsilon|\, dxdt + \iint_{t<\alpha} |w||\theta_\varepsilon - \theta||v_x^\varepsilon|\, dxdt.$$

The first integral on the right-hand side is estimated using (5.39). For the second integral, we have, since $\alpha < 1/n < 1/N$,

$$\begin{aligned}
\iint_{t<\alpha} |w||\theta_\varepsilon - \theta||v_x^\varepsilon|\, dxdt &\le 2M \cdot 2M_1 \iint_{t<\alpha} |v_x^\varepsilon|\, dxdt \\
&= 4MM_1 \int_0^\alpha \int_{-\infty}^{\infty} |v_x^\varepsilon|\, dxdt \\
&= 4MM_1 \int_0^\alpha V_t(v^\varepsilon)\, dt \\
&\le 4MM_1 C_{1/n}\alpha < \tilde{\varepsilon}/2
\end{aligned}$$

by our choice of α. Here we have made use of the estimate (5.37). Thus, we conclude that

$$\left| \iint_{t\ge 0} w\phi\, dxdt \right| < \tilde{\varepsilon}.$$

Since $\tilde{\varepsilon} > 0$ is arbitrary, we conclude that $\iint_{t\geq 0} w\phi \, dxdt = 0$ for all test functions ϕ, which are C^1 and have compact supports in $\mathbb{R} \times (0, \infty)$. Thus, $w = u_1 - u_2 = 0$ a.e. in $t > 0$, proving the uniqueness. □

5.5 RIEMANN PROBLEM

The classical Riemann problem is the IVP with initial function consisting of only two constant states u_ℓ and u_r, with one jump discontinuity. Solutions of the Riemann problem may, in turn, be used as building blocks for constructing solutions of the IVP in the class BV of functions of bounded variations. The main limitation of this approach is that it generally applies only when the initial data have sufficiently small total variation; see Glimm (1965) (for the deterministic version, see Liu 1977). Till date, it has remained a challenge to extend the result in Glimm (1965) to higher dimensions.

In the previous sections, we have analyzed the Burgers' equation with different initial conditions; in particular, discontinuous initial values u_0 taking only two constant values, namely 1 on the left of the origin, 0 on the right of the origin and vice-versa. In the first case where u_0 is decreasing, we obtained a discontinuous entropy solution with a shock discontinuity, and in the second case, the rarefaction phenomena appears. In the latter case, we could define a continuous solution satisfying the R–H condition and entropy condition. In this section, we will see that the phenomena is the same if we consider the general conservation laws with any two constants. It is also possible to construct solutions where the initial value takes finitely many constants (see for instance, Exercise 2). Consider the IVP (5.9), with u_0 given by

$$u_0(x) = \begin{cases} u_\ell \text{ if } x < 0 \\ u_r \text{ if } x > 0. \end{cases}$$

Here $u_\ell, u_r \in \mathbb{R}$ are two constants. Assume that $f \in C^2(\mathbb{R})$ and uniformly convex. Now, we consider the two cases separately, namely $u_l > u_r$ and $u_l < u_r$.

Case $u_\ell > u_r$ (discontinuous solution – shock discontinuity): The R–H condition immediately suggests that the discontinuity curve is the line $x = \sigma t$, where σ is the speed given by (see Figure 5.9)

$$\sigma = \frac{f(u_\ell) - f(u_r)}{u_\ell - u_r}.$$

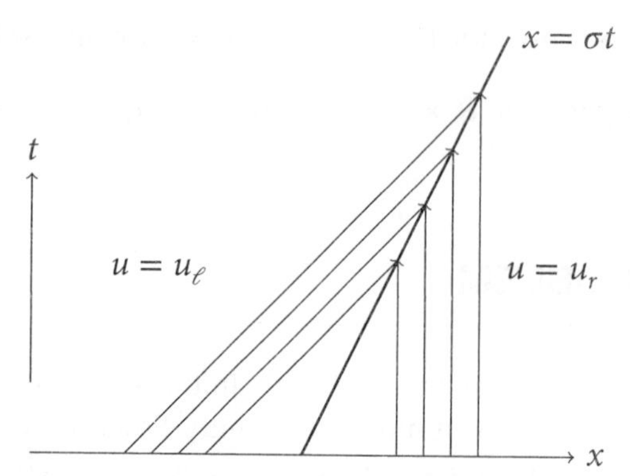

Figure 5.9 Solution of Riemann's problem when $u_\ell > u_r$

It is easy to verify that the function u defined by

$$u(x,t) = \begin{cases} u_\ell \text{ if } \frac{x}{t} < \sigma \\ u_r \text{ if } \frac{x}{t} > \sigma \end{cases} \tag{5.40}$$

is the discontinuous solution satisfying the entropy condition and hence unique. In fact, we have $f'(u_r) < \sigma < f'(u_\ell)$ that is the entropy inequality. The above solution can also be obtained from Lax–Oleinik formula.

Case $u_\ell < u_r$ (continuous solution – rarefaction): In this case, we have $f'(u_r) > f'(u_\ell)$ and thus, we get the diverging characteristics at the origin with slopes $G(u_\ell) = f'(u_\ell)^{-1}$ and $G(u_r) = f'(u_r)^{-1}$ (see Figure 5.10). These two characteristics will divide the upper half $x - t$ plane into 3 regions, namely I, II, III. Exactly as in the case of Burgers' equation, the solutions are constants u_ℓ and u_r, respectively in regions I and III. Indeed, if we define u in region II as

$$u(x,t) = G\left(\frac{x}{t}\right),$$

then,

$$u_t + f'(u)u_x = G'\left(\frac{x}{t}\right)\left(\frac{-x}{t^2}\right) + f'\left(G\left(\frac{x}{t}\right)\right)G'\left(\frac{x}{t}\right)\frac{1}{t} = 0$$

since $f'\left(G\left(\frac{x}{t}\right)\right) = \frac{x}{t}$. Further, on the separating line $x = f'(u_\ell)t$, we have $u(x,t) = G\left(\frac{x}{t}\right) = G\left(f'(u_\ell)\right) = u_\ell$. Similarly, on the line $x = f'(u_r)t$, we get $u(x,t) = u_r$ that proves the continuity across the regions. Thus, we have the following theorem:

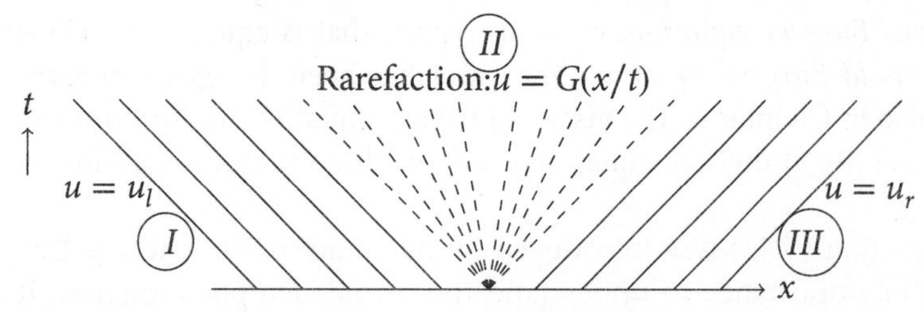

Figure 5.10 Solution of Riemann's problem when $u_\ell < u_r$

Theorem 5.12. Consider the Riemann problem with the conditions on f as above. Then, the following hold:

i. If $u_\ell > u_r$, the function u defined by (5.40) is the unique entropy solution with shock discontinuity curve $x = \sigma t$.
ii. If $u_\ell < u_r$, consider the function u defined by

$$u(x,t) = \begin{cases} u_l \text{ if } x < f'(u_\ell)t,\ t > 0 \\ G\left(\frac{x}{t}\right) \text{ if } f'(u_\ell)t < x < f'(u_r)t,\ t > 0 \\ u_r \text{ if } x > f'(u_r)t,\ t > 0. \end{cases} \tag{5.41}$$

Then, u is the unique (continuous) entropy solution to the Riemann problem. □

Proof We have already established the statement (1). For the statement (2), it is easy to verify, from the discussion prior to the statement of the theorem, that u defined by (5.41) is a continuous integral solution satisfying the R–H condition. It remains to show that u satisfies the entropy condition (5.25). We have already seen from Proposition 5.9 that G is Lipschitz continuous. Thus, for $x, x + z$ in region II, $z > 0$, we get

$$u(x+z,t) - u(x,t) = G\left(\frac{x+z}{t}\right) - G\left(\frac{x}{t}\right) \leq \frac{kz}{t},$$

where k is the Lipschitz constant of G. Thus, u satisfies the entropy inequality and hence is the unique solution. □

5.6 NOTES

The second-order PDE

$$u_t + uu_x = \nu u_{xx},\ \nu > 0 \tag{5.42}$$

is called *viscous Burgers' equation*. Its counterpart, that is equation (5.42) with $\nu = 0$, is also called *inviscid Burgers' equation*. The latter has been discussed in great detail in this chapter and also in Chapter 3. The viscous Burgers' equation has smooth solution whereas the limiting (as $\nu \to 0$) inviscid equation can have discontinuous solution as we have seen earlier.

Though equation (5.42) was introduced by Bateman in 1935, it was Burgers, in 1940, who stressed its importance to understand the turbulence phenomenon. It is somewhat surprising that this equation (of course, without the name Burgers' equation), in a more general form, has been discussed including its reduction to linear heat equation in a book by Forsyth in 1906. See Forsyth (1906),[4] Chapter XV, §229. The viscous Burgers' equation can be seen as one-dimensional approximation of the Navier–Stokes equations. Equation (5.42) was independently studied by Cole and Hopf, in 1950– (see Cole, 1951; Hopf, 1950). Using a non-linear change of variables, they succeeded in transforming the non-linear second-order equation (5.42) to the linear heat or diffusion equation:

$$\phi_t = \nu\phi_{xx}. \tag{5.43}$$

This transformation is called the *Cole–Hopf transformation*. We now briefly discuss this transformation and the solution of IVP associated with (5.42).

Let $u = v_x$. Then, (5.42) becomes

$$v_{xt} + v_x v_{xx} = \nu v_{xxx}.$$

Writing $v_x v_{xx} = \frac{1}{2}(v_x^2)_x$, an integration with respect to x gives

$$v_t + \frac{1}{2}v_x^2 = \nu v_{xx}.$$

Now put $v = -2\nu \log\phi$, $\phi > 0$. A simple computation then shows that ϕ satisfies the linear equation (5.43). If we impose an initial condition $u(x,0) = u_0(x)$, this transforms to an initial condition for ϕ: $\phi_0(x) \equiv \phi(x,0) = \exp\left(-\frac{1}{2\nu}\int_0^x u_0(\xi)\,d\xi\right)$. Therefore (see Chapter 8 on Heat Equation), the solution ϕ is given by

$$\phi(x,t) = (4\pi\nu t)^{-1/2}\int_{-\infty}^{\infty}\phi_0(y)\exp\left(-\frac{(x-y)^2}{4\nu t}\right)dy,\ t > 0. \tag{5.44}$$

[4]This reference was brought to our attention by our colleague K. T. Joseph. These volumes of Forsyth were reprinted by Dover Publications, New York, in a three-volume set in 1950.

This, in turn, gives the solution u of (5.42):

$$u(x,t) = \frac{\int_{-\infty}^{\infty} \frac{x-y}{t}\phi_0(y)\exp\left(-\frac{(x-y)^2}{4\nu t}\right) dy}{\int_{-\infty}^{\infty} \phi_0(y)\exp\left(-\frac{(x-y)^2}{4\nu t}\right) dy}, \ t > 0. \tag{5.45}$$

The limiting[5] case $\nu \to 0$ has been studied by Cole, Hopf and Burgers. These earlier works have been recorded in the book by Burgers (1974). The counterparts of the various solutions in the discontinuous theory (i.e., $\nu = 0$) discussed in this chapter can also be studied in the present improved theory. Except for very weak shocks (small $\nu > 0$), the only significant change is the smoothing of the shock into a thin transition layer. See Whitham (1974) for a detailed account.

We now briefly discuss the limiting case of $\nu \to 0$ in the work of Hopf. For the case of $f(u) = \frac{u^2}{2}$, we can trace the origins of the Lax–Oleinik formula and the entropy inequality here. To emphasize the dependence on ν, we write the solution u given in (5.45) as $u(x, t; \nu)$. Define the function F by

$$F(x,y,t) = \frac{(x-y)^2}{2t} + \int_0^y u_0(\eta)\, d\eta$$

for $x, y \in \mathbb{R}$ and $t > 0$ with u_0, an integrable function, as initial value. Then, F satisfies

$$\frac{F}{y^2} \to \frac{1}{2t} > 0 \text{ as } |y| \to \infty.$$

Hence, F attains its minimum value for one or more values of y. Denote the smallest and largest such y by y_* and y^*, respectively. Of course these are functions of x and t and $y_*(x,t) \le y^*(x,t)$. The following important inequality

$$\frac{x - y_*(x,t)}{t} \le \liminf_{\substack{\nu\to 0\\ \xi\to x\\ \tau\to t}} u(\xi,\tau;\nu) \le \limsup_{\substack{\nu\to 0\\ \xi\to x\\ \tau\to t}} u(\xi,\tau;\nu) \le \frac{x - y^*(x,t)}{t}. \tag{5.46}$$

[5]This procedure since then has been known as *vanishing viscosity method*.

is derived in Hopf (1950). In particular, if $y_* = y^*$, we get the important result that

$$\lim_{\substack{\nu \to 0 \\ \xi \to x \\ \tau \to t}} u(\xi, \tau; \nu) = \frac{x - y^*(x, t)}{t}.$$

Compare the above limit with Lax–Oleinik formula (5.22).

An extension of the works of Cole and Hopf, by Lax in 1957, to include strictly convex functions in place of $\frac{1}{2}u^2$ was a significant step in the theory of conservation laws. Here the important Lax entropy condition is introduced, which is an essential condition for the uniqueness of solutions. Further, Lax also derives the formula for a generalized solution, by converting the conservation law to a HJE; however, Lax does not mention the terminology of HJE. He also derives the entropy inequality for the case of strictly convex f.

Important results on the uniqueness and stability of solutions, due to Oleinik (1959), were followed. The uniqueness of a generalized solution for the conservation law

$$u_t + f(u)_x = 0$$

with f non-convex, still holds provided the Lax entropy condition is replaced by the following:

i. If $u_r < u_\ell$, then

$$f(\alpha u_r + (1 - \alpha)u_\ell) \leq \alpha f(u_r) + (1 - \alpha)f(u_\ell),\ 0 \leq \alpha \leq 1.$$

See Figure 5.11. Geometrically, the requirement is that the graph of the curve $f(u)$ lies below the chord joining the points $(u_r, f(u_r))$ and $(u_\ell, f(u_\ell))$ over the interval $[u_r, u_\ell]$.

ii. If $u_r > u_\ell$, then

$$f(\alpha u_r + (1 - \alpha)u_\ell) \geq \alpha f(u_r) + (1 - \alpha)f(u_\ell),\ 0 \leq \alpha \leq 1.$$

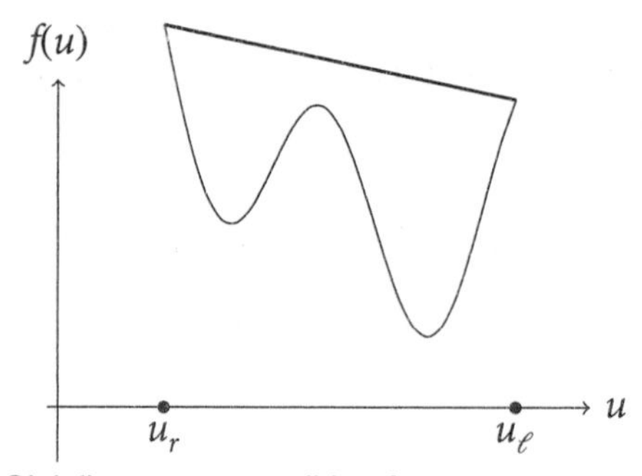

Figure 5.11 Oleinik entropy condition for $u_\ell > u_r$

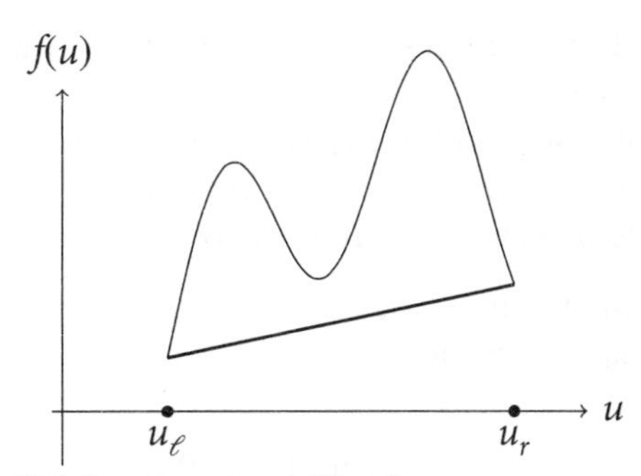

Figure 5.12 Oleinik entropy condition for $u_\ell < u_r$

See Figure 5.12. In this case, the graph of the curve $f(u)$ is required to lie above the chord joining the points $(u_\ell, f(u_\ell))$ and $(u_r, f(u_r))$ over the interval $[u_\ell, u_r]$.

These entropy conditions are due to Oleinik. But, now the Lax–Oleinik formula cannot be used for obtaining a solution, as f need not be convex. In this scenario, an existence result can be obtained by the vanishing viscosity method, by approximating the solution of the given conservation law by a second-order parabolic equation

$$u_t + f(u)_x = \nu u_{xx},$$

with $\nu > 0$, similar to the Burgers' equation. Unlike the Burgers' equation, we may not be able to reduce the above equation to a linear diffusion equation for a general f. However, it is possible to prove the existence of a solution and also, the existence of a limit as $\nu \to 0$. These are quite advanced topics and we refer the interested reader to a vast number of books on the subject (see, e.g., Dafermos, 2009; Lax, 1973; Morawetz, 1981; Smoller, 1994).

5.7 EXERCISES

1. Consider the Burgers' equation $u_t + uu_x = 0$ for $x \in \mathbb{R}$, $t > 0$ with the initial condition $u(x,0) = u_0(x)$, where u_0 is given by

$$u_0(x) = \begin{cases} 1 & \text{if } x \le 0 \\ 1 - x & \text{if } 0 \le x \le 1 \\ 0 & \text{if } x \ge 1 \end{cases}$$

a. Show that the characteristic curves do not meet till $t = 1$ and hence find the solution $u(x,t)$ for all $x \in \mathbb{R}$ and $0 < t < 1$.

b. Construct a curve of discontinuity $s(t)$ for $t \geq 1$ satisfying $s(1) = 1$. Construct a generalized solution that is smooth in the regions $x < s(t)$ and $x > s(t)$ for $t \geq 1$, and satisfying the Rankine–Hugoniot condition across the curve $x = s(t)$.

2. Consider the Burgers' equation $u_t + uu_x = 0$ for $x \in \mathbb{R}$, $t > 0$ with the initial condition $u(x,0) = u_0(x)$, where is u_0 is given by

$$u_0(x) = \begin{cases} 0 & \text{if } x \leq 0 \\ 1 & \text{if } 0 \leq x \leq 1 \\ 0 & \text{if } x \geq 1 \end{cases}$$

Construct a generalized solution.

3. Construct the characteristics of the generalized Burgers' equation

$$u_t + (f(u))_x = 0, \; x \in \mathbb{R}, \; t > 0,$$

satisfying the initial condition $u(x,0) = u_0(x)$ for $x \in \mathbb{R}$, where u_0 is a C^1 function and $f(u) = |u|^{p-2}u^2$, $p \geq 2$. Find a C^1 solution in a region in the upper half plane $t > 0$.

4. Consider the Burgers' equation in the upper half plane $t > 0$. Find the solution using the method of characteristics, in the appropriate regions in the upper half plane, satisfying the following initial conditions:

$$\text{(a) } u_0(x) = x^2 \text{ and (b) } u_0(x) = -x|x|, \; x \in \mathbb{R}.$$

(see Exercise 13 of Chapter 3).

5. Workout the details in Example 5.8, especially find the minimizer.

6. Consider the Burger's equation and find the solution using Lax–Olenik formula with the initial values as in Exercise 1, Exercise 2 and also with the initial value

$$u_0(x) = \begin{cases} 0 & \text{if } x \leq 0 \\ 1 & \text{if } x \geq 1 \end{cases}$$

In the last case $0 \leq u_0 \leq 1$ for $0 < x < 1$ and u_0 is strictly increasing.

7. Assume that the generalized solution of conservation law $u_t + (f(u))_x = 0$ has only shock discontinuities satisfying the entropy condition $f'(u_l) > f'(u_r)$, then any point in the upper half $x - t$ plane is connected by a backward characteristic.

CHAPTER 6

Classification of Second-Order Equations

6.1 INTRODUCTION

In this chapter, we discuss the classification of partial differential equations (PDE), concentrating mostly on linear equations or equations with linear principal part. In the modern theory of the subject, especially since the middle of the last century, the question of classification has taken altogether new directions, which we only mention very briefly in the section on Notes. Also, the classification of PDEs is in general not complete. The second-order equations in two variables have a fairly complete classification, which is our main topic of discussion in this chapter. We begin by a discussion on a Cauchy problem for a second-order linear PDE and highlighting certain subtle differences with a Cauchy problem for an ordinary differential equation (ODE). This naturally motivates towards the study of the classification of PDE. The general references for this chapter are Courant and Hilbert (1989), Rubinstein and Rubinstein (1998), Koshlyakov et al. (1964), John (1978), Mikhailov (1978), Ladyzhenskaya (1985), McOwen (2005), Renardy and Rogers (2004), Prasad and Ravindran (1996), Vladimirov (1979, 1984), and Hörmander (1976, 1984), among many others.

6.2 CAUCHY PROBLEM

We begin by describing a *Cauchy problem* or an *initial value problem* (IVP) for a linear PDE and comparing it with an IVP associated to a linear ODE. For simplicity of the exposition, we consider the second-order linear equation in a region Ω in $\mathbb{R}^n$:

$$\sum_{i,j=1}^{n} a_{ij}(x)\frac{\partial^2 u}{\partial x_i \partial x_j} + \sum_{i=1}^{n} a_i(x)\frac{\partial u}{\partial x_i} + a(x)u = f, \tag{6.1}$$

where the coefficients a_{ij}, a_i, a and the function f are given real (smooth) valued functions defined in Ω. The Cauchy problem for a second-order linear ODE

$$u'' + b(x)u' + c(x)u = g(x), \tag{6.2}$$

where the coefficients b, c and the function g are smooth functions defined in an interval in $\mathbb{R}$, consists in finding a solution u satisfying the initial conditions $u(x_0) = u_0$ and $u'(x_0) = u_1$ for some x_0 and *arbitrary* u_0, u_1.

A Cauchy problem (or an IVP) associated with the PDE (6.1), in which initial conditions for u and its normal derivative $\frac{\partial u}{\partial \nu}$ are assigned on an $(n-1)$-dimensional surface Γ in Ω. We will see that, in general, this may lead to a lack of existence or uniqueness depending on the nature of Γ. Thus consider an $(n-1)$-dimensional surface Γ of class C^k, $k \geq 2$ lying in Ω, given by the equation

$$F(x) = 0, \tag{6.3}$$

where F is a real valued $C^k(\Omega)$, $k \geq 2$, function such that $|\nabla F(x)| \neq 0$ for all $x \in \Gamma$. Let $x_0 \in \Gamma$ and choose $r > 0$ such that the ball $B = B_r(x_0) \subset \Omega$. Put $\Gamma_0 = \Gamma \cap B$. Let $u \in C^2(B)$ be a solution of (6.1). Define

$$u_0(x) = u(x),\ u_1(x) = \frac{\partial u}{\partial \nu}(x),\ x \in \Gamma_0. \tag{6.4}$$

In contrast with the Cauchy problem for an ODE, we will now show that the functions u_0 and u_1 cannot, in general, be arbitrary smooth functions.

Since $|\nabla F(x_0)| \neq 0$, we may assume that $\frac{\partial F}{\partial x_n}(x) \neq 0$ for all $x \in B$ (choose $r > 0$ smaller if necessary). By implicit function theorem, we may write (6.3) as

$$x_n = \phi(x'),\ x' = (x_1, \dots, x_{n-1}),$$

where ϕ is a smooth function. Consider the change of variables $x \mapsto y = y(x)$ given by

$$y_i = F_i(x),\ i = 1, 2, \dots, n, \tag{6.5}$$

where $F_i(x) = x_i - x_{0i}$ for $i = 1, 2, \dots, n-1$ and $F_n(x) = F(x)$. This is a one-one mapping of the region B onto a neighborhood $\widetilde{B}$ of the origin; origin being the image point of x_0. The inverse of this function is denoted by $x = x(y)$. Let

$$\Sigma = \widetilde{B} \cap \{(y', 0) : y' \in \mathbb{R}^{n-1}\}$$

and consider the function $v(y) = u(x) = u(x(y))$ defined on $\widetilde{B}$, where $y = y(x)$ is given by (6.5). Note that Σ lies in the hyper plane $y_n = 0$ and is referred to as a *flat surface*.[1] By chain rule, we have

$$\frac{\partial u}{\partial x_i} = \sum_{k=1}^{n} \frac{\partial v}{\partial y_k} \frac{\partial F_k}{\partial x_i}$$

[1]This procedure is referred to as flattening the surface Γ, locally.

$$\frac{\partial^2 u}{\partial x_i \partial x_j} = \sum_{k,m=1}^{n} \frac{\partial^2 v}{\partial y_k \partial y_m} \frac{\partial F_k}{\partial x_i} \frac{\partial F_m}{\partial x_j} + \sum_{k=1}^{n} \frac{\partial v}{\partial y_k} \frac{\partial^2 F_k}{\partial x_i \partial x_j}.$$

Thus, equation (6.1) in B may be transformed to an equation in $\widetilde{B}$, in the new variables as

$$\sum_{i,j=1}^{n} b_{ij}(y) \frac{\partial^2 v}{\partial y_i \partial y_j} + \sum_{i=1}^{n} b_i(y) \frac{\partial v}{\partial y_i} + b(y)v = f_1(y), \tag{6.6}$$

where $b_{ij}(y)$ are the elements of the matrix $\left[A(x(y))\nabla F_i(x(y)) \cdot \nabla F_j(x(y))\right]$ with $A(x)$ denoting the matrix $[a_{ij}(x)]$. In particular,

$$b_{nn}(y(x)) = A(x)\nabla F(x) \cdot \nabla F(x). \tag{6.7}$$

The initial conditions (6.4) respectively become

$$v = v_0(y'), \quad \nabla_y v \cdot \lambda(y) = \tilde{v}_1(y'), \text{ on } \Sigma, \tag{6.8}$$

where

$$v_0(y') = u_0(y', \phi(y')) \text{ and } \tilde{v}_1(y') = u_1(y', \phi(y')).$$

The vector λ is given by

$$\lambda(y(x)) = \left(\frac{\partial F_1}{\partial \nu}(x), \cdots, \frac{\partial F_n}{\partial \nu}(x) \right), \ x \in \Gamma_0.$$

We note that

$$\frac{\partial F_n}{\partial \nu} = \frac{\partial F}{\partial \nu} \neq 0 \text{ on } \Gamma_0.$$

The normal derivative prescribed in (6.4), may be replaced by a directional derivative along a direction ℓ, which need to satisfy the following condition: Suppose that $\ell(x) = (\ell_1(x), \ldots, \ell_n(x))$ is a non-vanishing C^1 vector field defined in Ω, which is *nowhere tangent* to Γ. This means, $|\ell(x)| \neq 0$ for all $x \in \Omega$ and

$$\left.\frac{\partial F}{\partial \ell}\right|_{\Gamma} \equiv \left.\frac{\ell \cdot \nabla F}{|\ell|}\right|_{\Gamma} \neq 0.$$

In particular $\frac{\partial F}{\partial l} \neq 0$ on Γ_0. For example, if $\Omega = \mathbb{R}^n$ and Γ is given by the hyperplane $x_n = 0$, we may take ℓ in the direction of the x_n co-ordinate.

We next show that the value of the vector ∇v at any point on the surface Σ is uniquely determined by v_0 and $\tilde{v}_1$. Indeed, the values of the derivatives $\frac{\partial v}{\partial y_i}$ on Σ for $i < n$ are determined by the first relation in (6.8):

$$\frac{\partial v}{\partial y_i} = \frac{\partial v_0}{\partial y_i} \text{ for } i < n, \text{ on } \Sigma.$$

The second relation in (6.8) gives

$$\frac{\partial v}{\partial y_n} = v_1(y') \text{ on } \Sigma, \tag{6.9}$$

where

$$v_1(y') = \left(\frac{\partial F}{\partial v}\right)^{-1}\left(\tilde{v}_1(y') - \sum_{i=1}^{n-1} \frac{\partial v_0}{\partial y_i}\frac{\partial F_i}{\partial v}\right).$$

Clearly the first condition in (6.4) and the condition (6.9) are equivalent to the conditions in (6.4).

We next proceed to determine the values of the second derivatives of $v(y)$ on Σ. We note that the first condition (6.8) and the condition in (6.9) uniquely determine the values of all second derivatives of v, except $\frac{\partial^2 v}{\partial y_n^2}$, on Σ. Using (6.6) and (6.7), we obtain that

$$A(x(y))\nabla_x F(x(y)) \cdot \nabla_x F(x(y))\frac{\partial^2 v}{\partial y_n^2} =$$

$$= f_1(y) - \sum_{i,j=1}^{n-1} b_{ij}\frac{\partial^2 v}{\partial y_i \partial y_j} - \sum_{i=1}^{n-1} b_{in}\frac{\partial^2 v}{\partial y_i \partial y_n} - \sum_{i=1}^{n-1} b_i \frac{\partial v}{\partial y_i} - bv. \tag{6.10}$$

If the function $A(x)\nabla F \cdot \nabla F \neq 0$ on the surface Γ_0, then the function $A(x(y))\nabla_x F(x(y)) \cdot \nabla_x F(x(y))$ does not vanish on Σ and therefore in $\widetilde{B}$ (choosing $r > 0$ small enough). Therefore, dividing by this non-vanishing coefficient through in (6.10) and setting $y_n = 0$, we obtain the value of $\frac{\partial^2 v}{\partial y_n^2}$ on Σ.

We therefore conclude that if $A(x)\nabla F \cdot \nabla F \neq 0$ on the surface Γ_0, all the derivatives of the solution u up to the second order are uniquely determined on Γ_0.

If on the other hand, $A(\bar{x})\nabla F(\bar{x}) \cdot \nabla F(\bar{x}) = 0$ for some $\bar{x} \in \Gamma_0$, then at the corresponding point $\bar{y}$ in Σ, we have

$$A(x(\bar{y}))\nabla_x F(x(\bar{y})) \cdot \nabla_x F(x(\bar{y})) = 0.$$

Then, at this point $\bar{y}$, (6.10) connects the known values of $v(\bar{y})$, $\frac{\partial v}{\partial y_i}(\bar{y})$ and $\frac{\partial^2 v}{\partial y_i \partial y_j}(\bar{y})$. This, in turn, implies that at the point $\bar{x}$, the functions u_0 and u_1 along with the corresponding derivatives are also connected, that is they cannot in general be arbitrary.

This leads to the following definition and, eventually, to a classification of PDE.

Definition 6.1. A point $\bar{x}$ on the surface Γ of class C^1, given by the equation $F(x) = 0$ (recall that F is a C^1 function with non-vanishing gradient on Γ), is called a *characteristic point* for the PDE (6.1) if

$$A(\bar{x})\nabla F(\bar{x}) \cdot \nabla F(\bar{x}) = 0.$$

The surface Γ is a *characteristic surface* for (6.1) if each of its points is a characteristic point of (6.1). The surface Γ is a *non-characteristic surface* of (6.1), if none of its points is a characteristic point of (6.1). □

Similar definitions for equations of higher order will be introduced in the next section.

Characteristic Cauchy Problem: We remark that if the surface Γ contains characteristic point(s), the Cauchy problem becomes more difficult. As was observed above, if the point $x^0 \in \Gamma$ is a characteristic point, then there are smooth functions u_0 and u_1 such that (6.1) may have no $C^2(B)$ solution in any neighborhood of this point satisfying the conditions in (6.4) on $\Gamma_0 = \Gamma \cap B$; see (6.10). It is further easy to see that if B^+ denotes one of the parts into which Γ_0 divides B, then there is no solution in $C^2(B^+ \cup \Gamma_0)$ also, which satisfies the conditions in (6.4) on Γ_0. If still a smooth solution exists, it may not be unique.

Here are some examples. Suppose $n = 2$ and consider the equation

$$\frac{\partial^2 u}{\partial x_1 \partial x_2} = f(x),$$

in $B = B_r(0)$. This is the wave equation in the *characteristic variables*, which is studied in detail in Chapter 9. The line $x_2 = 0$ is a characteristic. It is easy to see that for the existence of a $C^2(B)$ solution of the above equation satisfying the conditions

$$u(x_1, 0) = u_0(x_1) \text{ and } u_{x_2}(x_1, 0) = u_1(x_1),$$

it is necessary and sufficient that $u_1'(x_1) = f(x_1, 0)$. If this condition is satisfied, then the solution can be expressed as

$$u(x_1, x_2) = \int_0^{x_1} d\xi_1 \int_0^{x_2} f(\xi_1, \xi_2)\, d\xi_2 + u_0(x_1) + g(x_2),$$

where g is any C^2 function satisfying the conditions $g(0) = 0$ and $g'(0) = u_1(0)$. Clearly, the uniqueness is lost. The details are left as an exercise.

If Γ is a characteristic surface, then there may also be situations where equation (6.1) should be posed in analogy with the Cauchy problem for ODE, not of the second order but first order. For example, the heat equation (see Chapter 8 for its detailed study)

$$u_{x_1 x_1} - u_{x_2} = f(x), \; n = 2$$

has the line $x_2 = 0$ as the characteristic. Later we study the Cauchy problem in which the condition $u(x_1, 0) = u_0(x_1)$ is imposed on the characteristic and the solution is sought in the half-space $x_2 > 0$.

6.2.1 Non-characteristic Cauchy Problem

We now consider the Cauchy problem for (6.1) with the functions u_0 and u_1 (see (6.4)) given on a *non-characteristic surface* Γ. Consider equation (6.1) in the region Ω of $\mathbb{R}^n$, which contains the initial surface Γ given by (6.3). We assume that Γ is non-characteristic for (6.1), that is,

$$A(x)\nabla F \cdot \nabla F \neq 0 \text{ on } \Gamma. \tag{6.11}$$

It was shown above that all the derivatives of the solution u of (6.1) up to second order are uniquely determined on Γ in terms of u_0, u_1 and the coefficients in equation (6.1). We shall now assume that the data, namely the coefficients in (6.1), u_0, u_1, F, ν, and so on, are infinitely differentiable in their respective domains. Assuming that the Cauchy problem (6.1) and (6.4) has a C^∞ solution in B containing the point $x^0 \in \Gamma$, we will show that all the derivatives of u are uniquely determined on Γ in terms of the data.

To achieve this, we again use the transferred equation (6.6) via the change of variables given by (6.5). Since the functions F_i in (6.5) are C^∞, it suffices to show that all the derivatives of the function $v(y) = u(x(y))$ are uniquely determined on Σ in terms of the data. This is shown using an induction argument.

For any multi-index $\beta = (\beta_1, \dots, \beta_{n-1})$, the values of the derivatives $D^{(\beta,0)}v(y)$ and $D^{(\beta,1)}v(y)$ are directly determined by (6.8) and (6.9):

$$D^{(\beta,0)}v\Big|_\Sigma = D^\beta v_0, \; D^{(\beta,1)}v\Big|_\Sigma = D^\beta v_1.$$

Put

$$v_\alpha = \frac{1}{\alpha!}D^\alpha v(0), \; |\alpha| \geq 0. \tag{6.12}$$

Then, $v_{(\beta,0)}$ and $v_{(\beta,1)}$ are uniquely determined in terms of v_0 and v_1:

$$v_{(\beta,0)} = \frac{1}{\beta!}D^\beta v_0\Big|_{y'=0}, \quad v_{(\beta,1)} = \frac{1}{\beta!}D^\beta v_1\Big|_{y'=0}. \tag{6.13}$$

Since Γ is assumed to be non-characteristic, equation (6.10) can be written as

$$\frac{\partial^2 v}{\partial y_n^2} = \sum_{i,j=1}^{n-1} c_{ij}\frac{\partial^2 v}{\partial y_i \partial y_j} - \sum_{i=1}^{n-1} c_{in}\frac{\partial^2 v}{\partial y_i \partial y_n} - \sum_{i=1}^{n-1} c_i \frac{\partial v}{\partial y_i} + cv + h, \tag{6.14}$$

where the coefficients and the function h are C^∞. Denote by $H_1(y)$, the sums on the right hand side in (6.14). To compute the values of $D^{(\beta,2)}v(y)$ on Σ, we use (6.14). Differentiating (6.14) with respect y_j, $j < n$ and setting $y_n = 0$, we obtain

$$D^{(\beta,2)}v\Big|_\Sigma = D^{(\beta,0)}H_1\Big|_\Sigma.$$

Now $D^{(\beta,0)}H_1\Big|_\Sigma$ is a (linear) function (with known coefficients) of known quantities $D^{(\bar{\alpha},0)}v\Big|_\Sigma$ and $D^{(\bar{\beta},1)}v\Big|_\Sigma$ with multi-indices $\bar{\alpha}$ and $\bar{\beta}$ satisfying the conditions $0 \leq |\bar{\alpha}| \leq |\beta| + 2$, $0 \leq |\bar{\beta}| \leq |\beta| + 1$. Therefore, on Σ, all the derivatives $D^{(\beta,2)}v(y)$ are uniquely determined in terms of the data of the problem. In particular,

$$v_{(\beta,2)} = (2!\beta!)^{-1}\, D^{(\beta,0)}H_1(y)\Big|_{y=0}.$$

Now assume that all the derivatives $D^{(\beta,k-1)}v(y)$ have been uniquely determined by the data on Σ for some $k \geq 2$. We show that the same conclusion holds for $D^{(\beta,k)}v(y)$. For this, differentiate (6.14) β_i times with respect to y_i, $i < n$ and $k - 2$ times with respect to y_n, to obtain

$$D^{(\beta,k)}v\Big|_\Sigma = D^{(\beta,k-2)}H_1\Big|_\Sigma.$$

As before, the term $D^{(\beta,k-2)}H_1\Big|_\Sigma$ is a (linear) function (with known coefficients) of known quantities $D^{(\bar{\alpha},i)}v\Big|_\Sigma$ with $i < k$ and $\bar{\alpha}$ satisfying $0 \leq |\bar{\alpha}| \leq |\beta| + 2$ if $i < k - 1$ and $0 \leq |\bar{\alpha}| \leq |\beta| + 1$ if $i = k - 1$. Thus, all the derivatives $D^{(\beta,k)}v$ are uniquely determined on Σ in terms of the data. In particular,

$$v_{(\beta,k)} = (k!\beta!)^{-1}\, D^{(\beta,k-2)}H_1(y)\Big|_{y=0}. \tag{6.15}$$

This proves the assertion and completes the induction argument.

Conclusion: If the initial surface Γ is non-characteristic for equation (6.1), then the data of the problem uniquely determine all the derivatives of the infinitely differentiable solution u of (6.1) and (6.4), on Γ. Hence, the values of a function together with its derivatives along the curve Γ determine the function in a neighborhood of Γ, then the solution of the problem (6.1) and (6.4) is determined uniquely. One such class of functions is that of analytic functions. Later, in Chapter 11, we will show in detail the existence of a solution with analytic data. This is the celebrated *Cauchy–Kovalevsky Theorem*.

In contrast with the case of ODE, it should however be noted that mere C^∞ assumption on the data may not yield a solution. Hans Lewy (1957) produced the following example, with C^∞ coefficients, which has *no solution* in the class of C^∞ functions:

Example 6.2 (Han Lewy). The PDE

$$u_{x_1x_3} + iu_{x_2x_3} + 2i(x_1 + ix_2)u_{x_3x_3} = f(x_3)$$

does not have twice continuously differentiable solution in any neighborhood of the origin in $\mathbb{R}^3$ if the real-valued function f is not analytic. The details will be discussed in Chapter 8. □

6.3 CLASSIFICATION OF LINEAR EQUATIONS

Most of the linear PDE are classified based on the solvability of the Cauchy problem associated with it, by providing the Cauchy data on an initial surface. As discussed in the previous section, this, in general, depends on the notion of the surface being characteristic or non-characteristic with respect to the given PDE. However, in the modern treatment of the subject, especially after 1950s, the classification of linear PDE has undergone a major change; new concepts like, evolution and non-evolution PDE, hypo-elliptic and non-hypoelliptic equations have emerged.

In contrast, the linear ODE of arbitrary order (or systems) have a nice general theory and the related IVP is well-posed.

For second-order linear PDE in two dimensions with real coefficients, the classification is fairly complete. However, if we allow complex coefficients or go beyond two dimensions, a complete classification is not available. The traditional classification of PDE is based on the notion of what is known as *characteristics*. In some sense, the notion of characteristics is a measure of *strength* or *weakness* of a differential operator in certain directions. For the linear operator $L = \sum_{|\alpha|\leq k} a_\alpha(x)D^\alpha$, we define the *characteristic form*[2] $\chi_L(x, \xi)$ at a point x as a homogeneous polynomial in ξ:

$$\chi_L(x, \xi) = \sum_{|\alpha|=k} a_\alpha(x)\xi^\alpha, \ \xi \in \mathbb{R}^n, \tag{6.16}$$

where $\xi^\alpha = \xi_1^{\alpha_1} \cdots \xi_n^{\alpha_n}$.

A non-zero vector ξ is called a *characteristic* for L if $\chi_L(x, \xi) = 0$ and the *characteristic variety* is defined as

$$\text{char}_x(L) = \{\xi \in \mathbb{R}^n : \xi \neq 0, \chi_L(x, \xi) = 0\}.$$

The characteristic variety has the following interesting property: Suppose, we make a change of variable by a smooth invertible transformation, say $y = g(x)$. For a fixed $x \in \Omega$, if the

[2]This is also called the *principal symbol* of L.

Jacobian matrix $J(x) = \left[\frac{\partial y_i}{\partial x_j}\right]$ is non-singular, then $D_x = J(x)^T D_y$, where the superscript T denotes the matrix transpose. Therefore, the operator L transforms into

$$L' = \sum_{|\alpha|\le k} a_\alpha(g^{-1}(y))\left(J(g^{-1}(y))^T D_y\right)^\alpha$$

and

$$\chi_{L'}(y,\eta) = \sum_{|\alpha|=k} a_\alpha(g^{-1}(y))\left(J(g^{-1}(y))^T\eta\right)^\alpha.$$

Consequently, the characteristic variety $\text{char}_x(L)$ is the image of the characteristic variety $\text{char}_x(L')$ under the Jacobian of the transformation $J(x)^T$. Thus the nature of the PDE does not change.

Now suppose $\xi = (0,\cdots,0,\xi_j,0,\cdots,0) \ne 0$ is a coordinate vector. Then it is easy to see that $\xi \in \text{char}_x(L)$ if and only if the coefficient of ∂_j^k in L vanishes at x. Moreover, for any $\xi \ne 0$, by a rotation of coordinates, we can arrange ξ to lie in a coordinate direction. Combining the above remarks, we see that a vector $\xi \in \text{char}_x(L)$ means in some sense, *weakening* the nature of kth (highest) order of the PDE; that is L fails to be genuinely kth order in the ξ direction at x. In other words, a lack of information. This suggests that the PDE with empty characteristic variety may be easier to tackle as there is *full* information. However, as we will see later, the Cauchy problem associated with the Laplace's equation is not *well-posed*.

We say that L is *elliptic* at x if $\text{char}_x(L) = \phi$, the empty set, and it is elliptic in a domain if it is elliptic at all the points of that domain. We now give some examples.

i. $L = D_1 = \dfrac{\partial}{\partial x_1}$, $\text{char}_x(L) = \{\xi \in \mathbb{R}^n : \xi \ne 0, \xi_1 = 0\}$.

ii. $L = D_{12} = \dfrac{\partial^2}{\partial x_1 \partial x_2}$, $\text{char}_x(L) = \{\xi \in \mathbb{R}^n : \xi \ne 0, \xi_1 = 0 \text{ or } \xi_2 = 0\}$.

iii. Let $n = 2$. The Cauchy–Riemann operator, $L = \frac{1}{2}\left(\dfrac{\partial}{\partial x_1} + i\dfrac{\partial}{\partial x_2}\right)$, $i = \sqrt{-1}$ is elliptic. This has complex coefficients, but similar definition may be made.

iv. The Laplace operator, $L = \Delta = \sum_{i=1}^{n} \dfrac{\partial^2}{\partial x_i^2}$ is elliptic.

v. For the heat operator, $L = \dfrac{\partial}{\partial x_1} - \sum_{i=2}^{n} \dfrac{\partial^2}{\partial x_i^2}$,
$\text{char}_x(L) = \{\xi \in \mathbb{R}^n : \xi \ne 0, \xi_j = 0 \text{ for } j \ge 2\}$.

vi. For the wave operator, $L = \dfrac{\partial^2}{\partial x_1^2} - \sum_{i=2}^{n} \dfrac{\partial^2}{\partial x_i^2}$,
$\text{char}_x(L) = \{\xi \in \mathbb{R}^n : \xi \ne 0, \xi_1^2 = \sum_{i=2}^{n} \xi_j^2\}$.

Further Discussion and Reduction to Canonical Form: The classical theory of PDEs is concerned with the simplest and the most typical equations, such as Laplace and Poisson, heat or diffusion, and wave equations. These equations describe physical processes arising in very different contexts. For example, Laplace and Poisson equations appear in the theories of gravimetry, electrostatics and inviscid, incompressible fluid flow. The heat or diffusion equation arises in the theories of thermal conductivity, diffusion and percolation of elastic liquid through a porous medium. The wave equation describes the transversal and longitudinal vibrations in solids and liquids, and field propagation in electrodynamics. (See, e.g., Courant and Hilbert, 1989; Koshlyakov et al., 1964; Rubinstein and Rubinstein, 1998).

Even from purely mathematical point view, the diversity of phenomena described by the same type of second-order equation is not accidental; it reflects the essential properties of these equations. We will now demonstrate this statement through consideration of the possible classification of these equations and their reduction to canonical forms.

Consider a second-order equation with linear principal part (terms containing highest order derivatives):

$$\sum_{i,j=1}^{n} a_{ij}(x)u_{x_ix_j} + F(x, u, u_{x_1}, \dots, u_{x_n}) = 0, \tag{6.17}$$

in a region in $\mathbb{R}^n$, where $a_{ij} = a_{ji}$ are given functions. The term F contains at most first-order derivatives of the unknown function u; for most part of the discussion, it is immaterial whether F is linear in u and its first derivatives. As was done above with the characteristic variety, we begin with an approach towards the classification of (6.17) based on consideration of all possible non-singular change of co-ordinates:

$$y_j = \phi_j(x),\ j = 1, 2, \dots, n. \tag{6.18}$$

Here non-singular means the Jacobian matrix $\left[\frac{\partial \phi_i}{\partial x_j}\right]$ is non-singular at the points of the region under consideration. We have

$$\frac{\partial}{\partial x_j} = \sum_{k=1}^{n} \frac{\partial \phi_k}{\partial x_j}\frac{\partial}{\partial y_k}$$

$$\frac{\partial^2}{\partial x_i \partial x_j} = \sum_{k,m=1}^{n} \frac{\partial \phi_k}{\partial x_j}\frac{\partial \phi_m}{\partial x_i}\frac{\partial^2}{\partial y_k \partial y_m} + \sum_{k=1}^{n} \frac{\partial^2 \phi_k}{\partial x_i \partial x_j}\frac{\partial}{\partial y_k}.$$

Therefore, equation (6.17) may be written as

$$\sum_{i,j=1}^{n} \tilde{a}_{ij}(y)u_{y_iy_j} + \widetilde{F}(y, u, u_{y_1}, \dots, u_{y_n}) = 0, \tag{6.19}$$

where

$$\tilde{a}_{ij} = \sum_{k,m=1}^{n} a_{km} \frac{\partial \phi_i}{\partial x_k} \frac{\partial \phi_j}{\partial x_m}$$
$$\widetilde{F} = F + \sum_{i,j=1}^{n} a_{ij} \left(\sum_{k=1}^{n} \frac{\partial^2 \phi_k}{\partial x_i \partial x_j} \frac{\partial u}{\partial x_k} \right) \tag{6.20}$$

The principal part associated with equation (6.17) is the operator

$$L = \sum_{i,j=1}^{n} a_{ij}(x) \frac{\partial^2}{\partial x_i \partial x_j} \tag{6.21}$$

and the characteristic form associated with L is the quadratic form

$$Q(x, \xi) = \sum_{i,j=1}^{n} a_{ij}(x) \xi_i \xi_j, \tag{6.22}$$

where $\xi = (\xi_1, \ldots, \xi_n) \in \mathbb{R}^n$. Consequently, at a fixed point x in the region, the quadratic form associated with the transformed equation (6.19) is given by

$$\hat{Q}(\eta) = \sum_{i,j=1}^{n} \tilde{a}_{ij}(x) \eta_i \eta_j, \tag{6.23}$$

where $\tilde{a}_{ij}$ are given by (6.20). This suggests that the classification of the quadratic forms may be used for the classification of PDEs with principal linear part.

Quadratic forms are classified by reducing them to canonical forms and using the so-called law of inertia. Thus, the quadratic form (remember x is fixed) Q can be reduced by a non-singular (linear) transformation to

$$\hat{Q}(\eta) = \sum_{i=1}^{p} \lambda_i \eta_i^2 + \sum_{i=p+1}^{p+q} \lambda_i \eta_i^2, \tag{6.24}$$

where

$$\lambda_k = \begin{cases} +1, & \text{for } k = 1, 2, \ldots, p \\ -1, & \text{for } k = p + 1, \ldots, p + q \\ 0, & \text{for } k = p + q + 1, \ldots, p + q + r = n. \end{cases}$$

Here the non-negative integers p, q, r are invariants of the transformations. That is, they are independent of the choice of the transformation reducing the given quadratic form to its canonical form. We may also assume $p \geq q$. The numbers p, q, r also indicate the number of positive, negative, zero eigenvalues of the associated matrix, respectively.

Therefore, the PDE (6.17) reduces to its *canonical form*, namely

$$\sum_{i=1}^{p} \frac{\partial^2 u}{\partial y_i^2} - \sum_{i=p+1}^{p+q} \frac{\partial^2 u}{\partial y_i^2} + \tilde{F} = 0. \tag{6.25}$$

In what follows we consider only three simple cases:

(1) $p = n,\ q = 0,\ r = 0$.
(2) $p = n - 1,\ q = 1,\ r = 0$.
(3) $p = n - 1,\ q = 0,\ r = 1$.

In $\mathbb{R}^3$, the quadratic surface given by the equation

$$\lambda_1 x_1^2 + \lambda_2 x_2^2 + \lambda_3 x_3^2 = 1$$

with $\lambda_i = \pm 1$ or 0, is called an ellipsoid in case (1), a hyperboloid in case (2) and a paraboloid in case (3). This terminology is extended to the quadratic forms and then to the second-order PDE with linear principal part they represent. Accordingly, we say that equation (6.17) is *elliptic, hyperbolic* and *parabolic* if case (1), (2) and (3) holds, respectively.[3]

In the passing, we also mention that in case $p \geq q \geq 2$ and $r = 0$, thus requiring $n \geq 4$, equation (6.17) is termed as *ultrahyperbolic*. For example, the equation

$$u_{x_1x_1} + u_{x_2x_2} = u_{x_3x_3} + u_{x_4x_4}$$

in $\mathbb{R}^4$ is ultrahyperbolic. There is a good discussion on constant coefficient ultrahyperbolic equations in Courant and Hilbert (1989), somewhat parallel to that of wave equation. Since these equations perhaps do not describe any physical process, we do not find their mention in any subsequent literature.

Global Transformation: We have seen that equation (6.17) may be reduced to canonical form at any fixed point of the region, where the PDE is considered. We now look at the possibility of reducing an equation to canonical form not at a fixed point, but everywhere in the region under consideration. This means, we seek a non-singular change of variables so that the matrix $[\tilde{a}_{ij}]$, where $\tilde{a}_{ij}$ are as in (6.20), becomes a diagonal matrix with ± 1 and 0 as the only elements on the diagonal. Thus, the n functions ϕ_k, $1 \leq k \leq n$ must satisfy $n(n-1)$ differential equations; see (6.20). Even if we consider the symmetry of $[a_{ij}]$, which implies the symmetry of $\tilde{a}_{ij}$ as well, there are $n(n-1)/2$ equations to be satisfied by ϕ_k. If $n > 3$, then $n(n-1)/2 > n$ and the system of equations is overdetermined. If $n = 3$, we have $n(n-1)/2 = n$ and there may be a possibility to find such a transformation.

[3]However, a word of caution with regard to parabolicity; see the discussion in the case of two variables below.

However, it will be impossible to make $p + q$ diagonal elements of this matrix equal to ± 1, so we cannot expect the equation to be reducible to canonical form, in a whole region of $\mathbb{R}^3$, except when the equation has constant coefficients. If the equation has constant coefficients in its principal part, a suitable scaling of the independent variables does this job of reducing the given equation to canonical form. This implies that the problem of reducing the given equation to canonical form in a whole region must be restricted to two dimensions, which we discuss next.

6.3.1 Second-Order Equations in Two Variables

The most general form of linear second order PDE in two variables is given by

$$u_{xx} + 2bu_{xy} + cu_{yy} + du_x + eu_y + fu = g, \tag{6.26}$$

where we are using x and y to denote the independent variables. Here a, b, c, d, e, f, g are smooth functions of x, y and a, b, c do not vanish simultaneously, in a domain Ω in $\mathbb{R}^2$. If $g = 0$, we say equation (6.26) is *homogeneous*, otherwise it is *non-homogeneous*. Consider the operator L, the *principal part* of (6.26):

$$L = aD_x^2 + 2bD_xD_y + cD_y^2,$$

so that equation (6.26) can be written as

$$L(u) + \text{ lower order terms} = g. \tag{6.27}$$

The corresponding quadratic form is given by

$$Q(\xi, \eta) = a\xi^2 + 2b\xi\eta + c\eta^2 = \eta^2 q(\zeta), \tag{6.28}$$

where

$$q(\zeta) = a\zeta^2 + 2b\zeta + c, \quad \zeta = \xi/\eta.$$

It is simpler to classify the quadratic forms in two dimensions. The classification just depends on the *discriminant* δ defined by

$$\delta = ac - b^2.$$

Of course δ varies in Ω. However, when δ is either positive or negative at a point in Ω, then by continuity it retains its sign in a neighborhood of that point.

Definition 6.3. The quadratic form Q is said to be *definite, semi-definite* or *indefinite* according as $\delta > 0$, $= 0$ or < 0, respectively. □

Using the procedure described in the previous section, we transform L to

$$\tilde{L} = \tilde{a}\frac{\partial^2}{\partial \tilde{x}^2} + 2\tilde{b}\frac{\partial^2}{\partial \tilde{x}\partial \tilde{y}} + \tilde{c}\frac{\partial^2}{\partial \tilde{y}^2}, \tag{6.29}$$

using a non-singular change of variables:

$$\tilde{x} = \phi(x, y), \quad \tilde{y} = \psi(x, y). \tag{6.30}$$

The coefficients in (6.29) are given by (see (6.20))

$$\left.\begin{aligned} \tilde{a} &= a\phi_x^2 + 2b\phi_x\phi_y + c\phi_y^2 \\ \tilde{b} &= a\phi_x\psi_x + b(\phi_x\psi_y + \phi_y\psi_x) + c\phi_y\psi_y \\ \tilde{c} &= a\psi_x^2 + 2b\psi_x\psi_y + c\psi_y^2 \end{aligned}\right\} \tag{6.31}$$

A simple calculation shows that

$$\tilde{a}\tilde{c} - \tilde{b}^2 = (ac - b^2)(\phi_x\psi_y - \phi_y\psi_x)^2. \tag{6.32}$$

Also at a fixed point (x, y) in Ω, the quadratic form Q in (6.28) transforms into

$$\tilde{Q}(\tilde{\xi}, \tilde{\eta}) = \tilde{a}\tilde{\xi}^2 + 2\tilde{b}\tilde{\xi}\tilde{\eta} + \tilde{c}\tilde{\eta}^2, \tag{6.33}$$

where ξ, η and $\tilde{\xi}, \tilde{\eta}$ are connected by

$$\xi = \tilde{\xi}\phi_x + \tilde{\eta}\phi_y, \; \eta = \tilde{\xi}\psi_x + \tilde{\eta}\psi_y.$$

Thus, the quadratic form Q and $\tilde{Q}$ are of the same type, as their discriminants have the same sign under a non-singular change of variables (6.30).

Next, we wish to transform L to $\tilde{L}$ with the following choices:

- Case 1. $\tilde{a} = \tilde{c}$ and $\tilde{b} = 0$.
- Case 2. $\tilde{a} = -\tilde{c}$ and $\tilde{b} = 0$.
- Case 3. $\tilde{b} = \tilde{c} = 0$.

The possibility of choosing ϕ, ψ to obtain the required canonical form depends on the form Q, or geometrically speaking on the character of the quadratic curve $Q(\xi, \eta) = 1$ in the ξ, η plane, for fixed x, y. This curve is

(1) an ellipse if $ac - b^2 > 0$.
(2) a hyperbola if $ac - b^2 < 0$.
(3) a parabola if $ac - b^2 = 0$.

The corresponding canonical forms of the principal parts of the PDE are given by

$$\tilde{L}u = u_{\tilde{x}\tilde{x}} + u_{\tilde{y}\tilde{y}}$$
$$\tilde{L}u = u_{\tilde{x}\tilde{x}} - u_{\tilde{y}\tilde{y}}$$
$$\tilde{L}u = u_{\tilde{x}\tilde{x}}.$$

Definition 6.4. The operator L or the PDE (6.26) is called *elliptic, hyperbolic* and *parabolic* in Ω according as $\delta > 0$, $\delta < 0$ and $\delta = 0$, respectively. □

For a fixed (x, y) in Ω, such a transformation can always be obtained simply by the linear transformation that takes the form Q to the corresponding canonical form. However, assuming that the operator L is of the same type everywhere in Ω, we would like to find ϕ, ψ transforming L to canonical form.

First Consider the Hyperbolic Case: $ac - b^2 < 0$ in Ω. In this case, we want $\tilde{a} = \tilde{c} = 0$. Thus $\tilde{b}$ is non-zero. From (6.31), we are led to the quadratic equation

$$Q = a\xi^2 + 2b\xi\eta + c\eta^2 = 0 \tag{6.34}$$

for the ratio $\zeta = \xi/\eta$ of the derivatives ϕ_x/ϕ_y and ψ_x/ψ_y. If a and c are identically 0 in Ω, the operator L is already in canonical form. We may therefore assume that a or c is non-zero. In this case we may also assume that $\eta = 1$. The distinct real roots ζ_1 and ζ_2 of (6.34) are continuously differentiable functions of x, y. Thus, the functions ϕ, ψ are determined by the first-order PDEs:

$$\phi_x - \zeta_1\phi_y = 0, \quad \psi_x - \zeta_2\psi_y = 0.$$

Using the method of characteristics, these equations produce two families of curves $\phi(x, y) = \text{constant}$ and $\psi(x, y) = \text{constant}$, whose characteristics are given by the ODEs

$$y' + \zeta_1 = 0, \; y' + \zeta_2 = 0$$

where y is considered as a function along the characteristics. Since $\zeta_1 - \zeta_2 = \frac{2}{a}\sqrt{b^2 - ac} \neq 0$, the curves of these families cannot be tangent at any point of Ω and therefore $\phi_x\psi_y - \phi_y\psi_x \neq 0$.

The curves $\tilde{x} = \phi(x, y) = \text{constant}$ and $\tilde{y} = \psi(x, y) = \text{constant}$ are called the *characteristic curves* of the linear hyperbolic operator L.

We Next Consider the Parabolic Case[4] $b^2 - ac = 0$. Here we wish to obtain $\tilde{b} = \tilde{c} = 0$. Note that a and c cannot vanish simultaneously, as then $b = 0$, which is contrary to our

[4]This case is in a way a sensitive case. If $b^2 - ac = 0$ holds at a point in Ω, it may not be true that the condition holds in a neighborhood of that point. See the example and the comment below.

assumption. Suppose $a \neq 0$ in Ω. Again from (6.31), we are led to the first-order PDE for ϕ:

$$a\phi_x + b\phi_y = 0,$$

as there is a real double root $\zeta = -b/a$ of the corresponding quadratic equation (6.34) in this case. This also implies that ϕ satisfies the equation $b\phi_x + c\phi_y = 0$ as well as $b^2 - ac = 0$. The function ψ is now to be chosen such that the non-singularity condition is satisfied. It is not difficult to see that the choice of $\psi(x, y) = x$ will do the job. Therefore, the transformation $\tilde{x} = \phi(x, y)$, $\tilde{y} = x$ transforms L to canonical form $\tilde{L}u = u_{\tilde{x}\tilde{x}}$.

Finally, We Consider the Elliptic Case: $ac - b^2 > 0$. In this case, the quadratic equation (6.34) does not have any real root, but has complex conjugate roots. Assuming that the coefficients a, b, c are analytic,[5] and denoting the roots by $\zeta, \bar{\zeta}$, we obtain complex-valued analytic functions ϕ, $\psi = \bar{\phi}$ satisfying the first-order PDE: $\phi_x - \zeta\phi_y = 0$. Let $\tilde{x} = \phi(x, y)$ and $\tilde{y} = \psi(x, y) = \bar{\phi}(x, y) = \bar{\tilde{x}}$. Thus, $\tilde{x}$ and $\tilde{y}$ are complex conjugates.

Now put

$$X = \frac{\tilde{x} + \tilde{y}}{2}, \; Y = \frac{\tilde{x} - \tilde{y}}{2i}, \; i = \sqrt{-1}.$$

Then, X, Y are real variables and we wish to show that they are the required change of variables. Thus

$$\tilde{x} = \phi(x, y) = X + iY.$$

From (6.31), equating $\tilde{b}$ to zero, we obtain

$$a|\phi_x|^2 + b(\phi_x\bar{\phi}_y + \phi_y\bar{\phi}_x) + c|\phi_y|^2 = 0$$

as $\psi = \bar{\phi}$. Writing $\phi_x = \frac{\partial X}{\partial x} + i\frac{\partial Y}{\partial x}$ and so on, we obtain the following equation

$$a\left(\frac{\partial X}{\partial x}\right)^2 + 2b\frac{\partial X}{\partial x}\frac{\partial X}{\partial y} + c\left(\frac{\partial X}{\partial y}\right)^2 - a\left(\frac{\partial Y}{\partial x}\right)^2 - 2b\frac{\partial Y}{\partial x}\frac{\partial Y}{\partial y} - c\left(\frac{\partial Y}{\partial y}\right)^2 +$$
$$+ 2i\left(a\frac{\partial X}{\partial x}\frac{\partial Y}{\partial x} + b\left(\frac{\partial X}{\partial y}\frac{\partial Y}{\partial x} + \frac{\partial X}{\partial y}\frac{\partial Y}{\partial x}\right) + c\frac{\partial X}{\partial y}\frac{\partial Y}{\partial y}\right) = 0$$

Equating the real and imaginary parts in the above equation, we see from (6.31) (X, Y in place of ϕ, ψ) that $\tilde{a} = \tilde{c}$ and $\tilde{b} = 0$, thus accomplishing the reduction to canonical form in the elliptic case. The canonical form is given by

$$\tilde{L}u = u_{XX} + u_{YY}.$$

[5]For this reduction to canonical form, we need this assumption. It is also possible to work with only real-valued functions, but the procedure is complicated. The interested reader may refer to Courant and Hilbert (1989).

Comment on the Condition $b^2 - ac = 0$**:** If $b^2 - ac = 0$, the PDE (6.26) is usually referred to as *parabolic*, in the literature. However, we wish to elaborate on certain subtle issues present here. First of all, if $a = 0$ or $c = 0$, then $b = 0$ and hence the PDE (6.26) may degenerate into an ODE, which we certainly do not wish to classify as parabolic. Secondly, the PDE

$$u_{xx} + 2u_{xy} + u_{yy} = 0$$

satisfies the condition $b^2 - ac = 0$ and reduces to an ODE $u_{XX} = 0$, which certainly not in the form of the standard heat equation. Hence, the above PDE cannot be called parabolic. This equation is referred to as *weakly hyperbolic*. Thus, apart from the condition $b^2 - ac = 0$, the presence of certain lower-order term is essential to designate the PDE (6.26) as parabolic. See the notes at the end of the chapter. Also see Hörmander (1976).

Example 6.5. The Laplace equation $u_{xx} + u_{yy} = 0$ is elliptic, the heat equation $\frac{\partial u}{\partial y} - \frac{\partial^2 u}{\partial x^2} = 0$ is parabolic and the wave equation $\frac{\partial^2 u}{\partial x^2} - \frac{\partial^2 u}{\partial y^2} = 0$ is hyperbolic. □

Example 6.6. Consider the equation $x^2 u_{xx} - y^2 u_{yy} = 0$ in the first quadrant $x > 0, y > 0$. Let us discuss its type and reduction to canonical form. In this case, we have $b = 0$, $a = x^2$ and $c = y^2$. Therefore, $b^2 - ac = x^2y^2 > 0$ and the equation is hyperbolic. The roots of the corresponding quadratic are $\pm y/x$. The characteristic equations are given by

$$xy' + y = 0, \quad xy' - y = 0.$$

Therefore, we find that $\tilde{x} = \phi(x, y) = xy$ and $\tilde{y} = \psi(x, y) = y/x$. With these change of variables, the given equation reduces to its canonical form and is given by

$$u_{\tilde{x}\tilde{y}} - \frac{1}{2\tilde{x}} u_{\tilde{y}} = 0, \ \tilde{x} > 0, \ \tilde{y} > 0.$$

□

6.4 HIGHER-ORDER LINEAR EQUATIONS

Here we barely discuss the question of classification of higher-order PDE. A linear partial differential operator of order m is the expression

$$L(x, D) = \sum_{|\alpha| \le m} a_\alpha(x) D^\alpha, \tag{6.35}$$

where a_α are given smooth functions defined in a domain Ω in $\mathbb{R}^n$ and $\sum_{|\alpha|=m} |a_\alpha| \neq 0$ in Ω. If a_α are all constants, we write $L(D)$ in place of $L(x, D)$. The *principal part* of L

is defined by

$$L_m(x, D) = \sum_{|\alpha|=m} a_\alpha(x) D^\alpha, \tag{6.36}$$

consisting of only the highest order terms in L. The *characteristic form* or the *principal symbol* of L is the homogeneous polynomial of order m given by

$$Q_m(x, \xi) = \sum_{|\alpha|=m} a_\alpha(x) \xi^\alpha \tag{6.37}$$

and the *full symbol* of L is the polynomial defined by

$$Q(x, \xi) = \sum_{|\alpha|\le m} a_\alpha(x) \xi^\alpha \tag{6.38}$$

for $x \in \Omega$ and $\xi \in \mathbb{R}^n$. The classification problem for $m \geq 3$ is no longer simple. The only thing that is similar to the case of $m = 2$ is the following:

Definition 6.7. The operator L is called *elliptic* if $Q_m(x, \xi) \neq 0$ for all $x \in \Omega$ and $\xi \neq 0$. □

In particular, if L has constant coefficients and L is elliptic, it follows that m is even and

$$|Q_m(\xi)| \geq c|\xi|^m$$

for some $c > 0$ and for all $\xi \in \mathbb{R}^n$.

On the other hand, the definitions of hyperbolicity and parabolicity, even in the case of constant coefficients are in no way can be explained simply in terms of Q_m or Q. These definitions require a reference to a hyperspace in $\mathbb{R}^n$ (essentially a *direction*) and the solvability of a Cauchy problem. (See, e.g., Hörmander, 1976).

However, when there is a special variable (the time variable) t, the definition of hyperbolicity may be made somewhat easily, not only for higher-order equations, but also for systems of such equations. (See, e.g., Benzoni-Gavage and Serre, 2007; Kreiss and Lorenz, 2004). We will consider some simple definitions in Chapter 9 and Chapter 10.

Non-linear Equations: As we have classified the equations by analyzing the highest-order terms, the semi-linear equations (second order) of the form

$$a(x, y)u_{xx} + 2b(x, y)u_{xy} + c(x, y)y_{yy} + f(x, y, u, u_x, u_y) = 0 \tag{4.23}$$

can be classified in the same way.

But in the quasilinear case, where a, b and c depend on u, u_x, u_y as well, then the determinant $b^2 - ac$ also depends on u, u_x, u_y as well. Thus the classification, in general, depends upon the particular solution. For some solutions, it may be elliptic; parabolic and hyperbolic for some other. The type in general cannot be determined a-priori.

Example 6.8. Consider the PDE

$$u^2 u_{xx} + 3u_x u_y u_{xy} - u^2 u_{yy} = 0.$$

This is hyperbolic for every solution as $ac - b^2 = -\left(\frac{3}{2}u_x u_y\right)^2 - u^2 u^2 < 0$. On the other hand, the equation

$$(1 - u_x^2)u_{xx} - 2u_x u_y u_{xy} + (1 - u_y^2)u_{yy} = 0$$

is hyperbolic (respectively, elliptic) for those solutions u satisfying $u_x^2 + u_y^2 > 1$ (respectively, $u_x^2 + u_y^2 < 1$). □

6.5 NOTES

It was observed that the procedure of reducing to the canonical form of a given second-order linear equation with variable coefficients, is cumbersome and may also be not possible, except for the case $n = 2$. There is also a possibility of the type change of a given equation from one region to another, as was observed in the case of Tricomi equation. This makes it very difficult to develop a general existence and uniqueness theory for a general equation. However, in many applications, especially in mathematical physics, the equations do come with a special variable called the *time* variable and is denoted by t. Such equations (second-order linear) usually come in two forms:

$$u_{tt} - \sum_{i,j=1}^{n} a_{ij}(x,t)u_{x_i x_j} + \sum_{i=1}^{n} a_i(x,t)u_{x_i} + a_0(x,t)u_t = f(x,t) \tag{6.39}$$

$$u_t - \sum_{i,j=1}^{n} a_{ij}(x,t)u_{x_i x_j} + \sum_{i=1}^{n} a_i(x,t)u_{x_i} = f(x,t). \tag{6.40}$$

Here the coefficients $a_{ij} = a_{ji}$ satisfy the ellipticity condition:

$$\sum_{i,j=1}^{n} a_{ij}(x,t)\xi_i\xi_j \geq c|\xi|^2$$

for some constant $c > 0$, all $\xi \in \mathbb{R}^n$ and all (x, t) in a region in $\mathbb{R}^n \times \mathbb{R}$. Equation (6.39) is hyperbolic and equation (6.40) is parabolic.

Complete theories of existence and uniqueness have been developed for both these equations. (See, e.g., Benzoni-Gavage and Serre, 2007; Evans, 1998; Kreiss and Lorenz, 2004; Ladyzhenskaya, 1985; Ladyzhenskaya et al., 1968; Trèves, 2006; Vladimirov, 1979, 1984). These equations and many other evolution equations (e.g., Schrödinger's equation) can be put in the framework of an ODE:

$$u_t = Au$$

where A is a linear operator, possibly unbounded, in a Banach space. The theory has its origins in the development of semi-group of linear operators and the celebrated Hille–Yosida theorem (Brezis, 2011; Goldstein, 1985; Pazy, 1983; Yosida, 1974). The theory has been extended by Kato and others to include the case of *variable coefficients* $A = A(t)$ and also the non-linear equations $u_t = A(t, u)$. There are many excellent books written on the subject.

6.6 EXERCISES

1. Determine the types of the following equations, and reduce them to canonical form.
   a. $u_{xx} + 2e^{x+y}u_{xy} + e^{2y}u_{yy} = 0$.
   b. $u_{xx} + 2u_{xy} + 4u_{xz} + 5u_{zz} + u_x + 2u_y = 0$ (in $\mathbb{R}^3$).
   c. $u_{xx} - 2\sin x u_{xy} - \cos^2 x u_{yy} - \cos x u_y = 0$.
   d. $y^2u_{xx} + x^2u_{yy} = 0\,(x > 0,\ y > 0)$.
   e. $x^2u_{xx} + 2xyu_{xy} + y^2u_{yy} = 0\,(x > 0)$.
   f. $\frac{\partial}{\partial x}\left((1-x)^2\frac{\partial u}{\partial x}\right) = \frac{1}{a^2}(1-x)^2\frac{\partial^2 u}{\partial t^2}\ (a \neq 0)$.
   g. $\sin^4(2x)u_{xx} + 4\sin^4(2x)u = \frac{\partial^2 u}{\partial t^2}$.
2. Consider the Cauchy problem

$$\frac{\partial^2 u}{\partial x_1 \partial x_2} = f(x_1, x_2),$$

with initial conditions

$$u(x_1, 0) = u_0(x_1), \quad u_{x_2}(x_1, 0) = u_1(x_1).$$

Note that $x_2 = 0$ is a characteristic for the given equation. Observing that the general solution of the equation is of the form

$$u(x_1, x_2) = F(x_1) + G(x_2),$$

where F, G are arbitrary C^2 functions, show that the Cauchy problem has a solution if and only if $u_1'(x_1) = f(x_1, 0)$ for all x_1. In this case, write down the solution and conclude that there is non-uniqueness.

3. Consider the wave equation $u_{tt} - u_{xx} = 0$ with prescribed initial and normal derivative condition on the characteristic $x = t$. Use the characteristic variables $x_1 = x+t$ and $x_2 = x-t$ to reduce the problem to that in (2) above. Derive necessary and sufficient conditions on the initial data so that the Cauchy problem has a solution. Conclude that the solution is not unique.[6]

[6]It is in general true that for a constant coefficient hyperbolic equation of any order, there is always non-uniqueness to a characteristic Cauchy problem. (See Hörmander, 1976).

CHAPTER 7

Laplace and Poisson Equations

7.1 INTRODUCTION

The reasons for studying Laplace and Poisson equations are twofold. Primarily, these equations arise in a wide variety of physical contexts. Secondly, as mentioned earlier, Laplace operator is a prototype of a very general class of linear elliptic operators. In fact, the Laplace operator possesses many features of the general class of elliptic operators. Recall that the most general form of second-order linear partial differential equations (PDE) in n variables is given by

$$Lu \equiv \sum_{i,j=1}^{n} a_{ij}(x)D_{ij}u + \sum_{i=1}^{n} b_i(x)D_iu + c(x)u + d(x) = 0, \tag{7.1}$$

where $x \in \Omega$, an open set in $\mathbb{R}^n$, $a_{ij} = a_{ji}$. The operator L is said to be *uniformly elliptic* if there exists an $\alpha > 0$ such that $\sum_{i,j=1}^{n} a_{ij}(x)\xi_i\xi_j \geq \alpha \sum_{i=1}^{n} |\xi|^2$, for all $x \in \Omega$ and $\xi \in \mathbb{R}^n$. Recall that $\chi_L(x, \xi) = \sum_{i,j=1}^{n} a_{ij}(x)\xi_i\xi_j$ is the characteristic form associated with the operator L. The ellipticity condition here implies that the characteristic variety is an empty set at all points in Ω.

Thus, the condition requires the uniform positive definiteness of the symmetric matrix $[a_{ij}(x)]$. If we take $b_i = c_i = d = 0$ and

$$a_{ij}(x) = \delta_{ij} = \begin{cases} 1 & \text{if } i = j \\ 0 & \text{if } i \neq j, \end{cases}$$

there results the Laplace operator Δ. The ultimate interest is to study the *classical solutions* of the equation $Lu = f$ for a given data f. The study of Laplace equation $\Delta u = 0$ and Poisson equation $\Delta u = f$ *(potential theory)* gives a starting point for the general theory of $Lu = f$. The *Schauder theory* provides a general theory for $Lu = f$ when the coefficients are smooth, that is, Hölder continuous and it is essentially an extension of the potential theory. The crucial result is the derivation of estimates (known as *a-priori estimates*), say, of the form: Any $u \in C^2(\Omega)$, a solution of $\Delta u = f$ in a domain $\Omega \subset \mathbb{R}^n$ satisfies a uniform estimate:

$$\|u\|_{C^{2,\alpha}(\Omega')} \leq C\left(\sup_{\Omega} |u| + \|f\|_{C^{0,\alpha}(\overline{\Omega})}\right) \tag{7.2}$$

where $\Omega' \subset\subset \Omega, 0 < \alpha < 1$ and C is a constant depending only on α. Here $C^{k,\alpha}(\overline{\Omega})$ is the standard Hölder space.

Such estimates eventually (with delicate analysis) lead to the solvability of the equation. The above estimate is an *interior estimate* and we also require boundary/global estimates that depend on the smoothness of the boundary as well. The Schauder theory is not applicable for equations with non-smooth coefficients. An alternate and more general approach (and physically meaningful) is that the solutions can be obtained by Hilbert space methods based on weak solutions (generalized functions/distribution theory/Sobolev spaces).

7.1.1 Physical Interpretation

As remarked earlier, the Laplace and Poisson equations appear in a variety of physical problems. In particular, it is well known that the real and imaginary parts of an analytic function of one complex variable are harmonic; that is, they satisfy the (two-dimensional) Laplace equation. Below, we mention a few physical situations where these equations appear.

1. A very general interpretation is as follows: Suppose u denotes the density of certain quantity in equilibrium. Suppose Ω is a smooth domain in $\mathbb{R}^n$. If $V \subset \Omega$ is an arbitrary smooth region in Ω, then, the net flux through the boundary ∂V is zero; that is,

$$\int_{\partial V} F \cdot \nu d\sigma = 0,$$

where F is the flux density. By divergence theorem, we get $\int_V \text{div}F = 0$. Since V is an arbitrary sub-domain of Ω, we obtain $\text{div}F = 0$ in Ω. In many physical situations, the flux density is proportional to the gradient of a scalar function u so that $F = -a\nabla u$, $a > 0$. Consequently, $\text{div}\nabla u = \Delta u = 0$ in Ω.

2. As a specific case, let ϕ be the electric potential. Then, the electric field is given by $e = -\nabla\phi = -\left(\frac{\partial \phi}{\partial x_1}, \cdots, \frac{\partial \phi}{\partial x_n}\right)$. By Ohm's law, the electric current vector j satisfies $j = Ce$, where $C = [C_{ij}]$ is the general conductivity coefficient (matrix tensor). Substituting this in the continuity equation $\text{div}\, j = \nabla \cdot j = 0$, we obtain, the second-order equation

$$\sum_{i,j=1}^{n} \frac{\partial}{\partial x_i}\left(C_{ij}(x)\frac{\partial \phi}{\partial x_j}\right) = 0.$$

In particular, if the medium is isotropic, then, $C = \gamma I$, where γ is the conductivity and we see that ϕ satisfies the Laplace equation.

3. The Laplace/Poisson equation appears in many other situations as well. Some examples are indicated in the following table. In the presence of charges or external force, the Laplace equation becomes Poisson equation.

	j	e	ϕ	C
Thermal conduction	Heat current	Temperature gradient	Temperature	Thermal conductivity
Electrical conduction	Electrical current	Electrical field	Electric potential	Conductivity
Di-electrics	Displacement field	Electric field	Potential	Permittivity
Diffusion	Particle current	Gradient of concentration	Concentration	Diffusivity
Magnetism	Magnetic induction	Magnetic field intensity	Potential	Permeability
Stokes flow	Particle current	Pressure gradient	Pressure	Viscosity

7.2 FUNDAMENTAL SOLUTION, MEAN VALUE FORMULA AND MAXIMUM PRINCIPLES

The main idea of the study of these special concepts is that the general solution for the Dirichlet problem can be given by an integral representation via a specific or particular (singular) solution known as *fundamental solution/Green's function.* This is known as *Green's representation formula* and later, it can be used in the solvability of the equation. In the process, we also deduce some interesting properties – *mean value property* (MVP), *minimum and maximum principles.* Interestingly, the fundamental solution is not a solution of the Laplace equation as it has a singularity, but very useful in the construction of solutions.

One of the important features of Laplace equation $\Delta u = 0$ is its spherical symmetry; that is the equation is invariant under rotations about any point. This means in particular that if $u(x)$ is a solution of $\Delta u = 0$ and R is a rotation matrix, then $v(x) \equiv u(Rx)$ is also a solution of the Laplace equation (see Exercise 3). This suggests that we look for radial solutions u, that is u satisfying $u(x) = u(|x|)$. Putting $r = |x| = \left(\sum_{i=1}^{n} x_i^2\right)^{1/2}$, it is easy to derive that the function $v = v(r) := u(|x|)$ satisfies

$$\Delta u \equiv \ddot{v}(r) + \frac{n-1}{r}\dot{v}(r) = 0, \tag{7.3}$$

where $\cdot$ denotes the differentiation with respect to r. This ordinary differential equation (ODE) can easily be solved for $r > 0$ and we obtain

$$v(r) = \begin{cases} b \log r + C \text{ if } n = 2 \\ \dfrac{b}{r^{n-2}} + C \text{ if } n \geq 3. \end{cases} \tag{7.4}$$

for arbitrary constants b, C. We define the *fundamental solution* for $-\Delta$ as[1]

$$\phi(x) = \begin{cases} -\dfrac{1}{2\pi} \log |x|, \text{ if } n = 2 \\ \dfrac{1}{n(n-2)\omega_n |x|^{n-2}}, \text{ if } n \geq 3, \end{cases} \tag{7.5}$$

where ω_n, is the volume of the unit ball in $\mathbb{R}^n$.

It is readily verified that $\Delta\phi(x) = 0$ for all $x \in \mathbb{R}^n \backslash \{0\}$ and 0 is a singularity for ϕ. The functions satisfying $\Delta u = 0$ in a domain Ω are called *harmonic functions*. Thus, ϕ is harmonic in $\mathbb{R}^n \backslash \{0\}$. This singular solution plays a vital role in the future analysis. More generally, the singularity can be shifted to any other point $y \in \mathbb{R}^n$. That is, the function $x \mapsto \phi(x-y)$ is harmonic in $\mathbb{R}^n \backslash \{y\}$. Further, for $f : \mathbb{R}^n \to \mathbb{R}$, the function $x \mapsto \phi(x-y)f(y)$ is also harmonic in $\mathbb{R}^n \backslash \{y\}$ and so are their finite sums. This motivates us to define the *infinite sum* or *integral* and look for a solution u of the Laplace equation of the form:

$$u(x) = \int_{\mathbb{R}^n} \phi(x-y)f(y)dy = \phi * f(x). \tag{7.6}$$

The term on the right-hand side is the *convolution* of ϕ and f. Indeed, if it were true that $\Delta u(x) = \int_{\mathbb{R}^n} \Delta_x \phi(x-y)f(y)dy$, then $\Delta u = 0$ and we obtain a solution. To take the Δ inside the integral sign, we need the local integrability of the second derivatives of ϕ. But the function $D^2\phi(x-y)$ is not locally integrable in any neighborhood of y. However, $\phi(x-y)$ and $D\phi(x-y)$ are locally integrable (see Exercise 6).

Proposition 7.1. There exists a constant $C > 0$ such that

$$|D\phi(x)| \leq \frac{C}{|x|^{n-1}}, \quad |D^2\phi(x)| \leq \frac{C}{|x|^n}, x \neq 0. \tag{7.7}$$

[1]The choice of the constants will be clear from the analysis that follows.

Here $D\phi(x)$ is the gradient vector of ϕ and $D^2\phi(x)$ is the Hessian. In fact, $D\phi(x)$ and $D^2\phi(x)$ behave like $\frac{1}{|x|^{n-1}}$ and $\frac{1}{|x|^n}$ near the origin, respectively. Further, ϕ and $D\phi$ are locally integrable, but $D^2(\phi)$ is not. □

Proof Proof follows by direct differentiation and the fact that the function $\frac{1}{|x|^\alpha}$ is integrable in a neighborhood of the origin in $\mathbb{R}^n$ if and only if $\alpha < n$. The details are left as an exercise (see Exercise 6). □

Under certain assumptions, u defined by (7.6), in fact, solves Poisson equation. Equation (7.6), by changing variables: $y \to x - y$, can also be written as

$$u(x) = \int_{\mathbb{R}^n} \phi(x-y)f(y)dy = \int_{\mathbb{R}^n} \phi(y)f(x-y)dy.$$

Now assume $f \in C_c^2(\mathbb{R}^n)$. Then,

$$\frac{\partial u}{\partial x_i}(x) = \int_{\mathbb{R}^n} \phi(y)\frac{\partial f}{\partial x_i}(x-y)dy = \phi * \frac{\partial f}{\partial x_i}(x)$$

and similarly

$$\frac{\partial^2 u}{\partial x_i \partial x_j} = \phi * \frac{\partial^2 f}{\partial x_i \partial x_j}.$$

In the above computations, there is no problem of taking the differentiation inside the integral due to the smoothness of f and local integrability of ϕ. In particular,

$$\Delta u = \phi * \Delta f. \tag{7.8}$$

The convolution has this special property. To find the derivative of the convolution, it is enough to take the derivative of any one of the component functions and convolute it with the other. This is extremely useful for approximating non-smooth functions by smooth functions.

Now under the assumption that $f \in C_c^2(\mathbb{R}^n)$, we will show that $-\Delta u = -\phi * \Delta f = f$ in $\mathbb{R}^n$. Note that the integrand in

$$(\phi * \Delta f)(x) = \int_{\mathbb{R}^n} \phi(y)(\Delta f)(x-y)dy$$

has a singularity at the origin. Thus, we write

$$\Delta u(x) = I_\varepsilon + J_\varepsilon,$$

where $\varepsilon > 0$ and

$$I_\varepsilon = \int_{B_\varepsilon(0)} \phi(y)(\Delta f)(x-y)dy, \qquad J_\varepsilon = \int_{\mathbb{R}^n \setminus B_\varepsilon(0)} \phi(y)(\Delta f)(x-y)dy$$

It is easy to see, using the boundedness of Δf that $|I_\varepsilon| \longrightarrow 0$ as $\varepsilon \longrightarrow 0$. In fact, we can derive the following estimate (see Exercise 7): there exists a constant $C > 0$ such that

$$|I_\varepsilon| \leq \begin{cases} C\varepsilon^2 |\log \varepsilon|, & \text{if } n = 2 \\ C\varepsilon^2 & \text{if } n \geq 3. \end{cases}$$

An integration by parts gives

$$\begin{aligned} J_\varepsilon &= -\int_{\mathbb{R}^n \setminus B_\varepsilon(0)} \nabla\phi(y) \cdot \nabla f(x-y)dy - \int_{\partial B_\varepsilon(0)} \phi(y)\frac{\partial f}{\partial \nu}(x-y)d\sigma \\ &= K_\varepsilon + L_\varepsilon. \end{aligned}$$

Here, ν is the unit outward normal to $B_\varepsilon(0)$. Estimating L_ε is similar to that of I_ε, and we get

$$|L_\varepsilon| \leq \begin{cases} C\varepsilon |\log \varepsilon|, & \text{if } n = 2 \\ C\varepsilon, & \text{if } n \geq 3 \end{cases}$$

for some constant C. Thus, $|L_\varepsilon| \longrightarrow 0$ as $\varepsilon \longrightarrow 0$. To estimate K_ε, we do one more integration by parts to get

$$\begin{aligned} K_\varepsilon &= \int_{\mathbb{R}^n \setminus B_\varepsilon(0)} \Delta\phi(y) f(x-y)dy + \int_{\partial B_\varepsilon(0)} \frac{\partial \phi}{\partial \nu}(y) f(x-y)d\sigma(y) \\ &= \int_{\partial B_\varepsilon(0)} \frac{\partial \phi}{\partial \nu}(y) f(x-y)d\sigma(y), \end{aligned}$$

since $\Delta\phi(y) = 0$ for $y \neq 0$. On $\partial B_\varepsilon(0)$, $\nu = \dfrac{y}{|y|} = \dfrac{y}{\varepsilon}$ and so

$$\frac{\partial \phi}{\partial \nu}(y) = \nu \cdot \nabla\phi = -\frac{1}{n\omega_n \varepsilon^{n-1}} = -\frac{1}{|\partial B_\varepsilon(0)|}.$$

Therefore

$$\begin{aligned} K_\varepsilon &= -\frac{1}{|\partial B_\varepsilon(0)|} \int_{\partial B_\varepsilon(0)} f(x-y)d\sigma(y) \\ &= -\frac{1}{|\partial B_\varepsilon(x)|} \int_{\partial B_\varepsilon(x)} f(y)d\sigma(y) \\ &\to -f(x) \text{ as } \varepsilon \to 0. \end{aligned}$$

Thus, we have proved the following result:

Theorem 7.2. Let $f \in C_c^2(\mathbb{R}^n)$ and define u by

$$u(x) = \phi * f(x), \tag{7.9}$$

where ϕ is the fundamental solution of $-\Delta$. Then, $u \in C^2(\mathbb{R}^n)$ and u solves the Poisson equation $-\Delta u = f$ in $\mathbb{R}^n$. □

In the language of distributions, the function ϕ satisfies the equation $-\Delta\phi = \delta_0$ in the sense of distributions,[2] from which comes its nomenclature the *fundamental solution* of $-\Delta$.

Later, we show that the above theorem holds good when $f \in C^1(\overline{\Omega})$, but not for arbitrary $f \in C(\overline{\Omega})$. This is an important point to be noted that the continuity of f is not enough for the existence of a solution.

We now derive an important result for harmonic functions; more generally, for sub-harmonic and super-harmonic functions. *A* function $u \in C^2(\Omega)$ is said to be *sub-harmonic* if $\Delta u \geq 0$ in Ω and *super-harmonic* if $\Delta u \leq 0$ in Ω. These notions are fundamental in the study of uniqueness and existence results for Laplace equation via the *Perron's Method.*

7.2.1 Mean Value Formula

We now state and prove the following theorem:

Theorem 7.3. Let $u \in C^2(\Omega)$ be sub-harmonic. Then, for any ball $B = B_R(y) \subset\subset \Omega$, we have

$$u(y) \leq \frac{1}{n\omega_n R^{n-1}} \int_{\partial B} u(x)\, d\sigma(x) = \frac{1}{|\partial B|} \int_{\partial B} u\, d\sigma(x) \tag{7.10}$$

[2]The precise meaning of this is: $\int \phi(x)\Delta\psi(x)dx = -\psi(0)$ for every C_c^∞ function ψ.

and

$$u(y) \leq \frac{1}{\omega_n R^n} \int_B u(x)\,dx = \frac{1}{|B|} \int_B u(x)\,dx. \tag{7.11}$$

The above inequalities are reversed if u is super-harmonic. Finally, if u is harmonic, then, we have

$$u(y) = \frac{1}{|\partial B|} \int_{\partial B} u\,d\sigma = \frac{1}{|B|} \int_B u(x)\,dx. \tag{7.12}$$

□

Equation (7.12) is referred to as the *mean value formula* and the harmonic function u is said to satisfy the MVP.

Proof Assume $u \in C^2(\Omega)$ is sub-harmonic. We have, by divergence theorem, for $0 < \rho < R$,

$$\int_{\partial B_\rho(y)} \frac{\partial u}{\partial \nu}(x)\,d\sigma(x) = \int_{B\rho(y)} \Delta u(x)\,dx \geq 0.$$

Let $g(\rho) = \int_{\partial B_\rho(y)} \frac{\partial u}{\partial \nu}(x)\,d\sigma(x)$. Then,

$$g(\rho) = \rho^{n-1} \int_{\partial B_1(0)} \frac{\partial u}{\partial \rho}(y + \rho s)\,d\sigma(s) = \rho^{n-1} \frac{\partial}{\partial \rho} \int_{\partial B_1(0)} u(y + \rho s)\,d\sigma(s),$$

where $\rho = |x - y|, s = \frac{x-y}{\rho}$ are the radial and angular co-ordinates. Thus

$$g(\rho) = \rho^{n-1} \frac{\partial}{\partial \rho} \left(\rho^{1-n} \int_{\partial B_\rho(y)} u\,d\sigma \right) \geq 0,$$

which implies that

$$\frac{\partial}{\partial \rho} \left(\rho^{1-n} \int_{\partial B_\rho(y)} u\,d\sigma \right) \geq 0.$$

Fix ρ_1, ρ_2 such that $0 < \rho_1 < \rho_2 \le R$ and integrate the above inequality from ρ_1 to ρ_2 to obtain

$$\rho_1^{1-n} \int_{\partial B_{\rho_1}(y)} u\, d\sigma \le \rho_2^{1-n} \int_{\partial B_{\rho_2}(y)} u\, d\sigma.$$

But

$$\lim_{\rho_1 \to 0} \frac{1}{n\omega_n \rho_1^{n-1}} \int_{\partial B_{\rho_1}(y)} u(x)\, d\sigma(x) = u(y)$$

since the term inside the limit is the average of u on the sphere $\partial B_{\rho_1}(y)$. Now taking $\rho_2 = R$, we get the first inequality (7.10). We also have, by taking $\rho_2 = \rho$,

$$u(y) \le \frac{1}{|\partial B_\rho(y)|} \int_{\partial B_\rho(y)} u\, d\sigma.$$

This implies that

$$n\omega_n \rho^{n-1} u(y) \le \int_{\partial B_\rho(y)} u\, d\sigma$$

which is true for any $\rho \le R$. Now integrate with respect to ρ from 0 to R, to get the inequality (7.11).

If u is super-harmonic, observe that $-u$ is sub-harmonic. If u is harmonic, then both u and $-u$ are sub-harmonic. The other conclusions in the theorem will now readily follow. □

Remark 7.4. The proof above contains the following interesting result: If $u \in C^1(\Omega)$ and $\int_{\partial B} \frac{\partial u}{\partial \nu}\, d\sigma = 0$ for every ball $B \subset\subset \Omega$, then u satisfies MVP in Ω. □

Theorem 7.5 (Converse of MVP). Let $u \in C^2(\Omega)$ satisfy

$$u(x) = \frac{1}{|\partial B|} \int_{\partial B} u\, d\sigma = \frac{1}{|B|} \int_B u\, dx \tag{7.13}$$

for each ball $B = B_r(x) \subset\subset \Omega$, then u is harmonic. □

Proof First, we remark that the above theorem is also true with the assumption that $u \in C(\Omega)$. The present proof requires that $u \in C^2(\Omega)$. Let $h(r) = \frac{1}{|\partial B_r(x)|} \int_{\partial B_r(x)} u(y) d\sigma(y) = u(x)$, then on one hand $h'(r) = 0$ as h is independent of r using (7.13). On the other hand,

$$h(r) = \frac{1}{|\partial B_1(0)|} \int_{\partial B_1(0)} u(x + rz) d\sigma(z).$$

Therefore,

$$\begin{aligned} h'(r) &= \frac{1}{|\partial B_1(0)|} \int_{\partial B_1(0)} \nabla u(x + rz) \cdot z \, d\sigma(z) \\ &= \frac{1}{|\partial B_r(x)|} \int_{\partial B_r(x)} \nabla u(y) . \frac{y - x}{r} \, d\sigma(y) \\ &= \frac{1}{|\partial B_r(x)|} \int_{\partial B_r(x)} \frac{\partial u}{\partial r} \, d\sigma(y) \\ &= \frac{1}{|\partial B_r(x)|} \int_{B_r(x)} \Delta u(y) \, dy, \text{ using divergence theorem} \\ &= \frac{r}{n|B_r(x)|} \int_{B_r(x)} \Delta u(y) \, dy \end{aligned}$$

Thus, if $\Delta u(x) \neq 0$, say $\Delta u(x) > 0$, we can choose r small enough, so that $\Delta u(y) > 0$ in $B_r(x)$, leading to a contradiction. □

7.2.2 Maximum and Minimum Principles

Maximum and minimum principles are trademarks of second-order elliptic differential operators. Consider a C^2 function f defined on an interval (a, b). Suppose f has an interior strict local maximum. That is, there is a point $c \in (a, b)$ such that $f(c)$ is the maximum value of f in a neighborhood of c. Then, we know that $f'(c) = 0$ and $f''(c) < 0$. In other words, if $f'' \geq 0$ in a neighborhood of c, then f cannot have a strict local maximum at c. The operator Δ replaces the second derivative f'' in higher dimensions. More generally, it would be interesting to derive maximum principles for general second-order operators. This has far-reaching consequences in the study of uniqueness of solution, comparison principle, and so on, of differential operators. The maximum (minimum) principle, is useful in comparing a solution of, for example, $\Delta u = u^2$ in a domain Ω with the solution of v of $\Delta v = 0$ in Ω satisfying $v = u$ on $\partial\Omega$. The above discussion also indicates that the maximum principle is generally obtained for second-order elliptic operators, though some other operators also enjoy this property.

Theorem 7.6 (Strong Maximum Principle). Assume Ω is a bounded region and u is a bounded sub-harmonic function in Ω. If there is a $y \in \Omega$ such that $u(y) = \sup_\Omega u$, then u is a constant. □

That is, a bounded sub-harmonic function cannot assume an interior maximum unless it is a constant.

Proof Let $M = \sup_\Omega u$ and $\Omega_M = \{x \in \Omega : u(x) = M\}$. Then Ω is non-empty since $y \in \Omega_M$. By continuity of u, the set Ω_M is closed in Ω. Now $u - M$ is also sub-harmonic. Let $z \in \Omega_M$. Then by MVP, we get

$$0 = u(z) - M \leq \frac{1}{\omega_n R^n} \int_{B_R(z)} (u(x) - M)dx \leq 0$$

for $B_R(z) \subset\subset \Omega$. Since $u \leq M$ in Ω, we conclude that $u \equiv M$ in $B_R(z)$. Hence, $B_R(z) \subset \Omega_M$ and so Ω_M is open as well. Thus, by connectedness, $\Omega_M \equiv \Omega$ as Ω_M is non-empty. □

If u is super-harmonic, then by applying the above theorem to $-u$, we see that a non-constant super-harmonic function cannot assume an interior minimum and a non-constant harmonic function cannot assume both interior minimum and maximum. We state these observations in the following theorem:

Theorem 7.7. Assume Ω is a bounded region in $\mathbb{R}^n$. For $u \in C^2(\Omega) \cap C(\overline{\Omega})$, the following statements hold:

1. if u is sub-harmonic in Ω, then $\sup_\Omega u = \max_{\overline{\Omega}} u = \max_{\partial\Omega} u$.
2. if u is super-harmonic in Ω, then, $\inf_\Omega u = \min_{\overline{\Omega}} u = \min_{\partial\Omega} u$.
3. if u is harmonic in Ω, then, $\min_{\partial\Omega} u \leq u(x) \leq \max_{\partial\Omega} u$, for all $x \in \Omega$. □

The conclusions in Theorem 7.7 are referred as *weak maximum and minimum principles.* The maximum and minimum principles are also valid for more general second-order elliptic equations. We present a case here. Consider a second-order partial differential operator L of the form

$$L = \sum_{i,j=1}^{n} a_{ij} \frac{\partial^2}{\partial x_i \partial x_j} + \sum_{i=1}^{n} b_i \frac{\partial}{\partial x_i} + c, \tag{7.14}$$

where $a_{ij} = a_{ji}, b_i, c$ are real-valued functions defined on Ω and are assumed to be smooth. If $x_0 \in \Omega$, we say L is elliptic at x_0 if the matrix $[a_{ij}(x_0)]$ is non-negative definite. The operator L is elliptic in Ω if it is elliptic at every point in Ω.

Theorem 7.8. Suppose L is elliptic in Ω and $u \in C^2(\Omega)$ takes its supremum at some point $x_0 \in \Omega$. Then, $Lu(x_0) \leq c(x_0)u(x_0)$. □

Proof We recall the following result from Linear Algebra. Suppose $C = [c_{ij}]$ and $D = [d_{ij}]$ are symmetric, non-negative definite matrices, then $\sum_{i,j=1}^{n} c_{ij}d_{ij} \geq 0$. In fact, this follows since the above sum is the trace of the matrix CD and trace of a matrix is invariant under similarity transformations.

At the interior supremum point x_0, we have $\frac{\partial u}{\partial x_i}(x_0) = 0$ for $i = 1, \dots n$ and the Hessian $\left[\frac{\partial^2 u}{\partial x_i \partial x_j}(x_0)\right]$ of u is non-positive definite. Hence, it follows from the above observation that

$$\sum_{i,j=1}^{n} a_{ij}(x_0)\frac{\partial^2 u}{\partial x_i \partial x_j}(x_0) \leq 0.$$

This completes the proof. □

Thus, we have the following maximum principle for L:

Theorem 7.9. Let L be given by (7.14) be elliptic in Ω. Assume that either $Lu \geq 0$ and $c < 0$ or $Lu > 0$ and $c \leq 0$ in Ω. Then, u cannot take an interior maximum in Ω. □

Analogous statements can be made for the case of minimum as well. In general, the maximum principle is not true for higher-order operators. The maximum principle may not also hold in unbounded domains. However, with some additional hypothesis on u, we do obtain certain maximum principle for unbounded domains, as we see below.

Example 7.10. Consider the function $u(x) = x(x-1)(x+1)$ for $x \in (0, 1)$. Then, u satisfies the fourth-order equation $\frac{d^4u}{dx^4} = 0$. But u has a positive strict minimum at $x = 1/2$. □

Example 7.11. Consider the upper half plane $\Omega = \{(x_1, x_2) \in \mathbb{R}^2 : x_2 > 0\}$ and let $u(x_1, x_2) = x_2$ in Ω. Then, u is harmonic in Ω. Clearly u is zero on the boundary $\{x_2 = 0\}$ and hence the maximum principle does not hold. □

But, we have the following theorem:

Theorem 7.12. Let $\Omega = \{(x_1, x_2) \in \mathbb{R}^2 : x_2 > 0\}$ and suppose $u \in C^2(\Omega)$ is a bounded harmonic function. Then,

$$\sup_{\overline{\Omega}} |u| = \sup_{\partial\Omega} |u|.$$

□

Proof For a given $\varepsilon > 0$, consider the function $v(x) = u(x) - \frac{\varepsilon}{2}\log\left(x_1^2 + (x_2+1)^2\right)$ for $x \in \Omega$. Then, v is harmonic in Ω, since the second term on the right-hand side is essentially the fundamental solution. Now, we apply the maximum principle (Theorem 7.7) to v on the bounded domain

$$\Omega_a = \{(x_1, x_2) \in \Omega : x_1^2 + (x_2+1)^2 \le a^2\}$$

to get

$$\max_{\overline{\Omega}_a} |v| = \max_{\partial\Omega_a} |v|.$$

The boundary $\partial\Omega_a$ consists of two parts, namely the flat part $\Gamma_1 = \{(x_1, x_2) \in \partial\Omega_a : x_2 = 0\}$ and the curved part $\Gamma_2 = \{(x_1, x_2) \in \partial\Omega_a : x_2 > 0\}$.

Claim: The maximum of v on $\partial\Omega_a$ cannot be achieved on Γ_2.

If not, let $(x_1^0, x_2^0) \in \Gamma_2$ where v assumes its maximum, that is

$$v(x_1, x_2) \le v(x_1^0, x_2^0)$$

for all $(x_1, x_2) \in \overline{\Omega}_a$. In particular, taking $(x_1, x_2) \in \Gamma_1$, we have

$$u(x_1, 0) - \frac{\varepsilon}{2}\log\left(x_1^2 + 1\right) \le u(x_1^0, x_2^0) - \frac{\varepsilon}{2}\log a^2$$

for all $|x_1| \le a$. This implies, for all a large,

$$\varepsilon \log a \le 2C,$$

where $|u| \le C$. Taking a sufficiently large, we get a contradiction and hence the claim. Now letting $\varepsilon \to 0$, we see that the maximum for u in $\overline{\Omega}$ is achieved on the x_1 axis. Hence the theorem. □

7.2.3 Uniqueness and Regularity of the Dirichlet Problem

The maximum principle can be used to prove the following uniqueness result:

Theorem 7.13 (Uniqueness). The Dirichlet problem for Poisson equation

$$\begin{aligned} -\Delta u &= f \text{ in } \Omega \\ u &= g \text{ on } \partial\Omega \end{aligned} \tag{7.15}$$

has at most one solution $u \in C^2(\Omega) \cap C(\overline{\Omega})$. □

Proof If u_1 and u_2 are solutions of (7.15), apply the third conclusion in the Theorem 7.7 to the harmonic function $u = u_1 - u_2$. □

For the existence of a classical solution u of (7.15), we only ask for $u \in C^2(\Omega) \cap C(\overline{\Omega})$ under certain minimum smoothness assumption on f, g and $\partial\Omega$, collectively referred to as *data*. If the data possess more smoothness, then it is natural to expect more smoothness for the solution u. Such results are called regularity results. The regularity results are particularly useful in a general situation. In many instances, it may be difficult or even not possible to prove the existence of a classical solution to a given PDE, working with *only* spaces of smooth functions. However, it is possible to prove the existence of a *weak solution*, using the techniques of Hilbert space or more generally Functional Analysis. The regularity results (in a weak sense) will then show, under suitable conditions on the data, when a weak solution is actually a classical solution using Sobolev embedding theorems and so on.

Theorem 7.14 (Regularity). If $u \in C(\Omega)$ satisfies MVP (7.13) for each ball $B_r(x) \subset\subset \Omega$, then $u \in C^\infty(\Omega)$. □

Proof Proof is simple, but it is based on the concept of *mollifiers* $\rho_\varepsilon, \varepsilon > 0$ and convolution. If $x \in \Omega_\varepsilon = \{x \in \Omega : d(x, \partial\Omega) > \varepsilon\}$, then the convolution $u_\varepsilon(x) = \rho_\varepsilon * u(x)$ is in $C^\infty(\Omega_\varepsilon)$ as $\rho_\varepsilon \in C^\infty(\mathbb{R}^n)$. (see Chapter 2).

Claim: $u_\varepsilon(x) = u(x)$ for $x \in \Omega_\varepsilon$.

The above claim shows that $u \in C^\infty(\Omega_\varepsilon)$. Since any $x \in \Omega$ is in some Ω_ε for $\varepsilon > 0$ small, we conclude that $u \in C^\infty(\Omega)$.

Proof of the Claim: We use the special properties of mollifier ρ_ε namely, it is radial with compact support contained in $B_\varepsilon(0)$. Recall the mollifier

$$\rho_\varepsilon(y) = \rho_\varepsilon(|y|) = \frac{1}{\varepsilon^n}\rho\left(\frac{|y|}{\varepsilon}\right)$$

for $y \in \mathbb{R}^n$. For $x \in \Omega_\varepsilon$, the ball $B_\varepsilon(x) \subset\subset \Omega$. Then, we have

$$\begin{aligned}
u_\varepsilon(x) &= \frac{1}{\varepsilon^n}\int_{B_\varepsilon(x)} \rho\left(\frac{|x-y|}{\varepsilon}\right) u(y)\, dy \\
&= \frac{1}{\varepsilon^n}\int_0^\varepsilon \int_{\partial B_r(x)} \rho\left(\frac{r}{\varepsilon}\right) u(y) d\sigma(y)\, dr \\
&= \frac{1}{\varepsilon^n}\int_0^\varepsilon \rho\left(\frac{r}{\varepsilon}\right) n\omega_n r^{n-1} u(x)\, dr, \text{ applying MVP for } u \\
&= u(x)\frac{1}{\varepsilon^n}\int_0^\varepsilon \rho\left(\frac{r}{\varepsilon}\right) n\omega_n r^{n-1}\, dr,
\end{aligned}$$

$$= u(x)\frac{1}{\varepsilon^n}\int_{B_r(x)} \rho\left(\frac{x-y}{\varepsilon}\right) dy$$
$$= u(x).$$

This proves the claim and the proof of the theorem is complete. □

The above Theorem 7.14 together with the Theorem 7.5 shows that a continuous function satisfying the MVP is C^∞ and harmonic. To prove this, we have used the concept of mollifiers, but we can also give a direct proof of this fact (see Theorem 7.24 and Exercise 35).

In fact, a harmonic function u is analytic. This is proved by establishing finer estimates on u and its derivatives.

Estimates on the Derivatives: Let $x_0 \in \Omega$, u be harmonic in Ω and r be such that $B_r(x_0) \subset\subset \Omega$. Observe that $\Delta u = 0$ implies $\Delta\left(\frac{\partial u}{\partial x_i}\right) = 0$ in Ω, so that $\frac{\partial u}{\partial x_i}$ is also harmonic. Now, apply MVP (7.13) to $\frac{\partial u}{\partial x_i}$ in the ball $B_{r/2}(x_0)$ to deduce that

$$\left|\frac{\partial u}{\partial x_i}(x_0)\right| \leq \frac{2n}{r}\|u\|_{L^\infty(\partial B_{r/2}(x_0))}. \tag{7.16}$$

To derive the above estimate, use the fact $\int_{B_{r/2}(x_0)} \frac{\partial u}{\partial x_i} dx = -\int_{\partial B_{r/2}(x_0)} u\nu_i d\sigma$. The MVP (7.13) applied to u in $B_{r/2}(x) \subset B_r(x_0)$, we get

$$|u(x)| \leq \frac{1}{\omega_n}\left(\frac{2}{r}\right)^n \|u\|_{L^1(B_r(x_0))}$$

for all $x \in B_{r/2}(x_0)$. Combining, we get

$$\left|\frac{\partial u}{\partial x_i}(x_0)\right| \leq \left(\frac{2}{r}\right)^{n+1} \frac{n}{\omega_n}\|u\|_{L^1(B_r(x_0))}.$$

This, in turn, implies (7.16). Using an induction argument, we deduce that

$$\left|D^\alpha u(x_0)\right| \leq \frac{(2^{n+1}nk)^k}{\omega_n r^{n+k}}\|u\|_{L^1(B_r(x_0))} \tag{7.17}$$

if $|\alpha| = k$. This estimate immediately establishes the analyticity of harmonic functions, that is, for each $x_0 \in \Omega$, there exists $r > 0$ such that $u(x)$ is represented as a convergent power series

$$u(x) = \sum_{\alpha} \frac{D^{\alpha}u(x_0)}{\alpha!}(x - x_0)^{\alpha},$$

for $|x - x_0| < r$. Further, we have the following Liouville's Theorem:

Theorem 7.15 (Liouville's Theorem). Suppose u is a bounded harmonic function in $\mathbb{R}^n$. Then, u is a constant. □

Proof From the estimate (7.16), we have $\left|\frac{\partial u}{\partial x_i}\right| \leq \frac{C}{r}\|u\|_{L^{\infty}(\mathbb{R}^n)}$ that tends to 0 as $r \to \infty$. Hence, $\frac{\partial u}{\partial x_i} = 0$ for all $i = 1, \cdots, n$. Thus, u is a constant. □

We have seen that if $f \in C_c^2(\mathbb{R}^n)$, then u defined using the fundamental solution, namely $u(x) = \int \phi(x - y)f(y)dy$ solves the Poisson equation $-\Delta u = f$ in $\mathbb{R}^n$. In some sense the converse is also true and we have the following result:

Theorem 7.16 (Representation Formula). Let $f \in C_c^2(\mathbb{R}^n)$ and u be a *bounded* solution of $-\Delta u = f$ in $\mathbb{R}^n$, $n \geq 3$. Then, u is represented as

$$u(x) = \int_{\mathbb{R}^n} \phi(x - y)f(y)dy + C, \tag{7.18}$$

where C is a constant. □

Remark 7.17. The above result need not be true when $n = 2$, since $\phi(x) = -\frac{1}{2\pi}\log|x|$ is unbounded as $|x| \to \infty$, it may happen that $\int \phi(x - y)f(y)dy$ is unbounded. See Exercise 34. □

Proof By Theorem 7.2,

$$\tilde{u} = \int_{\mathbb{R}^n} \phi(x - y)f(y)dy$$

is a solution of $-\Delta\tilde{u} = f$. Now, for $n \geq 3$, $\phi(x) \to 0$ as $|x| \to \infty$, we see that $\tilde{u}$ is a bounded solution. Thus, if u is any other bounded solution, then $u - \tilde{u}$ is harmonic and bounded. Hence, $u - \tilde{u} \equiv$ constant, by Liouville's theorem. □

Point-wise comparison of a non-negative harmonic function away from the boundary is the theme of the Harnack's inequality, which we now discuss.

Theorem 7.18 (Harnack's Inequality). Let $V \subset\subset \Omega$ be a region and $u \geq 0$ be harmonic in Ω. Then, there exists a constant $C > 0$ depending only on V and n such that

$$\sup_V u \leq C \inf_V u. \tag{7.19}$$

In particular,

$$\frac{1}{C}u(y) \le u(x) \le Cu(y), \quad \text{for all } x, y \in V.$$

□

Proof Let $r = d(V, \partial\Omega)$, which is positive by assumption on V and $x, y \in V$ be such that $|x - y| \le r/4$. Then, by MVP (7.13),

$$\begin{aligned} u(x) &= \frac{1}{|B_{r/2}(x)|} \int_{B_{r/2}(x)} u(z)\, dz \\ &\ge \frac{2^n}{\omega_n r^n} \int_{B_{r/4}(y)} u(z)\, dz \\ &= \frac{1}{2^n |B_{r/4}(y)|} \int_{B_{r/4}(y)} u(z)\, dz \\ &= \frac{1}{2^n} u(y). \end{aligned}$$

By interchanging x and y, we also get $2^n u(y) \ge u(x)$. Hence for all $x, y \in V$, $|x - y| \le r/4$, we have

$$2^n u(y) \ge u(x) \ge \frac{1}{2^n} u(y).$$

By compactness of $\overline{V}$, we can cover $\overline{V}$ by finitely many balls, say N balls of radius $r/4$. This yields

$$2^{nN} u(y) \ge u(x) \ge \frac{1}{2^{nN}} u(y), \text{ for all } x, y \in V.$$

This completes the proof of the theorem. □

As an application, we prove the following result, *Hopf's Lemma,* for the ball, which gives an estimate for the normal derivative of the solution at the boundary:

Lemma 7.19 (Hopf's Lemma). Let $u \in C^2(B_R(x_0)) \cap C(\overline{B}_R(x_0))$ be harmonic in $B_R(x_0)$ and $x_* \in \partial B_R(x_0)$ be a strict minimum point of u in $\overline{B}_R(x_0)$. Then,

$$-\frac{\partial u}{\partial \nu}(x_*) \ge 2^{1-n} \left(\frac{u(x_0) - u(x_*)}{R} \right) > 0,$$

provided that the one-sided normal derivative

$$\frac{\partial u}{\partial \nu}(x_*) = \lim_{t \to 0+} \frac{u(x_* - t\nu) - u(x_*)}{t}$$

exists. □

Proof Observe that the inward unit normal to $\partial B_R(x_0)$ at x_* is $\nu = \frac{x_0 - x_*}{|x_0 - x_*|}$. The non-negative function $u - u(x_*)$ is harmonic in $B_R(x_0)$. Therefore, by Harnack's inequality, we have

$$\begin{aligned}\frac{u(x) - u(x_*)}{R - \rho} &\geq \left(\frac{R}{R+\rho}\right)^{n-2} \frac{u(x_0) - u(x_*)}{R + \rho} \\ &\geq 2^{1-n} \frac{u(x_0) - u(x_*)}{R},\end{aligned}$$

with $\rho = |x - x_0| < R$; here the simple estimate $(R+\rho)^{-1} \geq (2R)^{-1}$ is used. Now, let $x = x_* - t\nu$ for $0 < t < R$. Then, we have $R - \rho = t$. Letting $t \to 0$, the required estimate follows. □

7.2.4 Green's Function and Representation Formula

In Theorem 7.16, we have seen that if u is a solution to the Poisson equation in $\mathbb{R}^n$, then u is represented as in (7.18). We have then used this representation formula to show that u is indeed a solution to the Laplace equation. Thus, it is important to get a representation formula for the solution whenever it exists, which we will do it in this section. In the process, we will introduce the concept of a Green's function. Consider the Dirichlet problem

$$\begin{cases} -\Delta u = f \text{ in } \Omega, \\ u = g \text{ on } \partial\Omega. \end{cases} \tag{7.20}$$

Our aim in this section, towards solvability, is to obtain a representation formula for u in terms of the given data f, g and the fundamental solution. Now fix $x \in \Omega$. Then, the function $y \mapsto \phi(y - x)$ is harmonic except at $y = x$ and in particular, it is so in $V_\varepsilon = \Omega \backslash B_\varepsilon(x)$, where $B_\varepsilon(x) \subset\subset \Omega$. Applying Green's Formula to $u(\cdot)$ and $\phi(x - \cdot)$ in V_ε, we get

$$\begin{aligned}&\int_{V_\varepsilon} \left[u(y)\Delta\phi(y - x) - \phi(y - x)\Delta u(y)\right] dy \\ &\qquad = \int_{\partial V_\varepsilon} \left[u(y)\frac{\partial \phi}{\partial \nu}(y - x) - \phi(y - x)\frac{\partial u}{\partial \nu}(y)\right] d\sigma(y),\end{aligned}$$

where $\partial V_\varepsilon = \partial\Omega \cup \partial B_\varepsilon(x)$. We have

$$\left|\int_{\partial B_\varepsilon(x)} \phi(y - x)\frac{\partial u}{\partial \nu}\, d\sigma(y)\right| \leq \begin{cases} C\varepsilon \text{ if } n \geq 3 \\ C\varepsilon|\log\varepsilon| \text{ if } n = 2 \end{cases}$$

which tends to 0 as $\varepsilon \to 0$ and

$$\int_{\partial B_\varepsilon(x)} u(y)\frac{\partial \phi}{\partial \nu}\,d\sigma(y) = \frac{1}{|\partial B_\varepsilon(x)|}\int_{\partial B_\varepsilon(x)} u(y)\,d\sigma(y) \longrightarrow u(x)$$

as $\varepsilon \to 0$. Since ϕ is locally integrable, we have

$$\int_{V_\varepsilon} \phi(y-x)\Delta u(y)\,dy = -\int_{V_\varepsilon} \phi(y-x)f(y)\,dy \to -\int_{\Omega} \phi(y-x)f(y)\,dy$$

and using $\Delta\phi(y-x) = 0$ in V_ε, we arrive at the formula

$$u(x) = \int_{\Omega} f(y)\phi(y-x)\,dy - \int_{\partial\Omega} g(y)\frac{\partial \phi}{\partial \nu}(y-x)\,d\sigma(y) + \int_{\partial\Omega} \phi(y-x)\frac{\partial u}{\partial \nu}\,d\sigma(y). \qquad (7.21)$$

On the right-hand side, the normal derivative $\frac{\partial u}{\partial \nu}$ on $\partial\Omega$ is unknown. Hence, we would like to get rid of this term. This is achieved by a suitable *corrector function.* Fix $x \in \Omega$, the last term suggests that we look for a harmonic function whose boundary value is $\phi(\cdot - x)$. So introduce, $\phi^x = \phi^x(y)$ as the solution of

$$\begin{cases} -\Delta_y \phi^x = 0 \text{ in } \Omega \\ \phi^x(y) = \phi(y-x) \text{ on } \partial\Omega. \end{cases} \qquad (7.22)$$

Apply the Green's formula to the functions u and ϕ^x in Ω to arrive at

$$0 = \int_{\Omega} f(y)\phi^x(y)\,dy - \int_{\partial\Omega} g(y)\frac{\partial \phi^x}{\partial \nu}(y)\,d\sigma(y) + \int_{\partial\Omega} \phi(y-x)\frac{\partial u}{\partial \nu}\,d\sigma(y). \qquad (7.23)$$

Define, the *Green's function*

$$G(x,y) = \phi(x-y) - \phi^x(y),\ x \neq y,\ x,y \in \Omega. \qquad (7.24)$$

Now, subtracting equation (7.23) from (7.21), we get the following *Green's representation formula* for u as

$$u(x) = \int_{\Omega} f(y)G(x,y)\,dy - \int_{\partial\Omega} g(y)\frac{\partial G}{\partial \nu}(x,y)\,d\sigma(y),\ x \in \Omega. \qquad (7.25)$$

Indeed $G(x,\cdot)$ is harmonic in $\Omega\setminus\{y\}$ like $\phi(\cdot - x)$ and both have the same singularity. The difference is that G satisfies a boundary condition. Symbolically, $G(x,\cdot)$ solves the *measure valued* PDE

$$\Delta_y G = \delta_x \text{ in } \Omega$$
$$G = 0 \text{ on } \partial\Omega,$$

where δ_x is the Dirac-delta function concentrated at x. We will not go into a discussion of this topic here.

We will show that G is symmetric, that is $G(x,y) = G(y,x)$, which is not apparent in (7.24). To see this, fix $x, y \in \Omega$, $x \neq y$, define $v(z) = G(x,z)$, $w(z) = G(y,z)$. Apply Green's identity to v and w in $\Omega\setminus\left(B(x,\varepsilon)\cup B(y,\varepsilon)\right)$, with ε small. Then, it is easy to deduce that $v(y) = G(y,x)$ and $w(x) = G(x,y)$. See Exercise 31. Thus $G(x,y) = v(y) = w(x) = G(y,x)$.

If we can construct a Green's function for a given domain, there is a hope of answering the solvability question via the representation formula. But finding Green's function, in general, is as difficult as the solvability itself. However, there are some specific domains where this is possible, taking into account the geometrical structures enjoyed by these domains. Here, we present two cases, namely the upper half plane $\Omega = \mathbb{R}^n_+$ and the ball $\Omega = B_r(0)$. The corresponding Green's functions for the upper half plane and balls, are very important in the analysis of Poisson equation and it is presented below.

Due to the linearity of the problem (7.20), we can write u as $u = v + w$, where v solves the Poisson equation as in (7.20) with zero boundary condition, whereas w is harmonic and satisfies $w = g$ on $\partial\Omega$. We study these problems separately. The existence of w is obtained by the Perron's method that we will do it in the next section. For the existence of v, we need to analyze the Newtonian potential. This will be carried in the sections to follow.

Green's Function for the Upper Half Space: Consider the upper half space

$$\mathbb{R}^n_+ = \left\{x = (x_1, \cdots x_n) \in \mathbb{R}^n \;:\; x_n > 0\right\}.$$

Our goal is to obtain the corrector function $\phi^x = \phi^x(y)$ that solves

$$\Delta\phi^x = 0 \text{ in } \mathbb{R}^n_+$$
$$\phi^x = \phi(y - x) \text{ on } \partial\mathbb{R}^n_+.$$

The function $\phi(y - x)$ is singular only at $y = x$ and hence it satisfies the Laplace equation except at $y = x$. The geometry of $\mathbb{R}^n_+$ suggest to consider the reflection point $\tilde{x} = (x_1, \cdots x_{n-1}, -x_n) \in \mathbb{R}^n_-$ and hence $\phi(y - \tilde{x})$ is smooth in $\mathbb{R}^n_+$. Thus, $\phi^x(y) = \phi(y - \tilde{x})$ is harmonic in upper half space. Further, if $y = (y_1, \ldots, y_{n-1}, 0) = (y', 0)$ is on the boundary of the upper half space, then $\phi(y - \tilde{x}) = \phi(y - x)$. Thus, the required corrector function is ϕ^x and we have the Green's function

$$G(x, y) = \phi(y - x) - \phi(y - \tilde{x}),$$

$x, y \in \mathbb{R}^n_+,\ x \neq y$. A short calculation leads to

$$\frac{\partial G}{\partial \nu}(x, y) = -\frac{\partial G}{\partial y_n} = -\frac{2x_n}{n\omega_n |x - y|^n},$$

for $y \in \partial\mathbb{R}^n_+$. Thus, if $u \in C^2(\overline{\Omega})$ solves

$$\Delta u = 0 \text{ in } \mathbb{R}^n_+, \quad u = g \text{ on } \partial\mathbb{R}^n_+, \tag{7.26}$$

then, the representation formula is

$$u(x) = \frac{2x_n}{n\omega_n} \int_{\partial\mathbb{R}^n_+} \frac{g(y)}{|x - y|^n} d\sigma(y), \quad x \in \mathbb{R}^n_+. \tag{7.27}$$

Note that $d\sigma(y) = dy'$. The function

$$K(x, y) = \frac{2x_n}{n\omega_n |x - y|^n}, \ x \in \mathbb{R}^n_+, \ y \in \mathbb{R}^n_+$$

is known as *Poisson kernel* for $-\Delta$ in $\mathbb{R}^n_+$ and (7.27) is the *Poisson formula for the upper half space*. We, now verify that u defined by (7.27) indeed is the required solution, under the assumption g is a bounded continuous function in $\mathbb{R}^{n-1}$. Further, $\Delta_x K = 0$ for $x \neq y$ and $\int_{\partial\mathbb{R}^n_+} K(x, y) d\sigma(y) = 1$ for all $x \in \mathbb{R}^n_+$.

For $y \in \partial\mathbb{R}^n_+, x \in \mathbb{R}^n_+$, observe that K is a C^∞ function and if g is bounded, then we can differentiate u in (7.27) as many times as we wish by taking the differentiation under the integral sign. That is

$$D^\alpha u(x) = \int_{\partial\mathbb{R}^n_+} D^\alpha_x K(x, y) g(y) d\sigma(y).$$

Further, $\Delta u = 0$ in $\mathbb{R}^n_+$ as $\Delta_x K = 0$ in $\mathbb{R}^n_+$. Thus, we have the following theorem:

Theorem 7.20. Assume $g \in C_b(\mathbb{R}^{n-1})$, the space of continuous bounded functions, and define u by (7.27). Then, u satisfies the following:

1. $u \in C^\infty_b(\mathbb{R}^n_+)$.
2. $\Delta u = 0$ in $\mathbb{R}^n_+$.
3. $\lim\limits_{\substack{x \to x_0 \\ x \in \mathbb{R}^n_+}} u(x) = g(x_0)$ for $x_0 \in \partial\mathbb{R}^n_+ = \mathbb{R}^{n-1}$.

□

Proof Only part (3) need to be proved. Fix $x_0 \in \partial\mathbb{R}^n_+$. By continuity of g, given $\varepsilon > 0$, there is a $\delta > 0$ such that

$$|g(y) - g(x_0)| \leq \varepsilon$$

provided $|y - x_0| < \delta$ and $y \in \partial\mathbb{R}^n_+$. Now consider the ball $B_\delta(x_0)$ in $\mathbb{R}^n$ and let $A = \partial\mathbb{R}^n_+ \cap B_\delta(x_0)$, $B = \partial\mathbb{R}^n_+ \setminus B_\delta(x_0)$. Then, for $x \in B_{\delta/2}(x_0)$, we have

$$|u(x) - g(x_0)| \leq I + J,$$

where $I = \int_A K(x-y)|g(y) - g(x_0)|d\sigma(y)$ and $J = \int_B K(x-y)|g(y) - g(x_0)|d\sigma(y)$. Clearly, $I \leq \varepsilon$ as $\int_{\partial\mathbb{R}^n_+} K(x-y)d\sigma(y) = 1$. Now, if $y \in B, x \in B_{\delta/2}(x_0)$, then

$$|y - x_0| \leq |y - x| + |x - x_0| \leq |y - x| + \frac{\delta}{2} \leq |y - x| + \frac{1}{2}|y - x_0|.$$

Thus, $|y - x_0| \leq 2|y - x|$. Hence

$$J \leq \frac{2^{n+1}x_n\|g\|_0}{n\omega_n}\int_B \frac{1}{|y - x_0|^n}d\sigma(y)$$

which tends to zero as $x \to x_0$, that is as $x_n \to 0$. This completes the proof of the theorem. □

Green's Function for a Ball: First consider the unit ball $B = B_1(0)$. The idea is again to use the symmetry of B, to construct a corrector. Given $x \in \mathbb{R}^n \setminus \{0\}$, we take the inversion point $\tilde{x} = \frac{x}{|x|^2}$ so that if $x \in B_1(0)$, then $\tilde{x}$ is outside the ball.[3] Thus, the function $\phi(y - \tilde{x})$ is harmonic in $B_1(0)$. We need a certain normalization to fix the boundary values. We define, $\phi^x(y) = \phi(|x|(y - \tilde{x}))$. It is easy to check that

$$\Delta\phi^x = 0 \text{ in } B_1(0)$$
$$\phi^x = \phi(y - x) \text{ on } \partial B_1(0).$$

As before, define $G(x, y) = \phi(x - y) - \phi(|x|(y - \tilde{x}))$, $x, y \in B_1(0), x \neq y$. With a little further computation, we can see that, for $y \in \partial B_1(0)$,

$$\frac{\partial G}{\partial \nu}(x, y) = -\frac{1}{n\omega_n}\frac{1 - |x|^2}{|x - y|^n}.$$

[3]This is called the Kelvin's transform. For the ball $B_R(0)$, $\tilde{x} = \frac{R^2 x}{|x|^2}$.

Hence, if u solves the boundary value problem

$$\Delta u = 0 \text{ in } B_1(0)$$
$$u = g \text{ on } \partial B_1(0),$$

then, we get the *Poisson formula*

$$u(x) = \frac{1 - |x|^2}{n\omega_n} \int_{\partial B_1(0)} \frac{g(y)}{|x - y|^n} d\sigma(y). \tag{7.28}$$

If we consider the ball $B_r(0)$, $r > 0$, then, we get

$$u(x) = \frac{r^2 - |x|^2}{n\omega_n r} \int_{\partial B_r(0)} \frac{g(y)}{|x - y|^n} d\sigma(y) \tag{7.29}$$

for $x \in B_r(0)$. The function

$$K(x, y) = \frac{r^2 - |x|^2}{n\omega_n r} \frac{1}{|x - y|^n}$$

is known as the *Poisson kernel* for $-\Delta$ in the ball $B_r(0)$. We have the following theorem similar to Theorem 7.20, whose proof is also similar:

Theorem 7.21. Assume $g \in C(\partial B_r(0))$ and define u by (7.29). Then,

1. $u \in C^\infty(B_r(0))$.
2. $\Delta u = 0$ in $B_r(0)$.
3. $\lim_{\substack{x \to x_0 \\ x \in B_r(0)}} u(x) = g(x_0)$ for any $x_0 \in \partial B_r(0)$.

□

7.2.5 MVP Implies Harmonicity

We have earlier seen in Theorem 7.5 that a C^2 function satisfying MVP is harmonic and in Theorem 7.14, the same result was proved with just continuity assumption, making use of the mollifiers. Here we give a direct proof. We begin with a definition and an auxiliary result.

Definition 7.22. Let Ω be an open set in $\mathbb{R}^n$. A function $u \in C(\Omega)$ is said to be *sub-harmonic* if

$$u(x) \leq \frac{1}{|B_r(x)|} \int_{B_r(x)} u(y)\, dy$$

for all $x \in \Omega$ and $r > 0$ such that $B_r(x) \subset\subset \Omega$. A function $u \in C(\Omega)$ is said to be *super-harmonic* if $-u$ is sub-harmonic. A function $u \in C(\Omega)$ is said to satisfy the MVP if both u and $-u$ are sub-harmonic. □

Note that the above definition is weaker than the one defined earlier.

Proposition 7.23. Suppose Ω is a bounded open and connected subset of $\mathbb{R}^n$. If $u \in C(\overline{\Omega})$ is sub-harmonic, then the weak maximum principle holds for u, that is, $\max_{\overline{\Omega}} u = \max_{\partial\Omega} u$. □

Similarly, if u is super-harmonic, then the weak minimum principle holds for u. Finally, if u satisfies the MVP, then

$$\min_{\partial\Omega} u \leq u(x) \leq \max_{\partial\Omega} u$$

for all $x \in \overline{\Omega}$.

The proof is exactly the same as in Theorem 7.7 and is left as an exercise. We now come to the main result, namely the converse of the mean value theorem.

Theorem 7.24. Let Ω be an open set in $\mathbb{R}^n$ and $u \in C(\Omega)$ satisfies the MVP in Ω. Then, u is harmonic in Ω, that is $\Delta u(x) = 0$ for all $x \in \Omega$. Further, $u \in C^\infty(\Omega)$. □

Proof We only sketch a proof. The details are left as an exercise. Let $x_0 \in \Omega$. Choose $r > 0$ such that $B_r(x_0) \subset\subset \Omega$. Let v be a harmonic function in $B_r(x_0)$ such that $v = u$ on $\partial B_r(x_0)$; such a v is given by the Poisson formula (7.29), by shifting the origin to x_0. Now apply Proposition 7.23 to $\pm(u - v)$ to conclude that $u \equiv v$ in $B_r(x_0)$. This completes the proof. □

7.3 EXISTENCE OF SOLUTION OF DIRICHLET PROBLEM (PERRON'S METHOD)

We now want to address the question of existence of a solution to Dirichlet problem for the Laplace equation, in arbitrary bounded domains with prescribed boundary values. The method that we present here is known as *Perron's method* of sub-harmonic functions and requires certain assumptions on the regularity of the boundary of the domain. We sketch some of the ideas. The method is mainly based on the *maximum principle* and the *solvability* of the Dirichlet problem in a ball. This somewhat restricts the application of the method to general second order equations Another feature of this method is the separation of the *interior existence* from that of the *boundary behavior*.

Continuous Sub-harmonic and Super-harmonic Functions: Suppose $u \in C^2(\Omega) \cap C(\overline{\Omega})$ is sub-harmonic, that is $\Delta u \geq 0$ in Ω. If $v \in C^2(\Omega) \cap C(\overline{\Omega})$ is harmonic such that $u \leq v$ in

$\partial\Omega$, then, by maximum principle $u \leq v$ in Ω. More generally, for any ball $B \subset\subset \Omega$, if $u \leq v$ on ∂B, then $u \leq v$ in B. This, in fact, is a defining property of the sub-harmonic functions and hence the name sub-harmonic. This motivates us to define sub-harmonic and super-harmonic functions for continuous functions in a different way.

Definition 7.25. A function $u \in C(\Omega)$ is called *sub-harmonic* if the following condition holds: For any ball $B \subset\subset \Omega$ and for any harmonic function v in B satisfying $u \leq v$ on ∂B, the inequality $u \leq v$ is true in B. A function $u \in C(\Omega)$ is said to be super-harmonic if $-u$ is sub-harmonic. A continuous function u is *harmonic* if it is both sub-harmonic and super-harmonic. □

This definition of sub (super) harmonicity is due to F. Riesz and can be defined for upper (lower) semi-continuous functions. For continuous function, it coincides with the definition above. See DiBenedetto (2010) for an example. It is easy to check that the above definition of harmonicity coincides with the earlier definition via MVP.

Some properties of these functions are listed in the following theorem. Proof is not difficult; See Exercise 32.

Theorem 7.26. The following statements hold:

1. (**Comparison Principle**): If u is sub-harmonic in a connected bounded domain Ω, then, u satisfies the strong maximum principle; that is, if v is a super-harmonic function satisfying $u \leq v$ on $\partial\Omega$, then either $u < v$ in Ω or $u \equiv v$ in Ω.
2. (**Harmonic Lifting**): Suppose u is sub-harmonic in Ω and consider any ball $B_r \subset\subset \Omega$. Then, we can define $\tilde{u}$ in B_r via the Poisson integral using the boundary values of u on ∂B_r (see (7.27)). That is,

$$\tilde{u}(x) = \frac{r^2 - |x|^2}{n\omega_n r} \int_{\partial B_r} \frac{u(y)}{|x-y|^n} d\sigma(y) = \int_{\partial B_r} K(x,y)u(y)d\sigma(y). \tag{7.30}$$

   Thus, $\tilde{u}$ is harmonic in B_r and $\tilde{u} = u$ on ∂B_r. We define the *harmonic lifting* of u in B_r by

$$U(x) = \begin{cases} \tilde{u}(x) \text{ in } B_r \\ u(x) \text{ in } \Omega \backslash \bar{B}_r \end{cases} \tag{7.31}$$

   The function U is sub-harmonic in Ω.
3. Let $u_1, u_2, \ldots u_k$ be sub-harmonic in Ω. Then, $\max_i u_i$ is sub-harmonic in Ω. For super-harmonic functions, $\min_i u_i$ is super-harmonic. □

Now, given a continuous function g on $\partial\Omega$, the idea behind Perron's method in the solvability of $\Delta u = 0$ in Ω, $u = g$ on $\partial\Omega$ is to look for all continuous sub-harmonic functions v such that $v \leq g$ on $\partial\Omega$. Such a function v is called a *sub-function* relative to g. Then, hope that the maximizing sub-function will, in fact, be harmonic and will satisfy the boundary condition. Let S_g denote the set of all sub-functions relative to g. The set S_g is non-empty since any constant function $c \le \inf_{\partial\Omega} g$ is in S_g. A super harmonic function is a *super-function* relative to g if $v \ge g$ on $\partial\Omega$. The maximum principle shows that every sub-function is less than or equal to every super-function relative to g. We have the following theorem:

Theorem 7.27. Let $g \in C(\partial\Omega)$ and S_g be as above. Then, the function

$$u(x) = \sup_{v \in S_g} v(x) \tag{7.32}$$

is harmonic in Ω. □

Proof If $v \in S_g$, then, by maximum principle, $v(x) \le \sup_{\partial\Omega} g$ for all $x \in \Omega$. Thus, u is well-defined and $u(x) \le \sup_{\partial\Omega} g$, $x \in \Omega$. The proof that u harmonic is delicate. Indeed u is sub-harmonic by Theorem 7.26(3). Let B be an arbitrary ball in Ω, where any $v \in S_g$ can be harmonically lifted to B. Let $m = \inf_{\partial\Omega} g$ and $M = \sup_{\partial\Omega} g$. Clearly $u \le M$ by maximum principle. Let $x_1, x_2, \ldots$ be any sequence of points in B.

Claim: There is a sequence v_j in S_g such that $m \le v_j(x) \le M$ for all $x \in \Omega$ and $v_j(x_k)$ converges to $u(x_k)$ as $j \to \infty$ for all k.

Indeed by the definition of supremum, for each x_k, there exists a sequence $v_{j,k} \in S_g$, such that $v_{j,k}(x_k)$ converges to $u(x_k)$ as $j \to \infty$ for all k. Now define

$$\overline{v}_j(x) = \max\left\{v_{j,1}(x), \ldots v_{j,j}(x)\right\}.$$

Observe that $\overline{v}_j \in S_g$ and $v_{j,k}(x) \le \overline{v}_j(x) \le u(x)$. Thus $\overline{v}_j(x_k)$ converges to $u(x_k)$. Finally, take $v_j = \max\{\overline{v}_j, m\}$. Then, $v_j \in S_g$ and satisfies the required properties in the claim.

Now consider the lifting V_j of v_j as given in Theorem 7.26. Thus V_j is harmonic in B, $m \le V_j(x) \le M$ and $V_j(x_k)$ converges to $u(x_k)$. Since V_j is bounded, we see that $V_j(x)$ converges to $V(x)$ for some V that is harmonic in B, $m \le V(x) \le M$ and $V(x_k) = u(x_k)$ for all k. The delicate point is that the harmonic function V may depend on the choice of x_k.

For any arbitrary sequence x_k as above converging to a point x in B, without loss of generality, take $x_1 = x$. Then, $V(x_k)$ converges to $V(x)$ since V is continuous. Thus $u(x_k) \to V(x) = V(x_1) = u(x_1) = u(x)$. Thus u is continuous. Finally, choosing the sequence $\{x_k\}$ as a dense subset of B, we conclude that $u = V$ on a dense subset. By continuity of u, we now conclude that u is harmonic in a ball. Since the ball is arbitrary, we see that u is harmonic in Ω. □

Thus, we have shown the existence of a harmonic function in Ω. To complete the solvability, we need to study the boundary behavior of u. This requires the regularity of the boundary that is introduced via the *barrier function*. It eventually reduces to the *local solvability* of sub-harmonic functions at the boundary points, which preserves the negative sign.

Definition 7.28. Let $\xi \in \partial\Omega$. A continuous function $w \in C(\overline{\Omega})$ is called a *barrier* at ξ relative to Ω if w is sub-harmonic, $w < 0$ in $\overline{\Omega}\backslash\{\xi\}$ and $w(\xi) = 0$. □

Though w is defined in Ω, it is actually a local concept. In this direction, we define the concept of a local barrier. Once we get a local barrier, we can always produce a barrier relative to Ω. A function $\overline{w}$ is called a *local barrier* at $\xi \in \partial\Omega$ if there exists a neighborhood N of ξ such that $\overline{w}$ is a barrier at ξ relative to $\Omega \cap N$. We can, then, construct a barrier at ξ relative to Ω. Let B be a ball such that $\xi \in B \subset\subset N$ and $m = \sup_{N\backslash B} \overline{w}$, then the function w defined by

$$w(x) = \begin{cases} \max\{m, \overline{w}\}, \; x \in \overline{\Omega} \cap B \\ m, \; x \in \overline{\Omega}\backslash B \end{cases}$$

is a barrier at ξ relative to Ω.

Definition 7.29 (Barrier Function)**.** A boundary point is called *regular* if there exists a barrier at that point. A domain is called *regular* if all the boundary points are regular. □

Theorem 7.30. Let $g \in C(\partial\Omega)$ and u be the harmonic function constructed in Theorem 7.32. Then, $u(x) \to g(\xi)$ as $x \to \xi$ at every regular point $\xi \in \partial\Omega$. □

Proof Let $M = \sup_{\partial\Omega} |g|$ and w be a barrier at ξ. Given $\varepsilon > 0$, there exists $\delta > 0$ such that

$$|w(x)| < \varepsilon, \quad |g(x) - g(\xi)| < \varepsilon \text{ if } |x - \xi| < \delta$$

for $x \in \partial\Omega$. Further, we can find $k > 0$ such that

$$-kw(x) \geq 2M \text{ for } |x - \xi| \geq \delta.$$

The function $g(\xi) + \varepsilon - kw$ is a super-function and $g(\xi) - \varepsilon + kw$ is a sub-function and we have

$$g(\xi) - \varepsilon + kw(x) \leq u(x) \leq g(\xi) + \varepsilon - kw(x0.$$

Hence, we have

$$|u(x) - g(\xi)| \leq \varepsilon - kw(x).$$

Thus, $u(x) \to g(\xi)$ as $x \to \xi$ since $w(\xi) = 0$. □

This immediately gives the following existence theorem:

Theorem 7.31 (Existence and Uniqueness)**.** Let Ω be a bounded domain with regular boundary $\partial\Omega$ and $g \in C(\partial\Omega)$. Then, there exists a unique solution u to the Dirichlet boundary value problem

$$\begin{aligned} \Delta u &= 0 \text{ in } \Omega \\ u &= g \text{ on } \partial\Omega. \end{aligned} \tag{7.33}$$

Conversely, if the classical Dirichlet problem (7.33) is solvable for every continuous g, then the domain is regular. □

Proof Only the converse statement needs to be proved. For $\xi \in \partial\Omega$, consider the function $g(x) = -|x - \xi|$ that is continuous on $\partial\Omega$. Then, the solution of the Dirichlet problem with this g as the boundary values will be a barrier at ξ. □

The existence thus reduces to that of the geometric condition, namely local existence of subharmonic functions at the boundary. Therefore, we look for some sufficient conditions on the domain to guarantee the regularity of its boundary.

Example 7.32. The two-dimensional case $n = 2$ is simple. □

Let $z_0 \in \partial\Omega$ and assume without loss of generality that $z_0 = 0$. Now, use the polar coordinates r, θ and suppose, there is a neighborhood N of z_0 such that a single-valued branch of θ is defined on $\Omega \cap N$. Then, it can be verified that

$$w(z) = -\Re\left(\frac{1}{\log z}\right) = \frac{\log r}{\log^2 r + \theta^2}, \; z \neq 0, \; w(0) = \lim_{z\to 0} w(z)$$

is a barrier at 0. Here $\Re(z)$ is the real part of a complex number z.

For example, the boundary value problem in a domain in the plane is always solvable if its boundary values are accessible from the exterior of Ω by a simple arc. For example, the unit disk with a slit along an arc.

Example 7.33. In higher dimensions, the Dirichlet problem cannot be solved in such general domains, for example, domains with very sharp inward directed cusp. A simple sufficient condition can be given by *exterior sphere condition*; that is, there exists $R > 0$ such that for every $\xi \in \partial\Omega$, a ball B of radius R touches Ω at ξ, that is $\overline{B} \cap \overline{\Omega} = \{\xi\}$. In this case,

$$w(x) = \begin{cases} R^{2-n} - |x - \xi|^{2-n}, & n \geq 3 \\ \log \frac{|x-\xi|}{R}, & n = 2, \end{cases}$$

will be a barrier at ξ. □

7.4 POISSON EQUATION AND NEWTONIAN POTENTIAL

We now consider the general Poisson equation (7.20). Since, we have already studied the case with $f = 0$ and arbitrary g via Perron's method, we need to analyze the case when f is non-zero. Looking at the representation formula (7.25), the study reduces to that of understanding the first term on the right-hand side of (7.25). This is equivalent to understanding the first term with G replaced by the fundamental solution ϕ because G and ϕ have the same singularity. The difference between G and ϕ is the boundary values and we have already studied the problem with non-homogeneous boundary data. Thus, given an integrable function f in Ω, we define the *Newtonian potential* v of f by

$$v(x) = \int_{\Omega} \phi(x - y)f(y)dy, \; x \in \mathbb{R}^n. \tag{7.34}$$

Essentially, the study of Laplace equation reduces to the derivation of estimates on v and its derivatives. This involves a bit of regularization due to the singularity of ϕ. We will only do the bare minimum in this book to establish the unique existence of (7.20). More details are available in Gilberg and Trudinger (2001). The most important point to be made at this stage is that the Newtonian potential v need not be twice differentiable even for continuous f. Stated differently, we cannot expect, in general, to get solution of $\Delta u = f$ if f is just continuous. We need higher regularity of f. This is provided by the Hölder continuous functions. In some sense, Hölder continuity provides a quantitative measure of continuity. We will soon elaborate on this point.

First, we provide a counter-example to show that the continuity of f is not sufficient to obtain a classical solution. In this regard, we need the following proposition:

Proposition 7.34 (Removable Singularity). Suppose Ω is an open set in $\mathbb{R}^n$ and u be harmonic in $\Omega \setminus \{x_0\}$, $x_0 \in \Omega$. If $u(x) = o(\phi(x - x_0))$ as $x \to x_0$, where ϕ is the fundamental solution for $-\Delta$ in $\mathbb{R}^n$, then, u may be suitably defined at x_0 so that u becomes harmonic in Ω. □

Proof Choose $R > 0$ so that the closed ball $\overline{B_R(x_0)} \subset \Omega$. Let v be harmonic in $B_R(x_0)$ with boundary values $v = u$ on $\partial B_R(x_0)$; the existence of v follows from (7.29). Then, the function $w = u - v$ is harmonic in $B_R(x_0)\setminus\{x_0\}$ and $w = 0$ on $\partial B_R(x_0)$.

Claim: $w = 0$ in $B_R(x_0)\setminus\{x_0\}$.

Assuming the claim, we define $u(x_0) = v(x_0)$. Thus $u = v$ in $B_R(x_0)$. Since v is harmonic in $B_R(x_0)$, so is u. Thus, u is harmonic in Ω.

Proof of the Claim: We consider the case $n \geq 3$ and the arguments are similar for $n = 2$. For $\varepsilon > 0$, small, consider the functions z_+ and z_- defined by[4]

$$z_{\pm}(x) = \varepsilon|x - x_0|^{2-n} \pm w(x).$$

Then, $z_{\pm}$ are harmonic in $B_R(x_0)\setminus\{x_0\}$ and

$$z_{\pm} = \varepsilon R^{2-n} > 0$$

[4]If $n = 2$, use the corresponding fundamental solution.

on $\partial B_R(x_0)$. By hypothesis $u(x) = o(\phi(x - x_0))$ as $x \to x_0$ and v is continuous in $\overline{B}_R(x_0)$. Therefore, on $\partial B_\rho(x_0)$, we have

$$z_\pm(x) = \varepsilon\rho^{2-n} \pm w = \varepsilon\rho^{2-n} + o(\rho^{2-n}),$$

where $R > \rho > 0$ is sufficiently small. This implies that

$$z_\pm > 0$$

on $\partial B_\rho(x_0)$. Hence by maximum principle, $z_\pm > 0$ in the annulus $R > |x - x_0| > \rho$. Now, for any $x_1 \in B_R(x_0)\backslash\{x_0\}$, choose ρ small enough so that x_1 belongs to this annulus. Thus,

$$|w(x_1)| = \pm w(x_1) < \varepsilon|x_1 - x_0|^{2-n}.$$

As ε is arbitrary, we see that $w(x_1) = 0$ and thus $w = 0$ in $B_R(x_0)\backslash\{x_0\}$. This proves the claim and hence the proposition. □

Example 7.35. Consider the ball $B_R(0)$ in $\mathbb{R}^2$, where $0 < R < 1$. Define u by

$$u(x) = u(x_1, x_2) = \left(x_1^2 - x_2^2\right)\left(-\log|x|\right)^{1/2}.$$

Then, $u \in C^\infty\left(\overline{B}_R(0)\backslash\{0\}\right) \cap C\left(\overline{B}_R(0)\right)$ and satisfies

$$\Delta u(x) = \frac{x_2^2 - x_1^2}{2|x|^2}\left(\frac{n+2}{\left(-\log|x|\right)^{1/2}} + \frac{1}{2\left(-\log|x|\right)^{1/2}}\right)$$

in $\overline{B}_R(0)\backslash\{0\}$. Let us denote the right-hand side by $f(x)$. Then, f can be extended as a continuous function by defining $f(0,0) = 0$. Clearly u, is not a classical solution of $\Delta u = f$ in $B_R(0)$, as $\lim_{|x|\to 0} u_{x_1x_1} = \infty$. □

Claim: The equation $\Delta v = f$ with f given above, has no classical solution.

If not, assume $v \in C^2(B_R(0))$ is a classical solution, then $w = u - v$ is harmonic in $B_R(0)\backslash\{0\}$ with a possible singularity at the origin and $w(x) = o(\phi(x - x_0))$. In fact w is bounded. But, then w can be suitably redefined at the origin to make w harmonic in $B_R(0)$ by the above proposition. This shows that $u = w - v \in C^2(B_R(0))$ that is a contradiction.

But if we assume C^1 smoothness on f, then the existence of the solution is easy to prove and is given in the following theorem:

Theorem 7.36. Let $f \in C^1(\overline{\Omega})$ and define

$$v(x) = \int_{\Omega} \phi(x-y)f(y)dy,$$

where ϕ is the fundamental solution for $-\Delta$. Then, $v \in C^1(\overline{\Omega}) \cap C^2(\Omega)$ and satisfies $-\Delta v = f$ in Ω. □

Proof We have

$$\begin{aligned}\frac{\partial v}{\partial x_i}(x) &= \int_{\Omega} \frac{\partial \phi}{\partial x_i}(x-y)f(y)\, dy = -\int_{\Omega} \frac{\partial \phi}{\partial y_i}(x-y)f(y)\, dy \\ &= \int_{\Omega} \phi(x-y)\frac{\partial f}{\partial y_i}(y)\, dy - \int_{\partial\Omega} \phi(x-y)f(y)\nu_i(y)\, d\sigma(y).\end{aligned}$$

Since $\frac{\partial f}{\partial x_i} \in C(\overline{\Omega})$, the first term on the right-hand side of the above expression is differentiable. The second integral on the right-hand side is known as *single-layer potential* that is in $C^{\infty}(\mathbb{R}^n \backslash \partial\Omega)$ and it is harmonic in $\mathbb{R}^n \backslash \partial\Omega$. See Exercise 36. Thus $\frac{\partial v}{\partial x_i}$ is differentiable and hence $v \in C^1(\overline{\Omega}) \cap C^2(\Omega)$. To see v satisfies the Poisson equation, let $\psi \in C_c^2(\Omega)$. Then, we know that $\phi * \psi$ satisfies $-\Delta(\phi * \psi) = \psi$, that is, $-\int_{\Omega} \phi(x-y)\Delta\psi(y)dy = \psi(x)$ (see Theorem 7.2). Now applying Green's formula to v and ψ, we get

$$\begin{aligned}\int_{\Omega} \Delta v(x)\psi(x)dx &= \int_{\Omega} \Delta\psi(x)v(x)dx \\ &= \int_{\Omega} \Delta\psi(x)\left(\int_{\Omega} \phi(x-y)f(y)dy\right)dx \\ &= -\int_{\Omega} f(y)\left(\int_{\Omega} \phi(x-y)\Delta\psi(x)dx\right)dy, \text{ by Fubini's Theorem} \\ &= -\int_{\Omega} f(y)\psi(y)dy.\end{aligned}$$

Since ψ is arbitrary, this shows that $-\Delta v = f$ in Ω. This completes the proof. □

We remark that the assumption $f \in C^1(\overline{\Omega})$ is restrictive and on the other hand, continuity of f is not sufficient for existence. Thus, we look for a condition that is stronger than continuity, but weaker than differentiability. This is provided by the notion of Hölder continuity that is the topic of discussion in the section that follows.

7.4.1 Hölder Continuous Functions

Let $x_0 \in \Omega$, where Ω is a bounded domain in $\mathbb{R}^n$ and $0 < \alpha < 1$.

Definition 7.37 (Hölder Continuity). A function $f : \Omega \to \mathbb{R}$ is said to be *Hölder continuous* of order α at x_0 if there exists a constant $C > 0$ such that[5]

$$|f(x) - f(x_0)| \leq C|x - x_0|^\alpha \tag{7.35}$$

for all $x \in \Omega$. If $\alpha = 1$ in (7.35), then the function f is said to be *Lipschitz continuous*. □

The standard example is $f(x) = |x|^\alpha$, $0 < \alpha < 1$ that is Hölder continuous of order α.

Definition 7.38 (Uniform and Local Hölder continuity). A function $f : \Omega \to \mathbb{R}$ is said to be *uniformly Hölder continuous* of order α in Ω if there exists a constant $C > 0$ such that

$$|f(x) - f(y)| \leq C|x - y|^\alpha \tag{7.36}$$

for all $x, y \in \Omega$. The function f is locally Hölder continuous if f is uniformly Hölder continuous in every compact subset of Ω. When $\alpha = 1$, the function f is said to be *uniformly Lipschitz continuous.* It is also clear that a uniformly Hölder continuous function is also uniformly continuous. □

We denote by $C^{0,\alpha}(\overline{\Omega})$ the space of all uniformly Hölder continuous functions of order α in $\overline{\Omega}$ and define

$$\|f\|_{0,\alpha} = \|f\|_0 + \sup_{x,y \in \overline{\Omega},\, x \neq y} \frac{|f(x) - f(y)|}{|x - y|^\alpha}, \tag{7.37}$$

for $f \in C^{0,\alpha}(\overline{\Omega})$, where $\|f\|_0 = \sup_{x \in \overline{\Omega}} |f(x)|$ is the sup-norm. It is not difficult to verify that $C^{0,\alpha}(\overline{\Omega})$ is a Banach space equipped with the norm (7.37). We can also define the spaces $C^{k,\alpha}(\overline{\Omega})$, where $k \in \mathbb{N}$ as the space of all $C^k(\overline{\Omega})$ functions such that $D^\beta f \in C^{0,\alpha}(\overline{\Omega})$ for all $|\beta| = k$ and the norm is given by

$$\|f\|_{k,\alpha} = \|f\|_k + \sum_{|\beta|=k} \|D^\beta f\|_{0,\alpha} \tag{7.38}$$

[5] If $\alpha > 1$ and satisfies (7.35), it is an interesting exercise to show that f is a constant function.

with $\|f\|_k = \sup_{x\in\bar{\Omega},|\alpha|\le k} |D^\alpha f(x)|$. We now prove the following result that paves the way for solvability of the Poisson equation.

Proposition 7.39. Let Ω be a bounded domain in $\mathbb{R}^n$ and $f : \Omega \to \mathbb{R}$ be bounded and integrable. Consider the Newtonian potential v defined by (7.34). Then $v \in C^1(\mathbb{R}^n)$ and for $x \in \Omega$, we have

$$\frac{\partial v}{\partial x_i}(x) = \int_\Omega \frac{\partial \phi}{\partial x_i}(x-y)f(y)\, dy. \tag{7.39}$$

□

Proof Denote the integral on the right-hand side of (7.39) by $w_i(x)$. Recall that $\frac{\partial\phi}{\partial x_i}$ is locally integrable and satisfies $\left|\frac{\partial\phi}{\partial x_i}\right| \le \frac{1}{\omega_n|x-y|^{n-1}}$. It follows that the integral is well-defined, that is $|w_i(x)| < \infty$ for $x \in \Omega$. To show that v is differentiable and the equality in (7.39) holds, we proceed as follows: Choose $h \in C^1(\mathbb{R})$ be such that $h(t) = 0$ for $t \le 1$; $h(t) = 1$ for $t \ge 2$; $0 \le h(t) \le 1$ and $0 \le h'(t) \le a$ for some $a > 0$. For $\varepsilon > 0$, define $h_\varepsilon(x) = h(|x|/\varepsilon)$ and let

$$v_\varepsilon(x) = \int_\Omega \phi(x-y)h_\varepsilon(x-y)f(y)dy.$$

Note that the integration domain is actually $\Omega \cap \{y : |x-y| > \varepsilon\}$ and the integrand is smooth in this domain. Thus, we can differentiate under the integral sign to obtain

$$\frac{\partial v_\varepsilon}{\partial x_i}(x) = \int_\Omega \frac{\partial}{\partial x_i}\left(h_\varepsilon(x-y)\phi(x-y)\right) f(y)dy.$$

Now, compute

$$\begin{aligned} w_i(x) - \frac{\partial v_\varepsilon}{\partial x_i}(x) &= \int_{|x-y|\le 2\varepsilon} \frac{\partial}{\partial x_i}\left(\left[1-h_\varepsilon(x-y)\right]\phi(x-y)\right) f(y)dy \\ &= \int_{|x-y|\le 2\varepsilon} \left(-\frac{\partial h_\varepsilon}{\partial x_i}(x-y)\phi(x-y) + \left[1-h_\varepsilon(x-y)\right]\frac{\partial\phi}{\partial x_i}\right) f(y)dy \end{aligned}$$

Now use the estimates on h, h' and f to get

$$\begin{aligned} \left|w_i(x) - \frac{\partial v_\varepsilon}{\partial x_i}(x)\right| &\le \|f\|_0 \int_{|x-y|\le 2\varepsilon} \left(\frac{a}{\varepsilon}|\phi(x-y)| + \left|\frac{\partial\phi}{\partial x_i}(x-y)\right|\right) dy \\ &\le \begin{cases} C\varepsilon \text{ if } n > 2 \\ C\varepsilon(1 + |\log a\varepsilon|) \text{ for } n = 2. \end{cases} \end{aligned}$$

The last estimate follows from the estimates on the fundamental solution and its derivatives. Hence, $v_\varepsilon \to v$ and $\frac{\partial v_\varepsilon}{\partial x_i} \to w_i$ uniformly on compact subsets of $\mathbb{R}^n$. Thus $v \in C^1(\mathbb{R}^n)$ and $\frac{\partial v}{\partial x_i} = w_i$. This completes the proof of the proposition. □

Why Hölder Continuity? Heuristic Argument: Now we also need to consider the second derivatives of the fundamental solution and the corresponding integral. More precisely, we need to consider the integral of the form

$$\int_\Omega \frac{\partial^2 \phi}{\partial x_i \partial x_j}(x-y) f(y)dy.$$

Recall that for fixed x, $\frac{\partial^2 \phi}{\partial x_i \partial x_j}(x-y)$ is not locally integrable and hence the above integral is not meaningful. However, the second derivative satisfies the estimate

$$\left| \frac{\partial^2 \phi}{\partial x_i \partial x_j}(x-y) \right| \le \frac{1}{\omega_n |x-y|^n}.$$

If we have a slightly less singularity of the form $\frac{1}{|x-y|^{n-\alpha}}$ with $\alpha > 0$ on the right-hand side of the above expression, then there is local integrability. This is the α we are looking for, through the Hölder continuity of f by considering an expression of the form

$$\frac{\partial^2 \phi}{\partial x_i \partial x_j}(x-y)(f(y) - f(x)).$$

Suppose f is locally Hölder continuous of order α, $0 < \alpha < 1$. Then

$$|f(x) - f(y)| \le C|x-y|^\alpha$$

for all $x, y \in \Omega_0 \subset\subset \Omega$. Thus, for x fixed, we get

$$\left| \frac{\partial^2 \phi}{\partial x_i \partial x_j}(x-y)(f(y) - f(x)) \right| \le \frac{C}{\omega_n |x-y|^{n-\alpha}}.$$

That is, $\frac{\partial^2 \phi}{\partial x_i \partial x_j}(x-y)(f(y) - f(x))$ is locally integrable. Thus

$$\int_{\Omega_0} \frac{\partial^2 \phi}{\partial x_i \partial x_j}(x-y)(f(y) - f(x))\, dy$$

is well-defined. By incorporating a boundary integral, we now define the function u_{ij} for all $i, j = 1, \dots n$, by

$$u_{ij}(x) = \int_{B_R(x)} \frac{\partial^2 \phi}{\partial x_i \partial x_j}(x-y)(f(y)-f(x))\,dy - f(x) \int_{\partial B_R(x)} \frac{\partial \phi}{\partial x_i}(x-y)\nu_j(y)\,d\sigma(y) \tag{7.40}$$

which is well-defined for any $R > 0$ such that $B_R(x) \subset\subset \Omega$. Now, we can state the main theorem.

Theorem 7.40. Let f be bounded and locally Hölder continuous of order $\alpha \leq 1$ in Ω and v be the Newtonian potential defined as in (7.34). Then, $v \in C^2(\Omega)$ and satisfies $-\Delta v = f$. In fact, for any $x \in \Omega$, we have

$$\frac{\partial^2 v}{\partial x_i \partial x_j}(x) = u_{ij}(x). \tag{7.41}$$

Here u_{ij} is defined as in (7.40) with $R = 2\epsilon$, with $B_{2\epsilon}(x) \subset\subset \Omega$. □

Proof By (7.39), we have $w_i(x) = \frac{\partial v}{\partial x_i}(x)$. With h as in Proposition 7.39, let

$$w_{i,\epsilon}(x) = \int_{\Omega} \frac{\partial \phi}{\partial x_i}(x-y)h_\epsilon(x-y)f(y)\,dy.$$

Then, $w_{i,\epsilon} \in C^1(\Omega)$ and for ϵ small so that $B_{2\epsilon}(x) \subset\subset \Omega$, we get

$$\begin{aligned}
\frac{\partial w_{i,\epsilon}}{\partial x_j}(x) &= \int_{\Omega} \frac{\partial}{\partial x_j}\left(\frac{\partial \phi}{\partial x_i}(x-y)h_\epsilon(x-y)\right) f(y)\,dy \\
&= \int_{B_{2\epsilon}(x)} \frac{\partial}{\partial x_j}\left(\frac{\partial \phi}{\partial x_i}(x-y)h_\epsilon(x-y)\right)(f(y)-f(x))\,dy \\
&\quad + f(x) \int_{B_{2\epsilon}(x)} \frac{\partial}{\partial x_j}\left(\frac{\partial \phi}{\partial x_i}(x-y)h_\epsilon(x-y)\right) dy.
\end{aligned}$$

By divergence theorem, we have

$$\frac{\partial w_{i,\varepsilon}}{\partial x_j}(x) = \int_{B_{2\varepsilon}(x)} \frac{\partial}{\partial x_j}\left(\frac{\partial \phi}{\partial x_i}(x-y)h_\varepsilon(x-y)\right)(f(y)-f(x))\,dy$$
$$- f(x)\int_{\partial B_{2\varepsilon}(x)} \frac{\partial \phi}{\partial x_i}(x-y)\nu_j(y)\,d\sigma(y).$$

Thus, we get

$$u_{ij}(x) - \frac{\partial w_{i,\varepsilon}}{\partial x_j}(x) = \int_{|x-y|\le 2\varepsilon} \frac{\partial}{\partial x_j}\left(\left[1-h_\varepsilon(x-y)\right]\frac{\partial \phi}{\partial x_i}(x-y)\right)(f(y)-f(x))\,dy.$$

Now, use the Hölder continuity of f and estimates on the fundamental solution and its derivatives to obtain the estimate

$$\left|u_{ij}(x) - \frac{\partial w_{i,\varepsilon}}{\partial x_j}(x)\right| \le C\varepsilon^\alpha \|f\|_{0,\alpha}$$

for some constant $C > 0$ and $2\varepsilon < d(x, \partial\Omega)$. Hence $\frac{\partial w_{i,\varepsilon}}{\partial x_j}$ converges to u_{ij} uniformly on compact subsets of Ω. This together with the fact that $w_{i,\varepsilon}$ converges to $\frac{\partial v}{\partial x_i}$, we arrive at $u_{ij} = \frac{\partial^2 v}{\partial x_j \partial x_i}$. Using the fact that ϕ is harmonic except at $x = y$ (this is taken care of by the factor $f(x) - f(y)$), taking $j = i$ in (7.41) and summing over i, we get

$$\Delta v = \sum_{i=1}^{n} u_{ii} = -f(x)\int_{\partial B_{2\varepsilon}} \frac{\partial \phi}{\partial \nu}\,d\sigma(y) = -f(x).$$

This completes the proof. □

Now, to solve the Poisson equation (7.20), consider $w = u - v$, where v is the Newtonian potential. Thus, u solves (7.20) if and only if w solves the harmonic equation with the boundary data $w = g - v$ on $\partial\Omega$. This proves the existence and uniqueness of the Poisson equation. We state this in the following theorem:

Theorem 7.41 (Existence and Uniqueness). Let Ω be a regular bounded domain. Let f be bounded and locally Hölder continuous of order $\alpha \in (0, 1]$ in Ω and g is continuous on the boundary $\partial\Omega$. Then the Poisson equation (7.20) has a unique classical solution. □

7.5 HILBERT SPACE METHOD: WEAK SOLUTIONS

In Chapter 3 on first-order equations, we have seen that the conservation laws that do not admit smooth solutions even if we start with smooth initial data. We have observed the development of discontinuities or shocks as time evolves. This may be due to the physical nature of the problem. Thus, the lack of existence of smooth solutions does not give us freedom to neglect the problem, rather it is revealing interesting phenomena of physical problem via non-smooth functions as solutions. But our basic calculus will not allow us to differentiate non-smooth functions and hence we need to interpret non-smooth functions, but physically relevant, as solutions to a given PDE in a different sense. In fact, this is true with many PDE and we need to have new theory or theories. Though this discussion goes beyond the scope of the present book, we nevertheless, would like to present a few basic concepts of this theory in the context of Laplace operator.

Here, the basic step in finding the solution of the Dirichlet problem is that of putting it as an abstract problem in a suitable Hilbert space of functions. Then, use the powerful functional theoretic approach and the general available theorems. The Hilbert space would be a much bigger class than the space of smooth functions. This will immediately create the difficulty of understanding the concept of differentiation in such a class of functions. In this modern approach, we need to understand/introduce a weak notion of differentiability and this is achieved after the introduction of *generalized functions/distributions*. The Hilbert spaces introduced initially are known as *Sobolev spaces*. The method consists of the following steps:

i. Introduce appropriate Hilbert space.
ii. Give a suitable formulation of PDE so that the Hilbert space theory can be applied; this in literature is known as *weak formulation*.
iii. A solution to the weak formulation is known as *weak solution*; make sure that a weak solution that is smooth is indeed a classical solution and conversely, a classical solution is a weak solution as well.
iv. Prove the existence and uniqueness of a weak solution via the Hilbert space theory.
v. Prove then that the weak solution, thus obtained is smooth/regular and hence a classical solution by (iii). These are known as regularity results.

Even if the regularity results are unavailable, the weak solutions can be quite physical. In fact, in many physical situations the weak solutions are the actual physical solutions and we cannot expect the physical quantities to be always smooth. In such situations, the PDE may not be the right physical modelling and the PDE is the Euler equation arising from other problems like optimization with the additional assumption that the solution is smooth.

To elaborate further, consider an applied force f in an elastic body Ω and if v is the unknown displacement vector, associate the corresponding energy functional (for simplicity, we take $n = 2$):

$$
\begin{aligned}
F(v) &= \frac{1}{2}\int_\Omega \left[\left(\frac{\partial v}{\partial x}\right)^2 + \left(\frac{\partial v}{\partial y}\right)^2\right] - \int_\Omega fv \\
&= \frac{1}{2}\int_\Omega \nabla v \cdot \nabla v - \int_\Omega fv
\end{aligned}
\tag{7.42}
$$

At an equilibrium state, we look for a solution u that minimizes the energy functional. Thus, we consider the following problem: find u such that

$$
F(u) = \min_v F(v). \tag{7.43}
$$

Recall an elementary result from analysis. If $g : [a, b] \to \mathbb{R}$ is C^1, then at extremal points, we know that $g' = 0$. In (7.43), we are in fact looking for extremal (minimal) points, but the situation is that u is in a function space that is infinite-dimensional. The one-dimensional situation has a beautiful counterpart and can be achieved (with appropriate delicate analysis) similar results via the concept of *Fréchét derivative* of functions defined in a normed linear space.

Indeed, the problem (7.43) is a little vague as we have not specified the minimizing space. We have no intention to get into the finer details, but the aim is to project certain ideas in the modern theory of PDEs. Let us take the zero boundary condition and consider the space

$$
C_0^1 = \{v \in C^1(\overline{\Omega}) : v = 0 \text{ on } \partial\Omega\},
$$

where we are looking for solutions. In other words, we minimize F over C_0^1. The space is reasonable as we need differentiability of u to define F. Assuming $u \in C_0^1$ is a solution to (7.43), we can derive the following necessary condition: Take an arbitrary $v \in C_0^1$, then it is easy to see that

$$
F'(u)v \equiv \lim_{t\to 0} \frac{F(u + tv) - F(u)}{t} = \int_\Omega \nabla u \cdot \nabla v - \int_\Omega fv. \tag{7.44}
$$

In this novice approach, solving (7.43) reduces to that of finding extremal points. That is, we look for $u \in C_0^1$ such that $F'(u) = 0$ in the dual space. Equivalently

$$\begin{cases} \text{find } u \in C_0^1(\overline{\Omega}) \text{ such that} \\ \int_\Omega \nabla u . \nabla v - \int_\Omega fv = 0 \text{ for all } v \in C_0^1(\overline{\Omega}). \end{cases} \tag{7.45}$$

Connection to Dirichlet Problem: Recall the Poisson problem. Find $u \in C^2(\Omega) \cap C(\overline{\Omega})$ such that

$$\begin{cases} -\Delta u = f \text{ in } \Omega \\ u = 0 \text{ on } \partial\Omega. \end{cases} \tag{7.46}$$

Take $v \in C_0^1(\overline{\Omega})$, multiply (7.46) by v and integrate by parts, we can easily get

$$\int_\Omega \nabla u \cdot \nabla v = \int_\Omega fv, \quad \text{for all } v \in C_0^1(\overline{\Omega}).$$

Thus a classical solution of (7.46) indeed satisfies (7.45). A solution to the problem (7.45) is called a weak solution of the problem (7.46). This is justified by the following converse:

Conversely, if u is a weak solution and suppose that u is $C^2(\Omega)$, $f \in C(\overline{\Omega})$. Then, one can reverse the process of integration by parts in the above equation to get

$$\int_\Omega (-\Delta u - f)v = 0 \text{ for all } v \in C_0^1(\overline{\Omega}).$$

Since the above equation is true for all v, we deduce that $-\Delta u - f = 0$. Hence u is a classical solution.

Understanding the Problem (7.45): Recall the *Cauchy–Schwarz inequality:* if g and h are square integrable functions, then, the product gh is integrable and

$$\int_\Omega |gh| \leq \left(\int_\Omega |g|^2 \right)^{\frac{1}{2}} \left(\int_\Omega |h|^2 \right)^{\frac{1}{2}} = \|g\|_{L^2} \|h\|_{L^2}.$$

Here $\|h\|_{L^2}$ is the L^2 norm of the function h. Thus, the terms in (7.45) are all well-defined if $u, v, f, \frac{\partial u}{\partial x_i}, \frac{\partial v}{\partial x_i}$ are all square integrable. That is, they are in $L^2(\Omega)$. This is true since, $u, v \in C_0^1(\overline{\Omega})$. The integral formulation (7.45) is called the *weak formulation* of the problem (7.46).

The existence of the weak formulation can be obtained using the *Riesz representation theorem*; namely, for every bounded linear functional $L : H \longrightarrow \mathbb{R}$, there exists a unique $u \in H$ such that $\langle u, v \rangle = L(v)$, for all $v \in H$, where H is a Hilbert space with the inner product $\langle \cdot, \cdot \rangle$. We have the following situation: find u such that

$$\int_{\Omega} \nabla u \cdot \nabla v = \int_{\Omega} fv$$

for all $v \in C_0^1(\overline{\Omega})$. To apply Riesz representation theorem, left-hand side has to be interpreted as an inner product in $C_0^1(\overline{\Omega})$. If it is complete, then show that right-hand side defines a bounded linear functional. Indeed, left-hand side is an inner product, but the major difficulty is that $C_0^1(\overline{\Omega})$ is not complete under this inner product. We have to take the completion X of $C_0^1(\overline{\Omega})$ with respect to the inner product

$$\langle u, v \rangle := \int_{\Omega} \nabla u \cdot \nabla v = \int_{\Omega} \sum_{i=1}^{n} \frac{\partial u}{\partial x_i} \frac{\partial v}{\partial x_i}$$

and the norm is given by $\|v\|_X = \langle v, v \rangle^{\frac{1}{2}} = \|\nabla v\|_{L^2}$. But the biggest question is: what is X? Recall that the completion is defined in an abstract way via equivalence classes of Cauchy sequences and hence the elements in X are equivalence classes. But the bigger picture tells us that X can be identified with a class of functions, but they need not be differentiable and we are in a very delicate situation. This is the modern theory of distributions and Sobolev spaces: In a weak sense of differentiation, we can see that

$$X = \left\{ v \in L^2(\Omega) : \frac{\partial v}{\partial x_i} \in L^2(\Omega),\ 1 \leq i \leq n, v = 0 \text{ on } \partial\Omega \right\}. \tag{7.47}$$

The second step is to see that the linear functional $L : X \longrightarrow \mathbb{R}$ defined by $L(v) = \int fv$ is bounded. We have

$$|L(v)| \leq \|f\|_{L^2(\Omega)} \|v\|_{L^2(\Omega)} .$$

If we can show that

$$\|v\|_{L^2(\Omega)} \leq C \|v\|_X = \|\nabla v\|_{L^2(\Omega)} , \text{ for all } v \in X \tag{7.48}$$

for some constant $C > 0$, we are in the setup of Riesz representation theorem; that is, there exists $u \in X$ such that

$$\int_{\Omega} \nabla u . \nabla v = \int_{\Omega} fv, \text{ for all } v \in X. \tag{7.49}$$

Poincarè's Inequality: There is a constant $C > 0$ such that

$$\|v\|_{L^2} \leq C \|\nabla v\|_{L^2},$$

for all $v \in C_0^1(\overline{\Omega})$.

The proof is simple. Let $n = 2$ and $\Omega \subset K := [-a, a] \times [-a, a]$ for some $a > 0$ as Ω is bounded. For, $v \in C_0^1(\overline{\Omega})$, extend v to K by zero outside Ω. Then,

$$v(x, y) = \int_{-a}^{x} \frac{\partial v}{\partial t}(t, y) dt.$$

Apply Cauchy–Schwarz inequality, square it and integrate both sides with respect to x and y to show that

$$\|v\|_{L^2} \leq 2a \|\nabla v\|_{L^2}.$$

The proof is similar in any dimension.

Thus, in the Hilbert space approach, we have established the unique existence of a weak solution. To prove the weak solution is a classical solution, we need to establish smoothness of the solution and this is achieved by regularity results. In general, regularity may not be available and all these results can be part of a second course in PDE. Finally, we end this section by remarking that there are other concepts of weak solutions like viscosity solution, transposition solution, and so on.

However, we should bear in mind that the Dirichlet problem cannot be solved for all continuous boundary values by the Hilbert space approach.

Example 7.42. Let $\Omega = B_1(0)$ be the unit ball in $\mathbb{R}^2$ and $\Delta u = 0$ in Ω and $u = g$ on $\partial\Omega$. Introduce polar coordinates r, θ. Let $g(\theta) = \sum_0^\infty (a_k \cos k\theta + b_k \sin k\theta)$. Then, $u(r, \theta) = \sum_0^\infty (a_k \cos k\theta + b_k \sin k\theta) r^k$, $r < 1$. Now $\int_\Omega |\nabla u|^2 = \pi \sum_0^\infty k(a_k^2 + b_k^2)$. In particular, if we take $g(\theta) = \sum_1^\infty \frac{\cos(k^3\theta)}{k^2}$, then, $\int_\Omega |\nabla u|^2 = +\infty$. Thus, the energy term that defines the norm in the Hilbert space is infinite. □

7.5.1 Fourier Method

Here, we consider the Laplace equation in two dimensions in either rectangular or circular domains and discuss the representation of its solution satisfying a given boundary condition, in Fourier series. In this method, also known as the method of separation of variables, the Laplace equation is reduced to a couple of ODE.

Denoting the independent variables by x, y, we consider the Laplace equation

$$\Delta u = u_{xx} + u_{yy} = 0. \tag{7.50}$$

In the proposed method, the solution u is sought in the form

$$u(x,y) = X(x)Y(y).$$

Using (7.50), we then see that

$$\frac{X''}{X} = -\frac{Y''}{Y},$$

and thus the terms on the two sides must be a constant. The boundary condition that will be imposed on u now transfers to both X and Y and makes the possible constants to form a countable set, say $\{\lambda_n\}$. Since we are interested in non-trivial solutions X and Y, the boundary condition further imposes sign condition on these $\{\lambda_n\}$. Denoting the solutions corresponding to λ_n by X_n and Y_n, we then obtain the solution formally as

$$u(x,y) = \sum_n X_n(x)Y_n(y),$$

using the linear superposition principle. We now discuss the method in detail by considering different boundary conditions.

1. Consider the Laplace equation (7.50) in a strip

$$\{(x,y) : x \in [0,a],\, y \in [0,\infty)\},\ a > 0$$

with boundary conditions:

$$\begin{aligned} &u(0,y) = u(a,y) = 0,\ 0 \le y < \infty \\ &u(x,0) = A\left(1 - \frac{x}{a}\right),\ \lim_{y\to\infty} u(x,y) = 0,\ 0 \le x \le a. \end{aligned} \tag{7.51}$$

If we put $u(x,y) = X(x)Y(y)$, then the boundary conditions (7.51) imply that $X(0) = X(a) = 0$ and $\lim_{y\to\infty} Y(y) = 0$. We have $\frac{X''}{X} = -\frac{Y''}{Y} = \lambda$, where λ is a constant. Therefore,

$$\begin{aligned} X'' - \lambda X = 0 \\ Y'' + \lambda Y = 0. \end{aligned} \tag{7.52}$$

Since $X(0) = X(a) = 0$, it is straightforward to check that a non-zero solution X is possible only if $\lambda < 0$. In this case, the general solution is given by

$$X(x) = c_1 \cos(\sqrt{-\lambda}x) + c_2 \sin(\sqrt{-\lambda}x),$$

for arbitrary constants c_1 and c_2. The boundary condition $X(0) = 0$ gives $c_1 = 0$. And, for $c_2 \neq 0$, the other boundary condition implies that $\lambda = -\frac{n^2\pi^2}{a^2}$, for $n = 1, 2, \ldots$.

Thus, we get

$$X_n(x) = c_n \sin\left(\frac{n\pi x}{a}\right) \text{ and } Y_n(x) = d_n \exp\left(-\frac{n\pi y}{a}\right)$$

as the relevant solutions, where c_n and d_n are appropriate constants. Thus, we have

$$u(x, y) = \sum_{n=1}^{\infty} c_n d_n X_n(x) Y_n(y) = \sum_{n=1}^{\infty} c_n d_n \exp\left(-\frac{n\pi y}{a}\right) \sin\left(\frac{n\pi x}{a}\right).$$

Using the boundary condition $u(x, 0) = A\left(1 - \frac{x}{a}\right)$, we get

$$A\left(1 - \frac{x}{a}\right) = \sum_{n=1}^{\infty} c_n d_n \sin\left(\frac{n\pi x}{a}\right).$$

Now making the function $A\left(1 - \frac{x}{a}\right)$ as an *odd* periodic function of period $2a$ and expanding it in a Fourier sine series, we may choose $c_n = \frac{2A}{n\pi}$ and $d_n = 1$. Thus, the required solution is given by

$$u(x, y) = \frac{2A}{\pi} \sum_{n=1}^{\infty} \frac{1}{n} \exp\left(-\frac{n\pi y}{a}\right) \sin\left(\frac{n\pi x}{a}\right).$$

2. Consider the Laplace equation (7.50) in an annulus

$$\{(x, y) : R_1^2 < x^2 + y^2 < R_2^2\}$$

with boundary conditions:

$$\begin{aligned} \frac{\partial u}{\partial r} &= u_1 \text{ for } r = R_1 \\ u &= u_2 \text{ for } r = R_2. \end{aligned} \tag{7.53}$$

Here $0 < R_1 < R_2$ and u_1, u_2 are given constants; $r^2 = x^2 + y^2$.
Write equation (7.50) in polar co-ordinates (r, θ); $x = r\cos\theta$ and $y = r\sin\theta$:

$$\frac{\partial^2 u}{\partial r^2} + \frac{1}{r}\frac{\partial u}{\partial r} + \frac{1}{r^2}\frac{\partial^2 u}{\partial \theta^2} = 0 \tag{7.54}$$

or

$$r^2\frac{\partial^2 u}{\partial r^2} + r\frac{\partial u}{\partial r} + \frac{\partial^2 u}{\partial \theta^2} = 0.$$

We seek the solution in the separation of variables form:

$$u(r,\theta) = R(r)\Theta(\theta).$$

Proceeding as in the above example, we find that

$$r^2\Theta(\theta)R''(r) + r\Theta(\theta)R'(r) + \Theta''(\theta)R(r) = 0$$

or

$$\frac{\Theta''(\theta)}{\Theta(\theta)} = -\frac{r^2R''(r) + rR'(r)}{R(r)} = \lambda,$$

where λ is a constant. The boundary conditions (7.53) imply that Θ is a constant function and thus, u is a *radial* function. We may take $\Theta \equiv 1$. Using the boundary conditions (7.53), we see that R satisfies the boundary conditions

$$R'(r) = u_1 \text{ for } r = R_1 \text{ and } R(r) = u_2 \text{ for } r = R_2.$$

Thus, it follows that $\lambda = 0$ and R satisfies the second-order equation

$$r^2R''(r) + rR'(r) = 0.$$

The general solution of this equation is given by $R(r) = c_1 + c_2 \log r,\ r > 0$ for arbitrary constants c_1 and c_2. Using the given boundary conditions, we can easily determine the constants c_1 and c_2, and the solution is given by

$$u(x,y) = u_2 - u_1R_1 \log\left(\frac{R_2}{r}\right), \text{ for } R_1^2 < r^2 = x^2 + y^2 < R_2^2.$$

7.6 NOTES

1. Well-Posedness and Ill-Posedness: From applications, we understand that a PDE is always attached with a set of conditions in the form of boundary and/or initial conditions or both. The well-posedness to be defined soon, has to be understood in this setup. The initial or boundary conditions arise from physical situations and mathematically, it is not an easy task to give proper or appropriate conditions so that the problem is well-posed. Some type of conditions may be suitable for certain class of equations, whereas the same set of conditions may not work for other type of equations. In this direction, Jacques Hadamard (1902) proposed the following notion of well-posedness:

Definition 7.43. We say a PDE together with a set of conditions known as *data* from a certain class Y is said to be *well-posed* in a class X in the sense of Hadamard if the PDE has a unique solution in X satisfying the data and the solution depends continuously on the data. Here X and Y may be provided with suitable topologies. Problems that are not well-posed in the sense of Hadamard are termed as *ill-posed*. □

We remark that the last condition of continuous dependence is related to physical problems. Usually PDE are modelled by experimental data and physical laws. They are prone to errors and generally the data will also be an approximation. Thus, the solution obtained will be for an approximate data and we should like to know the solution obtained is also an approximation in the relevant topologies. The continuous dependence guarantees it. Inverse problems are often ill-posed. For example, the backward heat equation.

In Chapter 3, we have studied the initial value problem (IVP) for general first order equations, where the data, namely the initial values defined on a non-characteristic hypersurface. In the present chapter on Poisson and Laplace equations (more generally, elliptic equations), we have studied the problem with boundary data. We have indeed established the well-posedness of the boundary value problem, of course with the smoothness assumption on the domain with appropriate spaces like Hölder spaces. We now establish through an example that the IVP for Δ is ill-posed.

Example 7.44 (Ill-Posedness of IVP for Δ). Consider the problem

$$\frac{\partial^2 u}{\partial x_1^2} + \frac{\partial^2 u}{\partial x_2^2} = 0$$

with the initial data

$$u(x_1, 0) = 0, \quad \frac{\partial u}{\partial x_2}(x_1, 0) = ke^{-\sqrt{k}} \sin(kx_1),$$

where k is a positive integer. Since the operator is elliptic, the x_1-axis is indeed non-characteristic and the solution is given by

$$u(x_1, x_2) = e^{-\sqrt{k}} \sin(kx_1) \sinh(kx_2).$$

Now, observe that the Cauchy data $\frac{\partial u}{\partial x_2}(x_1, 0) = ke^{-\sqrt{k}} \sin(kx_1) \to 0$ as $k \to \infty$, whereas, for any $x_2 \neq 0$, we see that the solution $u(x_1, x_2) \to \infty$ as $k \to \infty$. That is the solution blows-up showing that it does not depend continuously on the data. □

2. Schauder Theory: The study in the present chapter on Poisson and Laplace equations, known as *potential theory*, is the natural starting point for the study of general second-order uniformly elliptic operators. This classical study is termed as *Schauder theory*. Then, we can of course go on to study non-uniformly elliptic operators as well. One of the classical example is the minimal surface equation given by

$$\sum_{i,j=1}^{n} D_j \left(\frac{D_i u}{(1 + |Du|^2)^{1/2}} \right) = 0,$$

which is a quasilinear non-uniformly elliptic operator. The Schauder theory essentially is an extension of the potential theory. One of the important observation is that the equation with Hölder coefficients can be locally treated as a perturbation of equations with constant coefficients by fixing the leading coefficient at a single value. Thus, as far as the highest order terms are concerned it is like a Laplacian locally. This allows us to derive the local (interior) estimates of the form

$$\|u\|_{C^{2,\alpha}(\Omega')} \leq C \left(\sup_{\Omega} |u| + \|f\|_{C^{0,\alpha}(\overline{\Omega})} \right)$$

as discussed in the beginning of the chapter. In addition to the interior estimates, we also require to obtain boundary and global estimates. We will not pursue this matter further in this book and the interested reader can refer to Gilberg and Trudinger (2001) and the references therein. The modern and general non-classical approach is based on suitable Hilbert spaces and we have briefly discussed this in Section 7.5. In this methodology, general data and coefficients can also be treated. This requires much more machinery like modern functional analysis, distribution theory, Sobolev spaces and so on. (see Brezis, 2001; Evans, 1998; Kesavan, 1989).

7.7 EXERCISES

1. Let $u, v \in C^2(\overline{\Omega})$. Using the divergence theorem, prove the following identities:

   a. $\int_\Omega \Delta u = \int_{\partial\Omega} \frac{\partial u}{\partial \nu}$.

   b. $\int_\Omega v\Delta u = -\int_\Omega \nabla u \cdot \nabla v + \int_{\partial\Omega} \frac{\partial u}{\partial \nu} v$.

   c. $\int_\Omega (v\Delta u - u\Delta) = \int_{\partial\Omega} \left(\frac{\partial u}{\partial \nu} v - u \frac{\partial v}{\partial \nu} \right)$.

   Here, $\frac{\partial u}{\partial \nu} = \nabla u \cdot \nu$ is the normal derivative and $\nabla = \left(\frac{\partial}{\partial x_1}, \cdots \frac{\partial}{\partial x_n}\right)$ is the *grad* operator. The above are known as *Green's identities.*
2. For the case $n = 2$, write the Laplace operator Δ in polar coordinates.
3. (Spherical Symmetry) Let R is a rotation matrix, that is $RR^t = I$ and u be harmonic in $\mathbb{R}^n$. Define v by $v(x) = u(Rx)$. Show that v is also harmonic in $\mathbb{R}^n$.
4. Let $v(r) = u(|x|)$ where $r = |x|$. Show that

$$\Delta u \equiv \ddot{v}(r) + \frac{n-1}{r}\dot{v}.$$

   Solve the equation to obtain the fundamental solution ϕ.

5. Let ϕ be the fundamental solution of $-\Delta$. Show that there exists a constant $C > 0$ such that

$$|D\phi(x)| \leq \frac{C}{|x|^{n-1}}, \; |D^2\phi(x)| \leq \frac{C}{|x|^n}, \quad x \neq 0.$$

6. Prove Proposition 7.1.
7. Let $f \in C_c^2(\mathbb{R}^n)$ and ϕ be the fundamental solution of $-\Delta$. Define $I_\varepsilon = \int_{B_\varepsilon(0)} \phi(y)(\Delta f)(x-y)\, dy\, \varepsilon > 0$. Show that there exists a constant $C > 0$ such that

$$|I_\varepsilon| \leq \begin{cases} C\varepsilon^2 |\log \varepsilon|, & \text{if } n = 2 \\ C\varepsilon^2 \text{ if } n \geq 3. \end{cases}$$

Also compute $\frac{\partial \phi}{\partial \nu}$ on $\partial B_\varepsilon(0)$.
8. Let Ω be a domain in $\mathbb{R}^2$ symmetric about the x-axis and let $\Omega^+ = \{(x, y) : y > 0\}$ be the upper part of Ω. Assume $u \in C(\overline{\Omega^+})$ is harmonic in Ω^+ with $u = 0$ on $\partial\Omega^+ \cap \{y = 0\}$. Define for $(x, y) \in \Omega$,

$$v(x, y) = \begin{cases} u(x, y) \text{ if } y \geq 0, \\ -u(x, -y) \text{ if } y < 0. \end{cases}$$

Show that v is harmonic.
9. Let $u \in C^2(\Omega) \cap C^0(\overline{\Omega})$ be a solution of

$$\Delta u + \sum_{k=1}^{n} a_k(x) \frac{\partial u}{\partial x_k} + c(x)u = 0 \text{ in } \Omega$$

with $c(x) < 0$ in Ω, $u = 0$ on $\partial\Omega$ and a_k's are smooth. Show that $u \equiv 0$.
10. Consider the PDE, $-\Delta u = \lambda u$ in Ω, $u = 0$ on $\partial\Omega$ where λ is a scalar and Ω is a bounded open set. If $\lambda \leq 0$, prove that $u \equiv 0$.
11. Let $u \in C^2(\overline{B_1(0)})$ solves $-\Delta u = f$ in $B_1(0)$, $u = 0$ on $\partial B_1(0)$. Show that there exists $C > 0$ such that

$$\max_{x \in B_1(0)} |u(x)| \leq C \max_{x \in B_1(0)} |f|.$$

(Hint: Consider the problem with $f = 1$ and $f = M$ where $M = \max_{x \in B_1(0)} |f|$.)
More generally, if u solves $-\Delta u = f$ in $B_1(0)$, $u = g$ on $\partial B_1(0)$, then

$$\max_{x \in B_1(0)} |u(x)| \leq C \left(\max_{x \in \partial B_1(0)} |g| + \max_{x \in B_1(0)} |f| \right).$$

12. Let u be a non-negative harmonic function in $\mathbb{R}^n$.

a. By using the Poisson's formula (7.29) for the ball and MVP, show that

$$\frac{R^{n-2}(R^2 - |x|^2)}{(R + |x|)^n} u(0) \leq u(x) \leq \frac{R^{n-2}(R^2 - |x|^2)}{(R - |x|)^n} u(0)$$

for any $R > 0$ and $|x| < R$.

b. By letting $R \to \infty$ in (a) above, conclude that u is a constant.

This gives the stronger form of the Liouville's theorem: If u is harmonic in $\mathbb{R}^n$ and bounded below (or above), then u is a constant function.

13. Let $A = [a_{ij}]$ be a real $n \times n$ matrix with zero trace and $a \in \mathbb{R}^n$. What can be said about a harmonic function in $\mathbb{R}^n$ that satisfies

$$u(x) \leq \sum_{i,j=1}^{n} a_{ij} x_i x_j + a \cdot x, \text{ for all } x \in \mathbb{R}^n?$$

Justify your answer. (Hint: Use Exercise 12.)

14. Let u be a harmonic function in $\mathbb{R}^n$. Describe the range of u.(Hint: Write $y \in \mathbb{R}^{n+1}$ as $y = (x, x_{n+1})$ with $x \in \mathbb{R}^n$. Define $U(y) = x_{n+1} - u(x)$. Suppose $y^0 = (x^0, x^0_{n+1})$ be such that $U(y^0) = 0$. What happens if $U(y) \leq x_{n+1} - x^0_{n+1} + \nabla u(x^0) \cdot (x - x^0)$ for all y? Use Exercise 12.)

15. If u is a harmonic function in $\mathbb{R}^n$ satisfying $|u(x)| \leq C(1 + |x|^s)$, for some non-negative real s and all $x \in \mathbb{R}^n$, show that u is a polynomial of degree at most $[s]$, where $[s]$ denotes the integer part of s.

16. Let Ω be an open, bounded set in $\mathbb{R}^n$. Suppose $u \in C^2(\Omega) \cap C^0(\overline{\Omega})$ satisfies $\Delta u = -1$ in Ω, $u = 0$ on $\partial\Omega$. Show that for $x \in \Omega$, $u(x) \geq \frac{1}{2n}(d(x, \partial\Omega))^2$. (Suggestion: For fixed $x_0 \in \Omega$, consider the harmonic function $u(x) + \frac{1}{2n}|x - x_0|^2$, $x \in \Omega$.)

17. If $x \in \mathbb{R}^n$, write $x = (x', x_n)$, $x' \in \mathbb{R}^{n-1}$. Let u be the unique solution of $\Delta u = 0$ in $B_1(0)$ and $u = \phi \in C(\partial B_1(0))$, on ∂B_1. If ϕ satisfies $\phi(x', x_n) = -\phi(x', x_n)$, show that $u(x', x_n) = -u(x', x_n)$. (Suggestion: *Uniqueness* is the key word.)

18. Let u be harmonic in $B_1^+ = x \in B_1(0) : x_n > 0$ and $u = 0$ on $x_n = 0$. Extend u to a harmonic function in $B_1(0)$. [Hint: Define $u(x', -x_n) = -u(x', x_n)$, $x_n > 0$. Then u is continuous in $B_1(0)$ and harmonic in B_1^+, B_1^-. Let $x^0 \in B_1(0)$ with $x^0_n = 0$ and $0 < r < 1$. Then

$$\int_{|x-x^0|=r} u(x)\, dS(x) = \int_{x_n>0} + \int_{x_n<0} = 0.$$

Thus, u satisfies the MVP.]

19. Prove that $u \in C(\Omega)$ is sub-harmonic if and only if for every open $\Omega' \subset \Omega$ and every harmonic function v in Ω' with $v = u$ on $\partial\Omega'$, the inequality $v \leq u$ holds.

20. Let Ω be an open connected (bounded or unbounded) set in $\mathbb{R}^n$. Suppose $u \geq 0$ is a harmonic function in Ω. Show that either $u \equiv 0$ in Ω or $u > 0$ in Ω.

21. Let Q be the rectangle with vertices $(0,0), (kr,0), (kr,2r), (0,2r)$ for some $r > 0$ and $k \in \mathbb{N}$. Let $P_0 = (r,r)$ and $P_* = ((k-1)r, r)$. If $u \geq 0$ is harmonic in Q, show that

$$2^{-2k}u(P_0) \leq u(P_*) \leq 2^{2k}u(P_0).$$

22. Consider the Neumann problem $\Delta u = 0$ in Ω, $\frac{\partial u}{\partial \nu} = 0$ on $\partial\Omega$. If $u \in C^2(\Omega) \cap C^1(\overline{\Omega})$ is a solution, show that u is identically a constant. (Hint: Apply Hopf's lemma to u and $-u$, if u were non-constant.)

23. If u is a smooth, non-negative solution of $\Delta u = u^3 + f$ with $f \leq 0$ in Ω and $u \geq 0$ on $\partial\Omega$, show that $u > 0$ in Ω.

24. Suppose Ω is a bounded, open subset of $\mathbb{R}^n$ and $u \in C^1(\overline{\Omega})$. If $\int_{\partial B} \frac{\partial u}{\partial \nu}\, dS = 0$ for every ball B with $\bar{B} \subset \Omega$, show that u is harmonic in Ω. (Hint: Consider the *spherical mean* of u:

$$M_u(x,r) = \frac{1}{\omega_n r^n} \int_{B_r(x)} u(y)\, dy,\ r > 0$$

and prove that u has the MVP.)

25. Let $u \in C^2(B_1(0)) \cap C^1(\overline{B_1(0)})$ be the unique solution of the mixed problem

$$\begin{aligned} \Delta u &= -1 \text{ in } B_1(0), \\ u &= 0 \text{ on } \partial B_1(0) \cap \{x_n > 0\}, \\ \frac{\partial u}{\partial \nu} &= -u \text{ on } \partial B_1(0) \cap \{x_n < 0\}. \end{aligned}$$

Show that $u \geq 0$ in $\overline{B_1(0)}$ and $u > 0$ on $\partial B_1(0) \cap \{x_n < 0\}$.

26. Give an example of a C^2 function u in $\mathbb{R}^n$ such that $u > 0$ and $\Delta u - u \geq 0$ in $\mathbb{R}^n$. Can such a function be bounded? Justify your answer. (Hint: If u has a (local) maximum at x_0, then $u(x_0) > 0$ and $\Delta u(x_0) \leq 0$, a contradiction. If u is bounded (above) and does not attain a maximum, consider the function $u/\cosh(\varepsilon|x|)$ for small ε.)

27. Let Ω be a bounded or unbounded domain in $\mathbb{R}^n$ and $u \in C(\overline{\Omega}) \cap C^2(\Omega)$ be bounded and satisfy $\Delta u - \lambda u \geq 0$ for some $\lambda > 0$. Prove that $u \leq \sup_{\partial\Omega} u_+$ in $\overline{\Omega}$. By convention, the sup is zero if $\partial\Omega$ is empty. Here and in the next exercise, $u_+ = \max\{u, 0\}$.

28. Let $\Omega = \mathbb{R}^n_+$, $u \in C(\overline{\Omega}) \cap C^2(\Omega)$ be bounded and $\Delta u - \lambda u \geq 0$ in Ω for some $\lambda > 0$. If $x_0 \in \partial\Omega$ is such that $u(x_0) = \sup_{\partial\Omega} u_+ > 0$ and $\frac{\partial u}{\partial \nu}(x_0)$ exists, show that $\frac{\partial u}{\partial \nu}(x_0) \leq -\sqrt{\lambda}u(x_0) < 0$. (Hint: Consider $v(x) = u(x) - u(x_0)\exp(-\sqrt{\lambda}x_n)$.)

29. Let u be harmonic in $B_R(0) \subset \mathbb{R}^n$, $n \geq 3$. For $x \in \mathbb{R}^n$, $|x| > R$, define

$$U(x) = \left(\frac{R}{|x|}\right)^{n-2} u\left(\frac{R^2}{|x|^2}x\right).$$

The function U is called the Kelvin transform of u. Show that U is harmonic in the region $|x| > R$ and that $U(x) \to 0$ as $|x| \to \infty$. Further, write down u in terms U.

30. Consider the ball $B_r(x_0)$ in $\mathbb{R}^3$. If $y \notin \bar{B}_r(x_0)$, compute $\int_{B_r(x_0)} \frac{dx}{|x-y|}$. (Hint: The function $x \mapsto |x-y|^{-1}$ is harmonic in $B_r(x_0)$.)

31. Let $G = G(x, y)$ be the Green's function for the $-\Delta$ in Ω. Show that G is symmetric in the sense that $G(x, y) = G(y, x)$.

32. Verify the statements in Theorem 7.26.

33. Let u be a C^2 function in a domain Ω in $\mathbb{R}^2$. Let $(x, y) \in \Omega$. Let $A(x-h, y-k)$, $B(x+h, y-k)$, $C(x+h, y+k)$, $D(x-h, y+k)$ be four points. Suppose $u(x, y) = \frac{1}{4}[u(A)+u(B)+u(C)+u(D)]$ for all $h, k > 0$ such that the rectangular region formed by A, B, C, D is in Ω. Then, show that u is harmonic in Ω. The converse need not be true.

34. If $f \in C_c^\infty(\mathbb{R}^n)$. Show that the convolution $\phi * f$ is an unbounded function in dimension $n = 2$ and is a bounded function for $n \geq 3$.

35. Provide the details of the proofs of Proposition 7.23 and Theorem 7.24.

36. Let Ω be a bounded open set in $\mathbb{R}^n$ with smooth boundary $\partial\Omega$ and $\rho_1, \rho_2 \in C(\partial\Omega)$. Define the functions u_1, u_2 by

$$u_1(x) = \int_{\partial\Omega} \phi(x-\xi)\rho_1(\xi)\, dS(\xi) \text{ and } u_2(x) = \int_{\partial\Omega} \rho_2(\xi)\frac{\partial \phi}{\partial \nu_\xi}(x-\xi)\, dS(\xi)$$

for $x \in \mathbb{R}^n \setminus \partial\Omega$. Here ν_ξ denotes the outward unit normal at $\xi \in \partial\Omega$. The functions u_1 and u_2 are referred to as *single-layer potential* and *double-layer potential* with *densities* ρ_1 and ρ_2, respectively. Show that $u_1, u_2 \in C^\infty(\mathbb{R}^n \setminus \partial\Omega)$ and harmonic in $\mathbb{R}^n \setminus \partial\Omega$.

CHAPTER 8

Heat Equation

8.1 INTRODUCTION

The heat or diffusion equation

$$u_t = a^2 \Delta u, \ t > 0, \ x \in \mathbb{R}^n, \tag{8.1}$$

models the heat flow in solids and fluids. It also describes the diffusion of chemical particles. It is also one of the fundamental equations that have influenced the development of the subject of partial differential equations (PDE) since the middle of the last century. *Heat and fluid flow* problems are important topics in fluid dynamics. Here the heat flow is combined with a fluid flow problem and the resulting equation is termed as *energy equation*. We begin with a derivation of one-dimensional heat equation, arising from the analysis of heat flow in a thin rod. Further, equation (8.1) is also a prototype in the class of parabolic equations and hence the importance of studying this equation.

8.1.1 Derivation of One-Dimensional Heat Equation

Consider a thin rod of length L and place it along the x-axis on the interval $[0, L]$. We assume that the rod is insulated so that its lateral surface is impenetrable to heat transfer. We also assume that the temperature is the same at all points of any cross-sectional area of the rod. Let ρ, c, k denote, respectively, the mass density, heat capacity and the coefficient of (internal) thermal conductivity, of the rod.[1] Let us analyze the heat balance in an arbitrary segment $[x_1, x_2]$ of the rod, with $\delta x = x_2 - x_1$ very small, over a time interval $[t, t + \delta t]$, δt small (see Figure 8.1).

Let $u(x, t)$ denote the *temperature* in the cross-section with abscissa x, at time t. According to *Fourier's law of heat conduction*, the rate of heat propagation q is proportional to $\frac{\partial u}{\partial x} S$, with

[1]Unless the rod is homogeneous, these quantities are functions of x and may also depend on the temperature. The dependence on temperature may be neglected if the variation of temperature is not too significant. However, in certain heat and mass transfer problems, these quantities are considered as functions of the temperature, in which case the resulting equation may become non-linear. The interested reader should consult books on *Thermodynamics* and *Heat and Mass Transfer*, for more details.

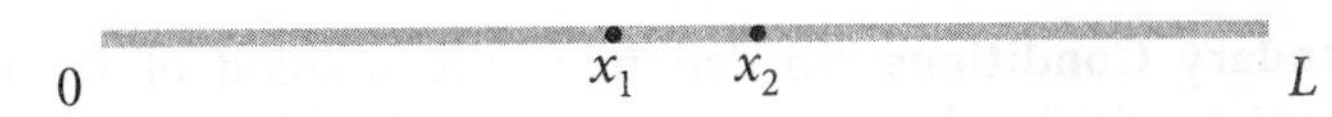

Figure 8.1 Temperature distribution in a rod

S denoting the area of the cross-section. Here,

$$q = \lim_{\delta t \to 0} \frac{\delta Q}{\delta t},$$

where δQ is the quantity of heat that has passed through a cross-section S during a time δt. Thus,

$$q = -k\frac{\partial u}{\partial x}S.$$

Applying Fourier's law at x_1 and x_2, we obtain

$$\delta Q_1 = \left(-k\frac{\partial u}{\partial x}\right)_{x=x_1} S\delta t \quad \text{and} \quad \delta Q_2 = \left(-k\frac{\partial u}{\partial x}\right)_{x=x_2} S\delta t.$$

Thus, the quantity of heat that has passed through the small segment $[x_1, x_2]$ of the rod, during time δt is given by

$$\begin{aligned} \delta Q_1 - \delta Q_2 &= \left(-k\frac{\partial u}{\partial x}\right)_{x=x_1} S\delta t - \left(-k\frac{\partial u}{\partial x}\right)_{x=x_2} S\delta t \\ &\approx \frac{\partial}{\partial x}\left(k\frac{\partial u}{\partial x}\right)\delta x\, S\delta t. \end{aligned} \tag{8.2}$$

This influx of heat during time δt was spent in raising or lowering the temperature of the rod by δu, say. According to the law of thermodynamics, this is expressed by

$$\delta Q_1 - \delta Q_2 = c\rho\delta x\, S\delta u \approx c\rho\delta x\, S\frac{\partial u}{\partial t}\delta t. \tag{8.3}$$

Note that the quantity $\rho\delta x\, S$ is the mass of the element $[x_1, x_2]$ of the rod. Comparing equations (8.2) and (8.3), in which the left sides represent the same quantity, we obtain

$$\frac{\partial}{\partial x}\left(k\frac{\partial u}{\partial x}\right)\delta x\, S\delta t = c\rho\delta x\, S\frac{\partial u}{\partial t}\delta t.$$

Or,

$$\frac{\partial u}{\partial t} = \frac{1}{c\rho}\frac{\partial}{\partial x}\left(k\frac{\partial u}{\partial x}\right) = \frac{k}{c\rho}\frac{\partial^2 u}{\partial x^2},$$

assuming k is a constant. This is precisely equation (8.1) with $a^2 = \frac{k}{c\rho}$.

Initial and Boundary Conditions: To determine the solution of (8.1) uniquely, the solution need to satisfy initial and boundary conditions. The initial condition

$$u(x, 0) = g(x),\ 0 < x < L, \tag{8.4}$$

representing the initial temperature distribution at all points of the rod at the initial instant of time $t = 0$. At the end points, $x = 0$ and $x = L$, different boundary conditions may be given. For example, consider the following conditions:

$$u(0, t) = h_1(t),\ u(L, t) = h_2(t) \tag{8.5}$$

for $t > 0$. These conditions represent that the rod is maintained at the prescribed temperatures at the end points. The resulting initial-boundary value problem is termed as the *Dirichlet problem*. If, instead, the heat flux at the end points are supplied, then u is replaced by $\frac{\partial u}{\partial x}$ in (8.5). The resulting problem is termed as the *Neumann problem*. If at one end of the rod temperature is prescribed and at the other end temperature flux, we arrive at a situation where the resulting problem is known as a *mixed problem*.

In three dimensions, the derivation of the heat equation is similar. In this case we only need to replace the operation of $\frac{\partial}{\partial x}$ in (8.2) by the gradient operator ∇. Similarly, in a boundary value problem, the Neumann condition is provided in terms of *normal derivative* in the direction of the outward unit normal to the boundary. In a *diffusion problem, Fick's law of diffusion* needs to be applied in place of Fourier's law of heat conduction.

8.2 HEAT TRANSFER IN AN UNBOUNDED ROD

Consider the case of an infinite rod in which the temperature is prescribed at all points of the rod at an initial instant of time $t = 0$. It is then required to determine the temperature distribution in the rod at instants of time $t > 0$. Physical problems reduce to that of heat transfer in an unbounded rod when the rod is so long that the conditions prescribed at the ends of the rod do not significantly influence the temperature in the interior points of the rod. Thus, an initial condition (8.4), now prescribed for all $x \in \mathbb{R}$, should suffice. Therefore, we consider the following initial value problem (IVP):

$$\begin{aligned} &u_t = a^2 u_{xx},\ x \in \mathbb{R}, t > 0, \\ &u(x, 0) = g(x),\ x \in \mathbb{R}. \end{aligned} \tag{8.6}$$

We apply the method of separation of variables, to find a solution of (8.6), that is, we assume that the solution u has the form

$$u(x, t) = X(x)T(t). \tag{8.7}$$

Substituting this into (8.6), we have $X(x)T'(t) = a^2X''(x)T(t)$ or

$$\frac{T'}{a^2T} = \frac{X''}{X}. \tag{8.8}$$

Here we have used $'$, as a convention, the derivative of any function of a *single* variable. Since neither of the quantities in (8.8) can be a function of x and/or t, both must be equal to a constant, say, $-\lambda^2$ with $\lambda > 0$.[2] Hence from (8.8) we obtain the following two ordinary differential equations (ODE):

$$T' + a^2\lambda^2 T = 0, \tag{8.9}$$

$$X'' + a^2\lambda^2 X = 0, \tag{8.10}$$

whose general solutions are given by

$$T(t) = Ce^{-a^2\lambda^2 t},$$
$$X(x) = A\cos\lambda x + B\sin\lambda x.$$

Substituting these expressions into (8.7), we therefore obtain a solution u_λ of (8.6) given by

$$u_\lambda(x,t) = e^{-a^2\lambda^2 t}[A(\lambda)\cos\lambda x + B(\lambda)\sin\lambda x], \tag{8.11}$$

for each $\lambda > 0$. Since equation (8.6) is linear, by *superposition*, we see that

$$u(x,t) = \int_0^\infty e^{-a^2\lambda^2 t}[A(\lambda)\cos\lambda x + B(\lambda)\sin\lambda x]\, d\lambda \tag{8.12}$$

is a solution of (8.6), provided that the integral in (8.12), its derivative with respect to t and its second derivative with respect to x all exist.

In order to determine the coefficients $A(\lambda)$ and $B(\lambda)$ in (8.12), we use the initial condition in (8.6):

$$u(x,0) = g(x) = \int_0^\infty [A(\lambda)\cos\lambda x + B(\lambda)\sin\lambda x]\, d\lambda. \tag{8.13}$$

[2] If the initial temperature g is bounded, we can expect the solution to be bounded, from physical considerations; hence the negative sign.

Suppose that the function[3] $g(x)$ satisfies the following integral identity:

$$g(x) = \frac{1}{\pi} \int_0^{\infty} \left(\int_{-\infty}^{\infty} g(y) \cos \lambda(y - x)\, dy \right) d\lambda,$$

or

$$g(x) = \frac{1}{\pi} \int_0^{\infty} \left[\left(\int_{-\infty}^{\infty} g(y) \cos \lambda y\, dy \right) \cos \lambda x + \left(\int_{-\infty}^{\infty} g(y) \sin \lambda y\, dy \right) \sin \lambda x \right] d\lambda. \tag{8.14}$$

Comparing the terms on the right-hand sides in (8.13) and (8.14), we obtain

$$A(\lambda) = \frac{1}{\pi} \int_{-\infty}^{\infty} g(y) \cos \lambda y\, dy \text{ and } B(\lambda) = \frac{1}{\pi} \int_{-\infty}^{\infty} g(y) \sin \lambda y\, dy. \tag{8.15}$$

Substituting these expressions into (8.12), we obtain the following expression for the solution:

$$\begin{aligned} u(x, t) &= \frac{1}{\pi} \int_0^{\infty} e^{-a^2 \lambda^2 t} \left[\left(\int_{-\infty}^{\infty} g(y) \cos \lambda y\, dy \right) \cos \lambda x + \left(\int_{-\infty}^{\infty} g(y) \sin \lambda y\, dy \right) \sin \lambda x \right] d\lambda \\ &= \frac{1}{\pi} \int_0^{\infty} e^{-a^2 \lambda^2 t} \left[\int_{-\infty}^{\infty} g(y)(\cos \lambda y \cos \lambda x + \sin \lambda y \sin \lambda x)\, dy \right] d\lambda \\ &= \frac{1}{\pi} \int_0^{\infty} e^{-a^2 \lambda^2 t} \left(\int_{-\infty}^{\infty} g(y) \cos \lambda(y - x)\, dy \right) d\lambda. \end{aligned}$$

[3]This condition appears to be very restrictive. However, the final formula we derive for the solution will remove this restriction.

If we interchange the order of integration in the last integral, we finally get

$$u(x,t) = \frac{1}{\pi} \int_{-\infty}^{\infty} \left[g(y) \left(\int_{0}^{\infty} e^{-a^2\lambda^2 t} \cos \lambda(y-x)\, d\lambda \right) \right] dy. \tag{8.16}$$

Let us analyze the inner integral in (8.16). By making the substitutions $a\lambda\sqrt{t} = z$ and $\eta = \frac{y-x}{a\sqrt{t}}$, we obtain that

$$\int_{0}^{\infty} e^{-a^2\lambda^2 t} \cos \lambda(y-x)\, d\lambda = \frac{1}{a\sqrt{t}} K(\eta),$$

where

$$K(\eta) = \int_{0}^{\infty} e^{-z^2} \cos \eta z\, dz. \tag{8.17}$$

Differentiating with respect to η inside the integral sign (which is easy to justify), we get

$$K'(\eta) = -\int_{0}^{\infty} e^{-z^2} z \sin \eta z\, dz,$$

which, upon an integration by parts, gives

$$K'(\eta) = -\frac{\eta}{2} \int_{0}^{\infty} e^{-z^2} \cos \eta z\, dz = -\frac{\eta}{2} K(\eta).$$

Solving this ODE, we get $K(\eta) = Ce^{-\eta^2/4}$ for some constant C. But,

$$K(0) = \int_{0}^{\infty} e^{-z^2}\, dz = \frac{\sqrt{\pi}}{2}.$$

Hence $C = \frac{\sqrt{\pi}}{2}$ and

$$K(\eta) = \frac{\sqrt{\pi}}{2} e^{-\eta^2/4}. \tag{8.18}$$

Reverting back to the original variables, we find that

$$\int_0^{\infty} e^{-a^2\lambda^2 t} \cos \lambda(y-x)\, d\lambda = \frac{1}{2a}\sqrt{\frac{\pi}{t}}e^{-\frac{(y-x)^2}{4a^2 t}}. \tag{8.19}$$

Putting back this expression into (8.16), we finally obtain the following expression for the solution:

$$u(x,t) = \frac{1}{2a\sqrt{\pi t}} \int_{-\infty}^{\infty} g(y) e^{-\frac{(y-x)^2}{4a^2 t}}\, dy. \tag{8.20}$$

This formula, valid for $t > 0$, is called the *Fourier–Poisson integral* or *Fourier–Poisson formula*. We note that the integral in (8.20) is well-defined for any function g, which is continuous (or even piece-wise continuous) and bounded.[4]

Physical Meaning of (8.20): Consider the initial function g defined by

$$g(x) = \begin{cases} 0, \text{ if } -\infty < x < x_0, \\ g_0(x), \text{ if } x_0 \le x \le x_0 + \delta x_0, \\ 0, \text{ if } x_0 + \delta x_0 < x < \infty, \end{cases} \tag{8.21}$$

for some x_0 and small δx_0 and continuous g_0; the discontinuity of g, if any, will not affect the integral in (8.20). Then, the function u_0 defined by

$$u_0(x,t) = \frac{1}{2a\sqrt{\pi t}} \int_{-\infty}^{\infty} g(y) e^{-\frac{(y-x)^2}{4a^2 t}}\, dy, \tag{8.22}$$

is a solution of the heat equation (8.6) satisfying the initial condition therein. We have

$$u_0(x,t) = \frac{1}{2a\sqrt{\pi t}} \int_{x_0}^{x_0+\delta x_0} g_0(y) e^{-\frac{(y-x)^2}{4a^2 t}}\, dy.$$

Applying mean-value theorem, we see that

$$u_0(x,t) = \frac{g_0(\xi)\delta x_0}{2a\sqrt{\pi t}} e^{-\frac{(\xi-x)^2}{4a^2 t}}, \tag{8.23}$$

[4]The readers who are familiar with Lebesgue integration will recognize that the assumption that g is a bounded measurable function will do.

for some $\xi \in (x_0, x_0 + \delta x_0)$. If for $t = 0$, the temperature is 0 everywhere except in a small interval $[x_0, x_0 + \delta x_0]$, where it is $g_0(x)$, then this formula gives the temperature at a point in the rod at any time $t > 0$. The "sum" of temperatures of form (8.23) is what yields the solution (8.22). We also notice that the heat in the element $[x_0, x_0 + \delta x_0]$ of the rod at $t = 0$ will be

$$\delta Q \approx g_0(\xi)c\rho\delta x_0, \tag{8.24}$$

where c is the specific heat and ρ is the density of the material. Now consider the expression

$$\frac{1}{2a\sqrt{\pi t}}e^{-\frac{(\xi-x)^2}{4a^2t}}. \tag{8.25}$$

Comparing this with the right side of (8.23) and taking into account (8.24), we may say that the term in (8.25) yields the temperature at any point of the rod at any instant of time $t > 0$ if, for $t = 0$ there was an instantaneous heat source with amount $Q = c\rho$ in the cross-section ξ (the limiting case of $\delta x_0 \to 0$). □

8.2.1 Solution in Higher Dimensions

We now consider the IVP for the heat equation in arbitrary space dimension $n \geq 1$:

$$\begin{aligned} &u_t = a^2\Delta u,\ t > 0,\ x \in \mathbb{R}^n, \\ &u(x, 0) = g(x),\ x \in \mathbb{R}^n. \end{aligned} \tag{8.26}$$

If we try to imitate the above procedure to obtain an expression for the proposed solution of the heat equation when $n > 1$, it is not immediately clear how to proceed with the separation of the variables. However, if we look at the Fourier–Poisson formula (8.20), we observe that it readily extends to higher dimensions. Thus, we consider the function

$$u(x, t) = \frac{1}{(4\pi a^2 t)^{n/2}} \int_{\mathbb{R}^n} g(y)e^{-\frac{|y-x|^2}{4a^2t}}\, dy, \tag{8.27}$$

for $x \in \mathbb{R}^n$ and $t > 0$, where g is as in (8.26). Another way to arrive at (8.27) is via certain invariant property the heat equation enjoys, resulting in some special solution of the heat equation.

The heat equation (8.26) is invariant under the dilation: $x \mapsto \lambda x$ and $t \mapsto \lambda^2 t$, for arbitrary $\lambda > 0$. This change of variables leaves the quotient $|x|^2/t$, $t > 0$ unchanged. This suggests to look for a solution of the heat equation[5] of the form $u(x, t) = v(|x|^2/t)$. A still more general approach is to look for a solution in *separation of variables* form:

$$u(x, t) = w(t)v(|x|^2/t),$$

[5] Recall the case of Laplace equation, where we looked for radial solutions using the rotational symmetry of Δ.

where the functions w and v are to be determined. Plugging this expression into the heat equation, we get

$$w'(t)v(|x|^2/t) = w(t)\left[\left(4a^2v''(|x|^2/t) + v'(|x|^2/t)\right)\frac{|x|^2}{t} + v'(|x|^2/t)\frac{2na^2}{t}\right]. \quad (8.28)$$

Choose v such that $4a^2v'' + v' = 0$. This yields a choice of v as $v(z) = \exp(-z/(4a^2))$. It then follows from (8.28) that $w' + \frac{n}{2t}w = 0$ and we choose $w(t) = t^{-n/2}$, $t > 0$ as a solution. Thus, we see that the function

$$t^{-n/2}\exp\left(-\frac{|x|^2}{4a^2t}\right)$$

is a solution of the heat equation for $x \in \mathbb{R}^n$ and $t > 0$. A suitable constant multiple of this function will be the fundamental solution of the heat equation, as we will see. The proposed solution u in (8.27) of IVP (8.26) is nothing but the convolution of this function (with a constant multiple) with the initial condition g, similar to the case of Laplace's equation. Yet another observation through the Fourier transform will also lead to the consideration of (8.27). This will be discussed in Section 8.5.

We now proceed to verify that the function u given by the Fourier–Poisson integral (8.27) indeed is a solution of (8.26). For this purpose, we define the *heat kernel* or *fundamental solution of the heat equation* by

$$K(x,t) = \begin{cases} (4\pi a^2 t)^{-n/2} e^{-\frac{|x|^2}{4a^2t}}, & \text{for } x \in \mathbb{R}^n,\ t > 0 \\ = 0, & \text{for } x \in \mathbb{R}^n,\ t < 0. \end{cases} \quad (8.29)$$

See (8.18). With this notation, we can write the Fourier–Poisson integral (8.27) as the *convolution* of K and g: $u(x,t) = (K(\cdot,t) * g)(x)$; see (8.30). For $n = 1$, some profiles of $K(\cdot,t)$ are shown in Figure 8.2, for different values of t. We can observe from these profiles that as t becomes smaller and smaller, the profile of K *concentrates* more and more near $x = 0$ and in order to maintain the integral over $\mathbb{R}$ for all t, the value of $K(0,t)$ becomes larger and larger as t becomes smaller and smaller. The family $\{K(\cdot,t),\ t > 0\}$ is also an example of an *approximate identity*.

Theorem 8.1. Let g be a continuous and bounded function defined on $\mathbb{R}^n$. Then, the function u defined by

$$u(x,t) = (K(\cdot,t) * g)(x) = (4\pi a^2 t)^{-n/2}\int_{\mathbb{R}^n} e^{-\frac{|x-y|^2}{4a^2t}} g(y)\,dy, \quad (8.30)$$

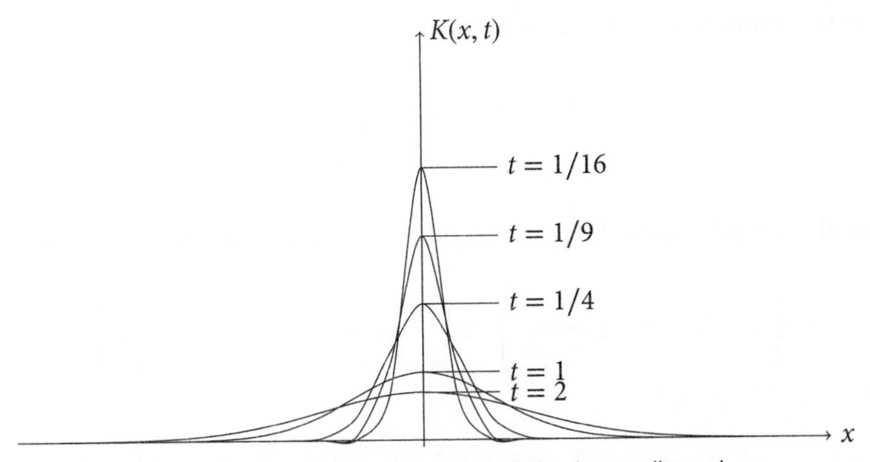

Figure 8.2 Profiles of the fundamental solution in one dimension

for $x \in \mathbb{R}^n$ and $t > 0$ is a C^∞ function and satisfies the heat equation (8.26) in the following sense: If we extend the function u to include $t = 0$ by $u(x, 0) = g(x)$, then u is continuous for $x \in \mathbb{R}^n$ and $t \geq 0$. □

The proof depends on the following properties of the heat kernel. For $x \in \mathbb{R}^n$ and $t > 0$, K satisfies:

1. $K(x, t) > 0$ and K is symmetric, that is $K(x, t) = K(-x, t)$.
2. $K(x, t)$ is a C^∞ function.
3. $\left(\partial_t - a^2 \Delta_x\right) K(x - y, t) = 0$, for all $y \in \mathbb{R}^n$.
4. $\int_{\mathbb{R}^n} K(x - y, t)\, dy = 1$.
5. For any $\delta > 0$, we have

$$\lim_{t \to 0+} \int_{|x-y| \geq \delta} K(x - y, t)\, dy = 0,$$

uniformly in $x \in \mathbb{R}^n$.

Properties (1)–(3) are easily verified from the definition of K. Property (4) follows from the fact that

$$\int_0^\infty e^{-z^2}\, dz = \frac{\sqrt{\pi}}{2}.$$

Next, by the substitution $x - y = 2a\sqrt{t}z$, we have

$$\int_{|x-y|\geq\delta} K(x-y,t)\,dy = \int_{|z|\geq\frac{\delta}{2a\sqrt{t}}} e^{-|z|^2}\,dz.$$

Note that the second integral does not depend on x and is estimated as follows: since

$$\left\{z \in \mathbb{R}^n : |z| \geq \frac{\delta}{2a\sqrt{t}}\right\} \subset \bigcup_{j=1}^{n} \left\{z \in \mathbb{R}^n : |z_j| \geq \frac{\delta}{2a\sqrt{nt}}\right\}$$

we get

$$\begin{aligned}\int_{|z|\geq\frac{\delta}{2a\sqrt{t}}} e^{-|z|^2}\,dz &\leq n\pi^{\frac{n-1}{2}} \cdot 2 \int_{\frac{\delta}{2a\sqrt{nt}}}^{\infty} e^{-\eta^2}\,d\eta \\ &\leq n\pi^{\frac{n-1}{2}} \frac{2a\sqrt{nt}}{\delta} \int_0^{\infty} (2\eta)e^{-\eta^2}\,d\eta \\ &\leq n\pi^{\frac{n-1}{2}} \frac{2a\sqrt{nt}}{\delta} \to 0 \text{ as } t \to 0.\end{aligned}$$

This proves Property (5).

It is also to be noted that for any fixed $T > 0$, the function $K(ix + iy, T - t)$, $i = \sqrt{-1}$ or a constant multiple of it, also satisfies Property (3) for $t < T$. This property will be used later to establish the uniqueness of solutions. Now to the proof of Theorem 8.1.

Proof (of Theorem 8.1) By repeated integrations under the integral sign,[6] it follows that $u \in C^\infty$ using Property (2) of K. Property (3) then proves that u satisfies (8.26) for $x \in \mathbb{R}^n$ and $t > 0$. For the stated continuity property of u, pick an $\eta \in \mathbb{R}^n$ and let $\varepsilon > 0$ be arbitrary. Using the continuity of g, we can find a $\delta > 0$ such that

$$|g(y) - g(\eta)| < \varepsilon \text{ for } |y - \eta| < 2\delta.$$

[6]Since the domain of integration is unbounded, a little care should be exercised in interchanging the limit and integral signs. Here the exponential factor in the integrand, which decays to 0 faster than any polynomial helps.

Denoting by $M = \sup_{y\in\mathbb{R}} |g(y)|$, we have for $|x-\eta| < \delta$ (which implies that $|y-\eta| < 2\delta$ if $|x-y| < \delta$),

$$\begin{aligned}
|u(x,t) - g(\eta)| &= \left| \int_{\mathbb{R}^n} K(x-y,t)(g(y) - g(\eta))\, dy \right| \\
&\leq \int_{|x-y|<\delta} K(x-y,t)|g(y) - g(\eta)|\, dy \\
&\quad + \int_{|x-y|\geq\delta} K(x-y,t)|g(y) - g(\eta)|\, dy \\
&\leq \int_{|y-\eta|<2\delta} K(x-y,t)|g(y) - g(\eta)|\, dy + 2M \int_{|x-y|\geq\delta} K(x-y,t)\, dy \\
&\leq \varepsilon \int_{\mathbb{R}^n} K(x-y,t)|g(y) - g(\eta)|\, dy + 2M \int_{|x-y|\geq\delta} K(x-y,t)\, dy \\
&< \varepsilon + 2M\frac{\varepsilon}{2M} < 2\varepsilon, \text{ if } t \text{ is sufficiently small.}
\end{aligned}$$

This completes the proof. □

Instead of the boundedness of g, if we assume that the initial function has exponential growth: $|g(x)| \leq Me^{bx^2}$ for some $M > 0$ and $b > 0$, the same proof shows that u given by the Fourier–Poisson integral is still a solution of the heat equation, but for a short duration of time: $0 < t < (4ba^2)^{-1}$. This restriction is to make the integral convergent.

Ir-reversibility, Infinite Speed of Propagation, Smoothing Effects: Before proceeding with uniqueness and other questions, we make some observations of the heat equation.

1. Unlike equations in mechanics, including the wave equation, the heat equation is *not* preserved under *time reversal,* that is the heat equation is not preserved under the transformation $t \mapsto -t$, as can be easily observed. This also explains the insolvability of the heat equation for $t < 0$, when the initial condition is prescribed at $t = 0$. This situation is the *ir-reversibility* of the physical problem. For example, we will not be able to predict the temperature in a rod some time back, given its temperature now. Such processes make distinction between the *future* and the *past.* This is a typical phenomenon of the parabolic problems.
2. Another observation of the heat equation is the propagation with *infinite speed.* If we look at the solution formula (8.30), we observe that any change in the initial condition g at a point x_0 or a small neighborhood around it, *instantly* felt at

any x for any $t > 0$. This makes the heat equation somewhat physically unrealistic. There have been attempts to rectify this shortcoming of the model. One such model is the so-called equation of the *porous media*. However, this is a non-linear equation.

3. The heat equation sits between the steady-state equation (Laplace's equation) and physically realistic wave equation. As such, mathematically, it enjoys good properties from both the sides. For example, the solution of the heat equation enjoys the maximum and minimum principles and a sort of mean value property, which are enjoyed by the solution of the Laplace's equation; it enjoys certain energy estimates similar to the solution of the wave equation. Some of these will be discussed in the following sections.
4. Another interesting property of the heat kernel is its *smoothing effect*. The function $u(x,t), t > 0$ given by (8.30) is C^∞ even though the initial function g is merely continuous; it may even be not continuous. In fact, it can be shown that for $t > 0$, the function $x \mapsto u(x,t)$ is analytic when extended to complex space, that is x is replaced by the complex variable z.

8.2.2 Uniqueness

Though the Fourier–Poisson integral gives us a solution for the IVP for the heat equation in the space $\mathbb{R}^n$, it is not immediately clear that whether the solution is unique. In fact, there is non-uniqueness unless u satisfies certain growth condition. Below we will give an example, following Tychonov, exhibiting non-uniqueness. We wish to mention here that there is a result due to Widder (1975), which states that there is at most one solution u to the heat equation such that $u(x,t) \geq 0$ for all $t \geq 0$ and for all x.

Consider the one-dimensional heat equation (8.6) with zero initial condition, that is $g \equiv 0$. Let

$$\psi(z) = \begin{cases} \exp\left(-1/z^2\right), \text{ for } z \in \mathbb{C}, z \neq 0, \\ 0, \text{ for } z = 0, \end{cases}$$

and define u by

$$u(x,t) = \begin{cases} \displaystyle\sum_{k=0}^{\infty} \psi^{(k)}(t)\frac{x^{2k}}{(2k)!}, \text{ for } x \in \mathbb{R}, t > 0, \\ 0, \text{ for } x \in \mathbb{R}, t = 0. \end{cases} \tag{8.31}$$

Here $\psi^{(k)}(t) = \frac{d^k\psi}{dt^k}(t)$. Assuming that the following formal arguments are valid, we show that u is a solution of the heat equation (8.6) satisfying $u(x,0) = 0$ for all $x \in \mathbb{R}$. Since $u \neq 0$ for $t > 0$ and identically zero function is also a solution, this proves non-uniqueness of the solution.

We have

$$\lim_{t\to 0+} u(x,t) = \sum_{k=0}^{\infty} \psi^{(k)}(0)\frac{x^{2k}}{(2k)!} = 0, \tag{8.32}$$

$$\begin{aligned}\frac{\partial^2 u}{\partial x^2} &= \sum_{k=1}^{\infty} \psi^{(k)}(t)(2k)(2k-1)\frac{x^{2k-2}}{(2k)!} \\ &= \sum_{k=1}^{\infty} \psi^{(k)}(t)\frac{x^{2(k-1)}}{(2(k-1))!} \\ &= \sum_{k=0}^{\infty} \psi^{(k+1)}(t)\frac{x^{2k}}{(2k)!} = \frac{\partial u}{\partial t}.\end{aligned} \tag{8.33}$$

We now show that the series in (8.31), (8.32) and (8.33) are uniformly convergent in a neighborhood of every point $(x,t), x \in \mathbb{R}, t > 0$. This justifies the above arguments of interchange of limit and summation and term-by-term differentiation in the infinite series, performed above.

The function $\psi(z)$ is analytic in $\mathbb{C}\backslash\{0\}$. Identify the t-axis as the real axis of the complex plane. For fixed $t > 0$, the circle

$$\Gamma = \{z \in \mathbb{C} : z = t + \frac{t}{2}e^{i\theta}\},\ 0 < \theta \le 2\pi$$

does not meet the origin. Hence by Cauchy formula,

$$\psi^{(k)}(t) = \frac{k!}{2\pi i}\int_{\Gamma} \frac{\psi(z)}{(z-t)^{k+1}}\,dz,\ k = 0, 1, 2, \ldots.$$

From this it follows that

$$|\psi^{(k)}(t)| \le \frac{k!}{2\pi}\int_{\Gamma} \frac{\exp(-\Re(z^{-2}))}{|z-t|^{k+1}}\,|dz| = \frac{k!}{2\pi}\left(\frac{2}{t}\right)^k \int_0^{2\pi} \exp(-\Re(z^{-2}))\,d\theta,$$

where $\Re(z)$ denotes the real part of a complex number $z \in \mathbb{C}$. For $z \in \Gamma$, we have $|z-t| = t/2$ and therefore,

$$z^2 = t^2\left(1 + \frac{t}{2}e^{i\theta}\right)^2.$$

It follows that

$$\Re(z^{-2}) = \frac{1}{t^2}\frac{\frac{1}{4}+\frac{1}{2}(1+\cos\theta)^2}{\left(\frac{1}{4}+(1+\cos\theta)\right)^2}.$$

Taking the minimum of the expression on the right, we get $\Re(z^{-2}) \geq \frac{1}{4t^2}$ and therefore

$$|\psi^{(k)}(t)| \leq k!\left(\frac{2}{t}\right)^k \exp\left(-\frac{1}{4t^2}\right),\ k = 0, 1, 2, \ldots.$$

Fix $a > 0$. Then for $|x| < a$, the series in (8.31) is majorized by the series

$$\exp\left(-\frac{1}{4t^2}\right)\sum_{k=0}^{\infty}\left(\frac{1}{t}\right)^k \frac{a^{2k}}{k!},$$

and the latter is uniformly convergent. Here use has been made of the inequality

$$\frac{2^k k!}{(2k)!} \leq \frac{1}{k!}.$$

Similar arguments hold for series in (8.32) and (8.33). □

8.2.3 Inhomogeneous Equation

We now consider the *inhomogeneous heat equation*:

$$\begin{aligned} &u_t - a^2\Delta u = f(x,t),\ x \in \mathbb{R}^n, t > 0 \\ &u(x,0) = g(x),\ x \in \mathbb{R}^n. \end{aligned} \tag{8.34}$$

Assume that the function f and its partial derivatives f_{x_j} are continuous in $x \in \mathbb{R}^n, t > 0$ and the function g is continuous and bounded. Owing to the linearity in the problem, the required solution can be written as some of two functions: one the solution of the homogeneous equation with initial condition g and another the solution of the inhomogeneous equation with zero initial condition. Thus it suffices to consider equation (8.34) with $g \equiv 0$ therein. A solution of this problem can be obtained via the *Duhamel's principle*. The idea is to solve the homogeneous equation at each level $t = s > 0$ and sum (integrate) all these solutions as follows: Fix $s \geq 0$ and consider the IVP for the heat equation:

$$\begin{aligned} &u_t - a^2\Delta u = 0,\ x \in \mathbb{R}^n, t > s \\ &u(x,s) = f(x,s),\ x \in \mathbb{R}^n. \end{aligned} \tag{8.35}$$

Denoting the solution by $v(x, t; s)$, we get

$$v(x, t; s) = \int_{\mathbb{R}^n} K(x - y, t - s)f(y, s)\, dy,\ t > s, \tag{8.36}$$

using (8.30) and the change of variable $t \mapsto t - s$. We can now write down the expression for a solution to (8.35) as:

$$u(x, t) = \int_0^t v(x, t; s)\, ds = \int_0^t \int_{\mathbb{R}^n} K(x - y, t - s)f(y, s)\, dyds. \tag{8.37}$$

Uniqueness of this solution is proved under the additional assumption that the solution u has certain restricted growth. However, care should be exercised in asserting that u is indeed the solution of the inhomogeneous equation, since K has a singularity at $t = 0$. The verification is left as an exercise.

8.3 MAXIMUM AND MINIMUM PRINCIPLES

We now consider the heat equation in a smooth bounded domain Ω in $\mathbb{R}^n$:

$$u_t - a^2\Delta u = f(x, t),\ x \in \Omega,\ t > 0 \tag{8.38}$$

with initial and boundary conditions

$$\begin{aligned} u(x, 0) &= g(x),\ x \in \Omega \\ u(x, t) &= h(x, t),\ x \in \partial\Omega,\ t > 0. \end{aligned} \tag{8.39}$$

Here $\partial\Omega$ is the boundary of Ω, f, g and h are given source functions.

In the physical problem of the heat conduction in a rod, if the sources are not present, the temperature in the rod can neither exceed the maximum of the initial and boundary temperatures nor can go below those temperatures. This is the content of the maximum (minimum) principle.

Fix any $T > 0$ and consider the domain

$$\Omega_T = \Omega \times (0, T]$$

and

$$\partial_p\Omega_T = \overline{\Omega}_T \setminus \Omega_T = \{\overline{\Omega} \times \{0\}\} \cup \{\partial\Omega \times [0, T]\}.$$

The partial boundary $\partial_p\Omega_T$ is called the *parabolic boundary* (see Figure 8.3).

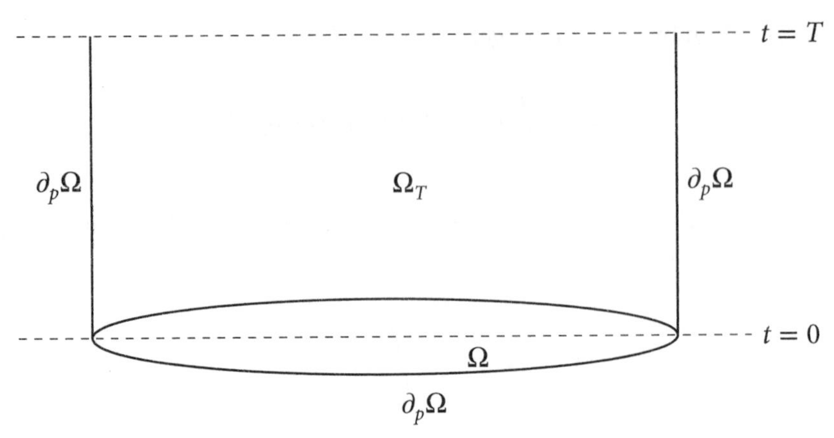

Figure 8.3 Parabolic boundary

Theorem 8.2 (Weak Maximum and Minimum Principles). Suppose the function u is continuous in $\overline{\Omega}_T$ such that the partial derivatives u_t and $u_{x_jx_k}$, $1 \leq j, k \leq n$ exist and are continuous in Ω_T. Then, the following statements hold: □

1. If $u_t - a^2\Delta u \leq 0$ in Ω_T, then $\max_{\overline{\Omega}_T} u = \max_{\partial_p\Omega_T} u$.
2. If $u_t - a^2\Delta u \geq 0$ in Ω_T, then $\min_{\overline{\Omega}_T} u = \min_{\partial_p\Omega_T} u$.
3. If $u_t - a^2\Delta u = 0$ in Ω_T, then $\max_{\overline{\Omega}_T} |u| = \max_{\partial_p\Omega_T} |u|$.

Proof It suffices to prove the first statement. The second statement follows from the first one, by applying it to $-u$ in place of u and the third statement follows by combining the first and second statements.

For $\varepsilon > 0$ small, consider the function

$$v(x, t) = u(x, t) - \varepsilon |x|^2.$$

Then,

$$v_t - a^2\Delta v = u_t - a^2\Delta u - 2\varepsilon \leq -2\varepsilon < 0. \tag{8.40}$$

If (x_0, t_0) is an interior point, $x_0 \in \Omega$, $0 < t_0 \leq T$, where v assumes a maximum, then, it follows that

$$\Delta v(x_0, t_0) \leq 0 \text{ and } v_t(x_0, t_0) \geq 0.$$

(If $t_0 < T$, then $v_t(x_0, t_0) = 0$.) But this contradicts (8.40) and shows that v can achieve its maximum only on $\partial_p \Omega_T$.

Since

$$u \leq v \leq u + C\varepsilon, \text{ in } \overline{\Omega}_T,$$

for some constant $C > 0$, which depends on Ω, the conclusion follows by letting $\varepsilon \to 0$. □

Remark 8.3. If u as in the above theorem satisfies condition (1) and attains its maximum at an interior point (x_0, t_0), then u is constant for all $t < t_0$, provided that we assume constant initial and boundary conditions. However, the solution may change for $t > t_0$ if we change the boundary condition for $t > t_0$, thus indicating that the solution responds to changes in the boundary condition only after the changes are made. This is in agreement with the view of taking the t variable as time variable. Similar conclusions can be made when u satisfies condition (2) or (3) in the above theorem. □

We have the following immediate consequences:

Corollary 8.4 (Comparison Theorem). Let u, v satisfy the heat equation (8.38). If $u \leq v$ on the parabolic boundary $\partial_p \Omega_T$, $T > 0$, then $u \leq v$ on $\overline{\Omega}_T$. □

Corollary 8.5 (Uniqueness Theorem). Let $f(x, t)$ be a continuous function on $\overline{\Omega} \times \{t \geq 0\}$, g be a continuous function on $\overline{\Omega}$ and h be a continuous function on $\partial\Omega \times \{t \geq 0\}$. Assume that the compatibility condition

$$g(x) = h(x, 0), \ x \in \partial\Omega \tag{8.41}$$

is satisfied. Then, the initial boundary value problem (8.38)–(8.39) has at most one solution. □

The compatibility condition is essential for the required continuity.

Proof Suppose u, v are two solutions of (8.38)–(8.39). Then, the difference $w = u - v$ is continuous for $x \in \overline{\Omega}$, $t \geq 0$ and satisfies (8.38)–(8.39) with f, g, h all identically zero. Theorem 8.2 then implies that $w = 0$ on $\overline{\Omega}_T$ for $T > 0$ arbitrary. This proves uniqueness. □

Backward Uniqueness for Heat Equation: At this point, we also discuss a curious property of the heat equation, namely the *backward in time uniqueness result*. This is somewhat surprising as the heat equation obeys irreversibility with respect to time, as discussed earlier. This is also an incident where we see that the heat equation enjoying an *energy* *estimate*, a property that is in the domain of wave equation (more generally, hyperbolic equation).

Theorem 8.6 (Backward Uniqueness for Heat Equation). Suppose $u, \tilde{u} \in C^2(\Omega_T)$ solve the following:

$$\begin{aligned} u_t &= \Delta u \text{ in } \Omega_T \\ u &= g \text{ on } \partial\Omega \times [0, T] \end{aligned} \tag{8.42}$$

and

$$\begin{aligned} \tilde{u}_t &= \Delta \tilde{u} \text{ in } \Omega_T \\ \tilde{u} &= g \text{ on } \partial\Omega \times [0, T] \end{aligned} \tag{8.43}$$

If $u(x, T) = \tilde{u}(x, T)$ for all $x \in \Omega$, then $u \equiv \tilde{u}$ in Ω_T. □

Here $T > 0$ is fixed. Note that no assumption has been made regarding the initial conditions of u and $\tilde{u}$.

Proof Let $w = u - \tilde{u}$. Then, w solves

$$\begin{aligned} w_t &= \Delta w \text{ in } \Omega_T \\ w &= 0 \text{ on } \partial\Omega \times [0, T] \end{aligned} \tag{8.44}$$

Consider the total energy of w at time t defined by

$$E(t) = \int_\Omega w^2(x, t)\, dx, \; t \in [0, T].$$

The conclusion of the theorem gets established once we prove that $E(t) = 0$ for all $t \in [0, T)$; by hypothesis, $E(T) = 0$.

Using (8.44) and integrating by parts once, we get

$$\frac{dE}{dt} = 2\int_\Omega w w_t\, dx = 2\int_\Omega w \Delta w\, dx = -2\int_\Omega |\nabla w|^2\, dx. \tag{8.45}$$

There is no boundary term here as $w = 0$ on $\partial\Omega$. Similar computation gives

$$\frac{d^2E}{dt^2} = -4\int_\Omega \nabla w \cdot \nabla w_t\, dx = 4\int_\Omega w_t \Delta w\, dx = 4\int_\Omega (\Delta w)^2\, dx. \tag{8.46}$$

Using (8.45) and Cauchy–Schwarz–Bunyakowski inequality, we get

$$\begin{aligned}\left(\frac{dE}{dt}\right)^2 &= 4\left(\int_\Omega |\nabla w|^2\,dx\right)^2\\ &= \left(-2\int_\Omega |\nabla w|^2\,dx\right)^2\\ &= \left(2\int_\Omega w\Delta w\,dx\right)^2\\ &\le \left(\int_\Omega w^2\,dx\right)\left(4\int_\Omega (\Delta w)^2\,dx\right)\\ &= E(t)\frac{d^2E}{dt^2}. \end{aligned} \tag{8.47}$$

Suppose, on the contrary, $E(t) \neq 0$ for some $t \in [0, T]$. Since $E(T) = 0$, by hypothesis, we see by continuity that there is a sub-interval $[t_1, t_2] \subset [0, T]$ such that $E(t) > 0$ for all $t \in [t_1, t_2)$ and $E(t_2) = 0$. Put $F(t) = \log E(t)$ for $t \in [t_1, t_2)$. Then,

$$\begin{aligned}\frac{d^2F}{dt^2} &= \frac{E\dfrac{d^2E}{dt^2} - \left(\dfrac{dE}{dt}\right)^2}{E^2}\\ &\ge 0, \text{ using (8.47)}.\end{aligned}$$

This shows that the function F is a convex function on the interval $[t_1, t_2)$. Therefore, for each $\alpha \in (0, 1)$ and a fixed $t \in (t_1, t_2)$, we have

$$F(\alpha t_1 + (1-\alpha)t) \le \alpha F(t_1) + (1-\alpha)F(t).$$

Or, in terms of E, we get

$$E(\alpha t_1 + (1-\alpha)t) \le E(t_1)^\alpha\, E(t)^{(1-\alpha)}.$$

Letting $t \to t_2$ in the last inequality and using $E(t_2) = 0$, we see that $E(t) = 0$ for all $t \in [t_1, t_2)$, a contradiction. This completes the proof. □

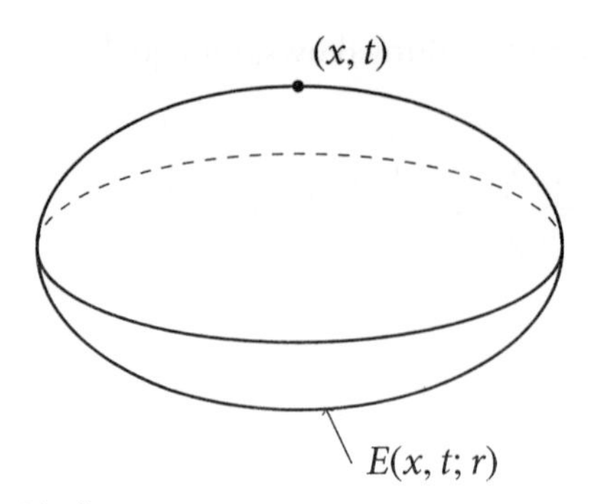

Figure 8.4 Heat ball

Heat Ball and Mean Value Property: Apart from sharing the maximum and minimum principles with the Laplace's equation, the heat equation also shares a mean value property, which we discuss now.

Fix $x \in \mathbb{R}^n$, $t > 0$ and $r > 0$. Define the space-time region $E(x, t; r)$ by

$$E(x, t; r) = \left\{ (y, s) \in \mathbb{R}^{n+1} \; : \; s \leq t, \; K(x - y, t - s) \geq \frac{1}{r^n} \right\}, \tag{8.48}$$

where K is the fundamental solution of the heat equation. The boundary of $E(x, t; r)$ is a level set of K. It is to be noted that the point (x, t) is at the centre of the top of $E(x, t; r)$. See Figure 8.4. The region $E(x, t; r)$ is sometimes referred to as the *heat ball.*

Theorem 8.7 (Mean Value Property). Let u be a smooth solution of the heat equation (with $a = 1$) in Ω_T. Then,

$$u(x, t) = \frac{1}{4r^n} \iint\limits_{E(x,t;r)} u(y, s) \frac{|y|^2}{s^2} \, dyds \tag{8.49}$$

for every $E(x, t; r) \subset \Omega_T$. □

The right side term in the above equation involves only the values of s that are below t. This is reasonable since $u(x, t)$ should not depend on future times. The condition that $K(x - y, t - s) \geq \frac{1}{r^n}$ determines the domain of integration with respect to the y-variable and further restricts the s-interval of integration. Since $\iint_{E(x,t;r)} \frac{|y|^2}{s^2} \, dyds = 4r^n$ (see (8.51)), the right-hand side term in (8.49) indeed describes the mean value of $u(y, s)$ over $E(x, t; r)$.

Proof By translating the variables, we may assume that $x = 0$ and $t = 0$. We write $E(r) = E(0,0;r)$, $r > 0$. Set

$$\chi(r) = \frac{1}{r^n} \iint\limits_{E(r)} u(y,s)\frac{|y|^2}{s^2}\, dyds$$
$$= \iint\limits_{E(1)} u(ry, r^2 s)\frac{|y|^2}{s^2}\, dyds.$$

The idea is to show that χ is a constant function of r. For this, consider

$$\frac{d\chi}{dr} = \iint\limits_{E(1)} \left(\sum_{i=1}^{n} \frac{\partial u}{\partial y_i}(ry, r^2 s) y_i + 2rs\frac{\partial u}{\partial s}(ry, r^2 s) \right) \frac{|y|^2}{s^2}\, dyds$$
$$= \frac{1}{r^{n+1}} \iint\limits_{E(r)} \left(\sum_{i=1}^{n} \frac{\partial u}{\partial y_i}(y,s) y_i \frac{|y|^2}{s^2} + 2\frac{\partial u}{\partial s}(y,s)\frac{|y|^2}{s} \right) dyds$$
$$= A + B, \text{ say.}$$

Consider the function ψ defined by

$$\psi(y,s) = -\frac{n}{2}\log(-4\pi s) + \frac{|y|^2}{4s} + n\log r, \tag{8.50}$$

which arises as a result of determining the domain of integration of the function χ, namely $s \leq 0$ and $K(y,-s) \geq \frac{1}{r^n}$. Note that $\psi = 0$ on the boundary $\partial E(r)$, on which $K(y,-s) = \frac{1}{r^n}$. We use ψ in writing B as

$$B = 2\frac{1}{r^{n+1}} \iint\limits_{E(r)} \frac{\partial u}{\partial s}\frac{|y|^2}{s}\, dyds$$
$$= 4\frac{1}{r^{n+1}} \iint\limits_{E(r)} \frac{\partial u}{\partial s} \sum_{i=1}^{n} \frac{\partial \psi}{\partial y_i} y_i\, dyds$$
$$= -4\frac{1}{r^{n+1}} \iint\limits_{E(r)} \left(n\frac{\partial u}{\partial s}\psi + \sum_{i=1}^{n} \frac{\partial^2 u}{\partial s \partial y_i} y_i \psi \right) dyds.$$

The last line follows by an integration by parts (with respect to the y- variable) and there are no boundary terms as $\psi = 0$ there. Now another integration by parts, with respect to the

s-variable this time, gives us

$$B = 4\frac{1}{r^{n+1}} \iint\limits_{E(r)} \left(-n\frac{\partial u}{\partial s}\psi + \sum_{i=1}^{n} \frac{\partial u}{\partial y_i} y_i \frac{\partial \psi}{\partial s} \right) dyds$$

$$= 4\frac{1}{r^{n+1}} \iint\limits_{E(r)} \left(-n\frac{\partial u}{\partial s}\psi + \sum_{i=1}^{n} \frac{\partial u}{\partial y_i} y_i \left[-\frac{n}{2s} - \frac{|y|^2}{4s^2} \right] \right) dyds$$

$$= -\frac{1}{r^{n+1}} \iint\limits_{E(r)} \left(4n\frac{\partial u}{\partial s}\psi + \frac{2n}{s} \sum_{i=1}^{n} \frac{\partial u}{\partial y_i} y_i \right) dyds - A.$$

Using the hypothesis that u solves the heat equation, we therefore obtain

$$\frac{d\chi}{dr} = A + B$$

$$= -\frac{1}{r^{n+1}} \iint\limits_{E(r)} \left(4n\psi\Delta u + \frac{2n}{s} \sum_{i=1}^{n} \frac{\partial u}{\partial y_i} y_i \right) dyds$$

$$= 2n\frac{1}{r^{n+1}} \sum_{i=1}^{n} \iint\limits_{E(r)} \left(2\frac{\partial u}{\partial y_i}\frac{\partial \psi}{\partial y_i} - \frac{1}{s}\frac{\partial u}{\partial y_i} y_i \right) dyds$$

$$= 0, \text{ as } \frac{\partial \psi}{\partial y_i} = \frac{y_i}{2s} \text{ (see (8.50))}.$$

Therefore, we conclude that χ is a constant function of r. Thus,

$$\chi(r) = \lim_{\rho \to 0} \chi(\rho) = u(0,0) \lim_{\rho \to 0} \frac{1}{\rho^n} \iint\limits_{E(\rho)} \frac{|y|^2}{s^2} dyds.$$

Now

$$\frac{1}{\rho^n} \iint\limits_{E(\rho)} \frac{|y|^2}{s^2} dyds = \iint\limits_{E(1)} \frac{|y|^2}{s^2} dyds.$$

In the integral on the right-hand side, we first determine the domain of integration of the y-variable using $K(y, -s) \geq 1$, and determine the s-interval of integration. It can then be shown that

$$\iint\limits_{E(1)} \frac{|y|^2}{s^2} dyds = 4. \tag{8.51}$$

This is left as an exercise. The proof is complete. □

Uniqueness Result: Using the weak maximum principle, we now prove a uniqueness result for the solution of the heat equation in $\mathbb{R}^n$, provided we assume a restricted growth on the solution.

Theorem 8.8. Let $T > 0$ be fixed and u be a continuous function on $\mathbb{R}^n \times [0, T]$ such that the functions u_t and Δu exist and are continuous in $\mathbb{R}^n \times (0, T)$ and satisfy the following:

$$\begin{aligned} &u_t - a^2 \Delta u \le 0,\ x \in \mathbb{R}^n,\ t \in (0, T) \\ &u(x, 0) = g(x),\ x \in \mathbb{R}^n, \\ &u(x, t) \le M e^{b|x|^2},\ x \in \mathbb{R}^n,\ t \in (0, T), \end{aligned} \tag{8.52}$$

for some positive constants b and M. Then,

$$u(x, t) \le \sup_z g(z),\ x \in \mathbb{R}^n,\ t \in (0, T). \tag{8.53}$$

□

The theorem in particular applies to the solution of the heat equation. For bounded g, therefore, the function given by the Fourier–Poisson integral is the only solution of the heat equation. We state this as a corollary.

Corollary 8.9 (Uniqueness). For bounded g, the function u defined by the Fourier–Poisson integral (8.30) is the only solution of (8.26). □

However, the assumption on the growth in (8.52) is essential in view of the Tychonov example discussed earlier.

Proof (of Theorem 8.8) Suffices to prove the theorem for T satisfying $4ba^2T < 1$. For the general case, we can divide the interval $[0, T]$ into subintervals of equal length $< \frac{1}{4ba^2}$ and repeat the arguments. Choose an $\varepsilon > 0$ such that $4ba^2(T + \varepsilon) < 1$.

For fixed $y \in \mathbb{R}^n$, consider the function

$$\begin{aligned} v_\mu(x, t) &= u(x, t) - \mu \frac{1}{(4\pi a^2(T + \varepsilon - t))^{n/2}} \exp\left(\frac{(x - y)^2}{4a^2(T + \varepsilon - t)}\right) \\ &= u(x, t) - \mu K(i(x - y), T + \varepsilon - t). \end{aligned} \tag{8.54}$$

Here $\mu > 0$, $i = \sqrt{-1}$ and $t \in [0, T]$. As noted earlier, the second function on the right satisfies the heat equation. Thus, for any $t > 0$, we have

$$\left(\frac{\partial}{\partial t} - \Delta\right) v_\mu = \left(\frac{\partial}{\partial t} - \Delta\right) u \le 0.$$

Applying the weak maximum principle with the domain $\Omega_T = \{|x-y| < \rho\} \times (0, T)$, with $\rho > 0$ to be chosen, we obtain that

$$v_\mu(y,t) \le \max_{\partial_p \Omega_T} v_\mu,$$

where the parabolic boundary is now given by

$$\partial_p \Omega_T = \{(x,0) : |x-y| \le \rho\} \cup \{(x,t) : |x-y| = \rho, t \in [0,T]\}.$$

On the bottom part of this boundary, namely $t = 0$, we have

$$v_\mu(x,0) \le u(x,0) \le \sup_z g(z),$$

as $K > 0$. On the other part of the parabolic boundary, we have $|x-y| = \rho$, $t \in [0,T]$ and therefore,

$$\begin{aligned} v_\mu(x,t) &\le Me^{b|x|^2} - \mu\left(4\pi a^2(T+\varepsilon-t)\right)^{-n/2} \exp\left(\frac{\rho^2}{4a^2(T+\varepsilon-t)}\right) \\ &\le Me^{b(|y|+\rho)^2} - \mu\left(4\pi a^2(T+\varepsilon)\right)^{-n/2} \exp\left(\frac{\rho^2}{4a^2(T+\varepsilon)}\right). \end{aligned}$$

Here we have used the hypothesis on the growth condition of u. We now show that the last term becomes negative and tends to $-\infty$ as $\rho \to \infty$, using the assumption that $4ba^2(T+\varepsilon) < 1$. To see this, take out the term $e^{b\rho^2}$ from the first term, as a common factor. This makes the exponential in the second term as

$$\exp\left(\frac{\rho^2(1-4ba^2(T+\varepsilon))}{4a^2(T+\varepsilon)}\right).$$

Since $1-4ba^2(T+\varepsilon) > 0$ by assumption, this exponential can be made as large as we please by choosing ρ large. Therefore, the second term dominates and since $\mu > 0$, the entire term becomes negative for sufficiently large ρ. In particular, this term is bounded by $\sup_z g(z)$, by letting $\rho \to \infty$. Thus,

$$\max_{\partial_p \Omega} v_\mu \le \sup_z g(z).$$

Hence,

$$v_\mu(y,t) = u(y,t) - \mu\frac{1}{2a\sqrt{\pi(T+\varepsilon-t)}} \le \sup_z g(z).$$

Finally letting $\mu \to 0$, we arrive at the inequality $u(y,t) \le \sup_z g(z)$. Since y is arbitrary, (8.53) follows. This completes the proof. □

8.4 HEAT EQUATION ON A FINITE INTERVAL: FOURIER METHOD

In this section, we see how the Fourier method can be used to obtain the solution of the IVP for the heat equation

$$\begin{aligned} &u_t - a^2 u_{xx} = 0,\ 0 < x < L, t > 0 \\ &u(x,0) = g(x),\ 0 < x < L \end{aligned} \tag{8.55}$$

under different boundary conditions. First consider the following simplest boundary conditions:

$$u(0,t) = 0,\ u(L,t) = 0,\ t > 0. \tag{8.56}$$

We seek a solution u of (8.55) in the variable separation form, similar to the procedure adapted in Section 8.2:

$$u(x,t) = X(x)T(t). \tag{8.57}$$

Using (8.55), we then get

$$\frac{T'}{a^2 T} = \frac{X''}{X} = -\lambda^2,$$

with $\lambda \geq 0$. Thus, X satisfies the equation

$$X'' + \lambda^2 X = 0$$

whose general solution is given by

$$X(x) = c_1 \cos \lambda x + c_2 \sin \lambda x.$$

But X needs to satisfy the boundary conditions $X(0) = X(L) = 0$. Thus, in order to obtain non-trivial X, that is X is not an identically zero function, λ must be equal to $\frac{n\pi}{L}$, for $n = 1, 2, \ldots$. Thus, we have

$$X_n(x) = \sin\left(\frac{n\pi}{L}x\right),$$

for $n = 1, 2, \ldots$ and consequently

$$T_n(t) = C_n \exp\left(-\frac{n^2\pi^2 a^2}{L^2}t\right),$$

for some arbitrary constant C_n. Owing to the linearity of equation (8.55), we thus formally obtain the solution as (see (8.57))

$$u(x,t) = \sum_{n=1}^{\infty} C_n \exp\left(-\frac{n^2\pi^2 a^2}{L^2}t\right) \sin\left(\frac{n\pi}{L}x\right). \tag{8.58}$$

We will now determine the constants C_n using the initial condition in (8.55). We assume that the function g is continuous in $[0, L]$, has piece-wise continuous derivative in $(0, L)$ and $g(0) = g(L) = 0$. Then, from the theory of Fourier series (*Dirichlet theorem*), it is known that

$$g(x) = \sum_{n=1}^{\infty} g_n \sin\left(\frac{n\pi}{L}x\right),$$

where the series converges absolutely and uniformly. The Fourier coefficients g_n of g are given by

$$g_n = \frac{2}{L}\int_0^L g(x) \sin\left(\frac{n\pi}{L}x\right) dx, \tag{8.59}$$

for[7] $n = 1, 2, \ldots$. Therefore, we have

$$u(x, 0) = g(x) = \sum_{n=1}^{\infty} C_n \sin\left(\frac{n\pi}{L}x\right)$$

implying that $C_n = g_n$. Therefore, the function u is given by

$$u(x, t) = \sum_{n=1}^{\infty} g_n \exp\left(-\frac{n^2\pi^2a^2}{L^2}t\right) \sin\left(\frac{n\pi}{L}x\right), \tag{8.60}$$

where g_n are defined as in (8.59). We will now verify that u given by (8.60) indeed is the required solution.

Since $0 < \exp\left(-\frac{n^2\pi^2a^2}{L^2}t\right) \leq 1$ for all $t \geq 0$, it readily follows that the series in (8.60) converges absolutely and uniformly. Also, we have for $t > 0$,

$$n^k \exp\left(-\frac{n^2\pi^2a^2}{L^2}t\right)$$

is bounded by a constant for $k = 0, 1, 2, \ldots$ if we choose n large enough. This shows that the series obtained from (8.60) by differentiating term-by-term both with respect to x and t is also absolutely and uniformly convergent. This, in turn, proves that u is in fact a C^∞ function for $0 < x < L$ and $t > 0$. It is now straightforward to verify that u satisfies the IVP (8.55) and the boundary conditions (8.56).

[7] If g is assumed to be a C^2 function, then it immediately follows that $n^2 g_n$ are bounded.

8.4.1 Prescribed Non-zero Boundary Conditions

We now change the boundary conditions and impose the following

$$u(0,t) = u_1,\ u(L,t) = u_2,\ t > 0, \tag{8.61}$$

where u_1 and u_2 are given constants. Physically, these conditions represent that the ends of the rod are maintained at given temperatures.

Let $u(x,t) = w(x) + v(x,t)$, where

$$w(x) = u_1\left(1 - \frac{x}{L}\right) + u_2\frac{x}{L}.$$

Then, v satisfies the IVP (8.55) with the homogeneous boundary conditions (8.56). This implies that the solution u we are now seeking satisfies the inhomogeneous boundary conditions (8.61). From (8.60), we get that

$$v(x,t) = \sum_{n=1}^{\infty} g_n \exp\left(-\frac{n^2\pi^2a^2}{L^2}t\right)\sin\left(\frac{n\pi}{L}x\right).$$

Thus, we obtain the following representation for the solution u:

$$u(x,t) = u_1\left(1 - \frac{x}{L}\right) + u_2\frac{x}{L} + \sum_{n=1}^{\infty} g_n \exp\left(-\frac{n^2\pi^2a^2}{L^2}t\right)\sin\left(\frac{n\pi}{L}x\right). \tag{8.62}$$

It is possible to write down the functions x/L and $1 - x/L$, $0 \le x \le L$ in the form of a sine series and cosine series, by suitably extending these functions to $\mathbb{R}$, periodically, as odd and even functions, respectively. However, care should be exercised to see that the resulting sine or cosine series converges absolutely and uniformly on $[0, L]$. We leave the details to the reader.

Remark 8.10. When u_1 and u_2 are not constants, but functions of t, we can still look for a solution of the form $u(x,t) = w(x,t) + v(x,t)$, where w satisfies the inhomogeneous boundary conditions. However, the function v now satisfies an inhomogeneous heat equation, with homogeneous boundary conditions. Under suitable conditions on the functions $u_1(t)$ and $u_2(t)$, it is possible to obtain a representation of the solution u similar to (8.62). □

8.4.2 Free Exchange of Heat at the Ends

We now discuss the situation of mixed boundary conditions and replace the conditions (8.61) by

$$u_x(0,t) - hu(0,t) = 0,\ u_x(L,t) + hu(L,t) = 0,\ t > 0, \tag{8.63}$$

for some $h \geq 0$. Physically these boundary conditions represent free exchange of heat in the rod-taking place with the surrounding medium. When $h = 0$, we get the Neumann boundary conditions.

Again, seeking a solution in the variable separable form $u(x, t) = X(x)T(t)$, we now find that the function X satisfies the ODE

$$X'' + \lambda^2 X = 0$$

and the boundary conditions

$$X'(0) - hX(0) = 0, \; X'(L) + hX(L) = 0.$$

The solution is then given by

$$X(x) = c_1 \cos \lambda x + c_2 \sin \lambda x,$$

with c_1 and c_2 satisfying the algebraic equations

$$\begin{cases} -hc_1 + \lambda c_2 = 0, \\ (-\lambda \sin \lambda L + h \cos \lambda L)c_1 + (\lambda \cos \lambda L + h \sin \lambda L)c_2 = 0. \end{cases} \tag{8.64}$$

For X to be non-trivial, that is c_1 and c_2 are *not* both zero, the coefficient matrix in (8.64) must be singular. Thus, λ satisfies the equation

$$2h \cos \lambda L = (\lambda - h^2) \sin \lambda L.$$

For $h = 0$, we then have $\lambda = \frac{n\pi}{L}$ for $n = 0, 1, 2, \ldots$. For $h > 0$, put $\mu = \lambda L$ and $b = hL$. Then, the above equation reduces to

$$2 \cot \mu = \frac{\mu}{b} - \frac{b}{\mu}.$$

Looking at the graphs of the functions $\mu \mapsto 2 \cot \mu$ and $\mu \mapsto \frac{\mu}{b} - \frac{b}{\mu}$ (see Figure 8.5), we see that this trigonometric equation has infinitely many roots.

Denote the roots by μ_n, $n = 1, 2, \ldots$, and thus $\lambda_n = \frac{\mu_n}{L}$, and the corresponding solutions by $X_n(x)$. It is not difficult to see that these solutions are *mutually orthogonal*, that is

$$\int_0^L X_n(x)X_m(x)\, dx = 0, \text{ if } n \neq m.$$

This is owing to the fact that the boundary value problem satisfied by X_n's is a regular Sturm–Liouville problem (see for instance Nandakumaran *et al.*, 2017). We also have that for the

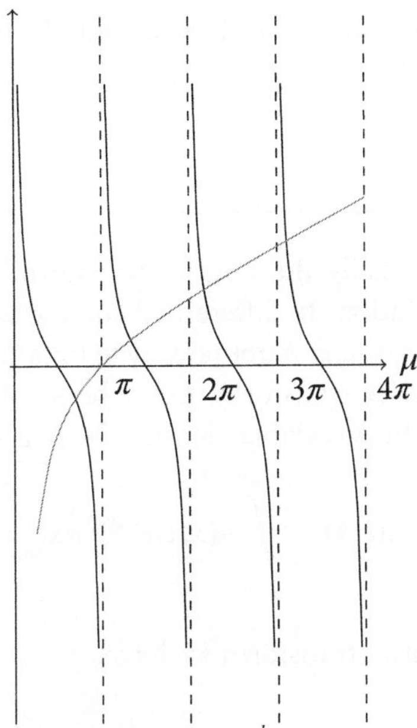

Figure 8.5 Graphs of $2\cot\mu$ and $\frac{\mu}{b} - \frac{b}{\mu}$, $b > 0$

initial function g with stated assumptions,

$$g(x) = \sum_{n=1}^{\infty} g_n X_n(x),$$

where the series converges absolutely and uniformly. The coefficients g_n, which are called the *Fourier coefficients* with respect to the orthogonal family $\{X_n\}$, suitably normalized, are given by

$$g_n = \int_0^L g(x)X_n(x)\,dx.$$

The corresponding T_n's are now given by

$$T_n(t) = c_n \exp\left(-\frac{a^2\lambda_n^2}{L^2}t\right),$$

with $c_n = T_n(0) = g_n$, the Fourier coefficients of the initial function g. Finally, the solution u is given by

$$u(x,t) = \sum_{n=1}^{\infty} T_n(t)X_n(x) = \sum_{n=1}^{\infty} g_n \exp\left(-\frac{a^2\lambda_n^2}{L^2}t\right)X_n(x). \tag{8.65}$$

We leave it to reader to verify that u is indeed a solution to the initial-boundary value problem under consideration.

8.5 NOTES

The transform techniques, especially the Fourier transform and Laplace transform, play important roles in obtaining solutions to differential equations, by converting the ODE into algebraic equations and PDE into ODE. Another way to obtain the Fourier–Laplace formula is making use of (partial) Fourier transform. Let u be a solution of (8.26). The Fourier transform of u with respect to the x variable, denoted by $\tilde{u}$, is defined as

$$\tilde{u}(\xi, t) = \int_{\mathbb{R}^n} u(x, t)e^{-ix\cdot\xi}\, dx.$$

Using the properties of the Fourier transform we have,

$$\begin{aligned}\tilde{u}_t(\xi, t) &= \int_{\mathbb{R}^n} u_t(x, t)e^{-ix\cdot\xi}\, dx \\ &= a^2 \int_{\mathbb{R}^n} \Delta u(x, t)e^{-ix\cdot\xi}\, dx, \text{ since } u \text{ satisfies the heat equation} \\ &= -\, a^2|\xi|^2\tilde{u}(\xi, t).\end{aligned}$$

Thus, $\tilde{u}$ satisfies the ODE

$$\frac{d\tilde{u}}{dt}(\xi, t) + a^2|\xi|^2\tilde{u}(\xi, t) = 0,$$

with ξ playing the role of a parameter. Further $\tilde{u}(\xi, 0) = \hat{g}(\xi)$, where $\hat{g}$ is the Fourier transform of g. Therefore,

$$\tilde{u}(\xi, t) = \exp(-a^2|\xi|^2 t)\hat{g}(\xi). \tag{8.66}$$

In order to recover u from $\tilde{u}$, we need to use the inverse Fourier transform:

$$u(x, t) = (2\pi)^{-n} \int_{\mathbb{R}^n} \tilde{u}(\xi, t)e^{ix\cdot\xi}\, d\xi. \tag{8.67}$$

It is well known that the Fourier transform of the function $e^{-|x|^2/2}$ is the function $\pi^{n/2}e^{-|\xi|^2/2}$. Therefore, using appropriate scaling, we see that the Fourier transform of the function

$$f_\lambda(x) = e^{-\lambda|x|^2/2},\ \lambda > 0,$$

is given by

$$\hat{f}_\lambda(\xi) = \lambda^{-n}\pi^{n/2}e^{-\frac{|\xi|^2}{2\lambda}}, \ \lambda > 0.$$

Therefore, if we take

$$K(x,t) = \lambda^n \pi^{-n/2} e^{-\frac{|x|^2}{4a^2t}}, \ \lambda = \frac{1}{2a^2t}$$

then,

$$\tilde{K}(\xi,t) = \lambda^n \pi^{-n/2} e^{-a^2|\xi|^2 t}, \ \lambda = \frac{1}{2a^2t}.$$

Thus, using (8.66) and (8.67), we get

$$u(x,t) = (2\pi)^{-n}\int_{\mathbb{R}^n} \tilde{K}(\xi,t)\hat{g}(\xi)e^{ix\cdot\xi}\, d\xi. \tag{8.68}$$

It is another important property of the Fourier transform that the Fourier transform of the convolution of two functions is the product of their Fourier transforms:

$$\widehat{f*g} = \hat{f}\,\hat{g}.$$

This gives that the inverse Fourier transform of the product of the Fourier transform of two functions is their convolution. Therefore, using (8.68), we get

$$u(x,t) = (K*g)(x,t),$$

which is precisely the Fourier–Laplace formula derived earlier.

In fact, this (inverse Fourier transform) is the procedure used for proving that every constant coefficient *partial differential operator* (PDO) possesses a *fundamental solution.* This is the celebrated *Malgrange–Ehrenpreis Theorem.*

Let $\Omega \subset \mathbb{R}^n$ be an open domain and $P(D) = \sum_{|\alpha|\le m} a_\alpha D^\alpha$ be a constant coefficient PDO of order m; the coefficients a_α are assumed to be real. The *(formal) adjoint* of $P(D)$ is the PDO given by $P'(D) = \sum_{|\alpha|\le m}(-1)^{|\alpha|}a_\alpha D^\alpha$. Let

$$p(\xi) = \sum_{|\alpha|\le m} a_\alpha \xi^\alpha \text{ and } p_m(\xi) = \sum_{|\alpha|=m} a_\alpha \xi^\alpha,$$

$\xi \in \mathbb{R}^n$ be, respectively the *full symbol* and *principal symbol* of $P(D)$; see Chapter 6.

A fundamental solution of $P(D)$ in Ω is a *(tempered) distribution* E satisfying $P(D)E = \delta$, where δ is the *Dirac distribution*. More precisely,

$$\langle E, P'(D)g\rangle = g(0),$$

for all C^∞ functions g with compact support in Ω; here $\langle E, g\rangle$ is the action of E on a test function g. If E is a fundamental solution of $P(D)$, it follows that $p(\xi)\hat{E}(\xi) = 1$, where $\hat{E}$ is the Fourier transform of E; this is quite technical as the definition of Fourier transform needs to be extended to tempered distributions. The major difficulty in the proof of Malgrange–Ehrenpreis theorem is the Fourier inversion of the inverse of a polynomial, which may have complex roots (see Nirenberg, 1976; Vladimirov, 1979, 1984).

The main advantage of knowing a fundamental solution is the following: If f is a suitable function or distribution, then a solution of the inhomogeneous equation $P(D)u = f$ is given by the convolution: $u = E * f$, where E is a fundamental solution. It is possible to include initial or boundary conditions in the inhomogeneous term. This is the reason why the solutions of Laplace, heat and wave equations are all given in terms of convolution. The fundamental solutions for Laplace and heat equations have been discussed; for the wave equation it is more complicated.

For a long time, there was no success story for the case of linear operators with variable coefficients. Then there was a surprising shock when Hans Lewy (1957) produced an example of a PDE with variable coefficients with *no solution*. This example eventually led Hörmander to systematically classify operators that have no solutions.

Example 8.11 (H. Lewy). The PDE, considered in $\mathbb{R}^3$,

$$u_{x_1x_3} + iu_{x_2x_3} + 2i(x_1 + ix_2)u_{x_3x_3} = f(x_3) \tag{8.69}$$

does not have C^2 solutions in any neighborhood of the origin in $\mathbb{R}^3$, unless the real-valued function $f(x_3)$ is analytic. □

Note that the above equation has variable complex coefficients. To prove the statement made in the example, it suffices to prove that the following equation (obtained from the above through an integration with respect to x_3 variable) $u_{x_1} + iu_{x_2} + 2i(x_1 + ix_2)u_{x_3} = f(x_3)$ does not possess any C^1 solution in any neighborhood of the origin. Suppose, on the contrary, that in the cylinder

$$\Omega = \{(x_1, x_2, x_3) : x_1^2 + x_2^2 < R^2, \ |x_3| < H\},$$

for some positive R and H, there is a solution $u \in C^1(\bar{\Omega})$ with a real-valued function $f(x_3)$ being non-analytic in the interval $(-H, H)$. Then, using the polar co-ordinates, we see that the function

$$u_1(\rho, \theta, x_3) = u(\rho\cos\theta, \rho\sin\theta, x_3)$$

satisfies the equation

$$u_{1\rho}e^{i\theta} + i\frac{u_{1\theta}}{\rho}e^{i\theta} + 2i\rho e^{i\theta}u_{1x_3} = f(x_3),$$

in $D = \{(\rho, \theta, x_3) : \rho \in (0, R), \theta \in (0, 2\pi), |x_3| < H\}$. Further, $u_1 \in C^1(\overline{D})$ satisfies $u_1(\rho, 0, x_3) = u_1(\rho, 2\pi, x_3)$. Keeping ρ and x_3 fixed, an integration with respect to θ shows that the function

$$u_2(\rho, x_3) = \int_0^{2\pi} u_1(\rho, \theta, x_3)e^{i\theta}\, d\theta$$

lies in $C^1(\overline{D}_1)$ and satisfies the equation

$$u_{2\rho} + \frac{u_2}{\rho} + 2i\rho u_{2x_3} = 2\pi f(x_3),$$

where $D_1 = \{(\rho, x_3) : \rho \in (0, R), |x_3| < H\}$. Hence, the function

$$v(r, x_3) = \sqrt{r}u_2(\sqrt{r}, x_3)$$

belonging to $C^1(D_2) \cap C(\overline{D}_2)$ is a solution of the equation

$$v_r + iv_{x_3} = \pi f(x_3),$$

where $D_2 = \{(r, x_3) : r \in (0, R^2), |x_3| < H\}$. Thus, if we put

$$w(r, x_3) = v(r, x_3) + i\pi \int_0^{x_3} f(\xi)\, d\xi$$

we see that w satisfies the Cauchy-Riemann equation $w_r + iw_{x_3} = 0$ in the complex variable $r + ix_3$. Therefore, $w(r, x_3) = g(r + ix_3)$, where g is an analytic function in D_2 and continuous in $\overline{D}_2$. Since $\Re g = 0$ for $r = 0$, by the principle of symmetry, the function g can be continued analytically into the rectangle $D_3 = \{(r, x_3) : |r| < R^2), |x_3| < H\}$. In particular, g is analytic on the line segment $\{r = 0, |x_3| < H\}$. But, $g = i\pi \int_0^{x_3} f(\xi)\, d\xi$ for $r = 0$. Consequently, the function $f(x_3)$ is also analytic in $|x_3| < H$, contradicting the assumption. □

Weak Solution: A continuous function $u \in C(\Omega)$ is called a *weak solution*[8] of $P(D)u = f$, where f is a given function, if

$$\int_\Omega u(x)P'(D)g(x)\, dx = \int_\Omega f(x)g(x)\, dx$$

for all *test functions* g, that is functions which are in $C^\infty(\Omega)$, having compact support in Ω. An operator $P(D)$ is said to be *hypoelliptic* in Ω if, whenever $f \in C^\infty(\Omega)$ and u is a weak

[8]The notion of a weak solution may be defined for a much larger class; for example u may be a locally square integrable function or even a *distribution*. However, we are not introducing these concepts in this book.

solution of $P(D)u = f$, then $u \in C^\infty(\Omega)$. The hypoellipticity of the Laplace operator Δ was established by H. Weyl and this result is known as *Weyl's lemma*, which plays an important role in the modern treatment of potential theory. There are many extensions of Weyl's lemma and the following result due to Hörmander is very far-reaching:

Theorem 8.12 (Hörmander). An operator $P(D)$ is hypoelliptic if and only if

$$\frac{p^{(\alpha)}(-i\xi)}{p(-i\xi)} \to 0 \text{ as } |\xi| \to \infty,$$

for all α, $|\alpha| \geq 1$, where

$$p^{(\alpha)}(\xi) = D_\xi^\alpha p(\xi).$$

□

It is interesting to note that the hypoellipticity of a constant coefficient operator is given by an algebraic condition in terms of its symbol. It is readily verified that the Laplace operator Δ and the heat operator $\partial_t - \Delta$ are hypoelliptic in any open domain of $\mathbb{R}^n$ and $\mathbb{R}^{n+1}$, respectively.

The necessary and sufficient condition for hypoellipticity of $P(D)$ can also be given in terms of a *fundamental solution* of $P(D)$. If $P(D)$ possesses a fundamental solution that is C^∞ except at the origin, then $P(D)$ is hypoelliptic and conversely. We have seen that Δ possesses a fundamental solution $\phi(x) = c|x|^{-n+2}$, $n \geq 3$, which is C^∞ except at the origin. Similarly, the heat operator $\partial_t - \Delta$ possesses a fundamental solution $K(x, t) = (4\pi a^2 t)^{-n/2} e^{-\frac{|x|^2}{4a^2 t}}$, $x \in \mathbb{R}^n$, $t > 0$ and $= 0$ for $t < 0$, which is C^∞ except at the origin.

The heat kernel also plays an important role in the *index theorem* (Gilkey, 1984) and in the study of *diffusion processes* (see Strook and Varadhan, 1979).

8.6 EXERCISES

1. Let $u \in C^2(\mathbb{R} \times (0, \infty))$ be a solution of the equation $u_t = a^2 u_{xx} + bu_x + cu + f(x, t)$, where a, b, c are real constants and f is a given function. Define the function v by $v(x, t) = e^{-ct} u(x - bt, t)$ for $x \in \mathbb{R}$ and $t > 0$. Show that v satisfies the non-homogeneous equation $v_t = a^2 v_{xx} + e^{-ct} f(x, t)$.
2. Solve the IVP

$$u_t = a^2 u_{xx} + bu_x + cu + f(x, t),\ x \in \mathbb{R}, t > 0 \text{ and } u(x, 0) = u_0(x),\ x \in \mathbb{R}$$

with the following data:

a. $f(x, t) = t \sin x$, $u_0 \equiv 1$, $a = c > 0$, $b = 0$.
b. $f(x, t) = h(t) \in C^1([0, \infty))$ and u_0 is a bounded continuous function.

3. For an arbitrary $s \geq 0$, let $u(x, t; s)$ be a C^2 solution the IVP

$$u_t = a^2 \Delta u,\ x \in \mathbb{R}^n, t > s \text{ and } u(x, s; s) = f(x, s),\ x \in \mathbb{R}^n.$$

Define v by $v(x, t; s) = \int_s^t u(x, t; \tau)\, d\tau$. Show that v satisfies the IVP

$$v_t = a^2 \Delta v + f(x, t),\ x \in \mathbb{R}^n, t > s \text{ and } v(x, s; s) = 0,\ x \in \mathbb{R}^n.$$

Further, show that the converse also holds.
Thus, the homogeneous problem with non-homogeneous initial data can be transformed into a non-homogeneous problem with zero initial data and vice-versa.

4. Let $u_0 : \mathbb{R}^n \to \mathbb{R}$ be such that $u_0(x) = \prod_{j=1}^n u_{0j}(x_j)$, where for each $j = 1, 2, \ldots, n$, $u_{0j} : \mathbb{R} \to \mathbb{R}$ is a bounded continuous function. If u_j solves the *one*-dimensional heat equation $\frac{\partial u_j}{\partial t} = a^2 \frac{\partial^2 u_j}{\partial x_j^2}$ with initial condition $u_j(x_j, 0) = u_{0j}(x_j)$ for $j = 1, 2, \ldots$, show that the solution u of the IVP $u_t = a^2 \Delta u$, $u(x, 0) = u_0(x)$ is given by $u(x, t) = \prod_{j=1}^n u_j(x_j, t)$.
5. Verify that the function u given by the expression in (8.37) satisfies the inhomogeneous equation (8.34).
6. Prove the formula (8.52).
7. Verify that the functions u given by the expressions (8.62) and (8.65) satisfy the respective initial-boundary value problems cited in the text.
8. Suppose $u_0 \in C(\mathbb{R}^n)$ satisfies the condition that $|u_0(x)| \leq Me^{-\delta |x|^2}$ for all $x \in \mathbb{R}^n$ and for some constants $M > 0$, $\delta \geq 0$. Show that the solution u of the heat equation $u_t = a^2 \Delta u$ with initial data u_0 satisfies the estimate

$$|u(x, t)| \leq M \left(1 + 4a^2 \delta t\right)^{-n/2} \exp\left(-\frac{\delta |x|^2}{1 + 4a^2 \delta t}\right)$$

for all $x \in \mathbb{R}^n$ and $t \geq 0$.

CHAPTER 9
One-Dimensional Wave Equation

9.1 INTRODUCTION

The wave equation

$$\Box_c u \equiv u_{tt} - c^2 u_{xx} = 0,\ t > 0,\ x \in \mathbb{R}, \tag{9.1}$$

models many real-world problems: small transversal vibrations of a string, the longitudinal vibrations of a rod, electrical oscillations in a wire, the torsional oscillations of shafts, oscillations in gases, and so on. It is one of the fundamental equations, the others being the equation of heat conduction and Laplace (Poisson) equation, which have influenced the development of the subject of partial differential equations (PDE) since the middle of the last century.

We shall now derive equation (9.1) in the case of transverse vibrations of a string. Physically, a string is a flexible and elastic thread. The tensions that arise in a string are directed along a tangent to its profile. We assume that the string is placed on the x-axis, with its end points at $x = 0$ and $x = L$ (not shown in the figure); see Figure 9.1.

We consider small transversal vibrations of the string, so that the motion of the points of the string is described by a function $u(x, t)$, which gives the amount that a point of the string with abscissa x has moved at time t. We also assume that the length of element MM' of the string corresponding to x and $x + \Delta x$, Δx being very small, is equal to[1] Δx. We also assume that the tension of the string is uniform and denote it by T.

Consider a small element of the string corresponding to the abscissa points x and $x + \Delta x$. Forces T act at the end points of this element along the tangents to the string. Let the tangents make angles ϕ and $\phi + \Delta\phi$ with x-axis, at M and M', respectively. Then, the projection on the u-axis of the forces acting on this element will be equal to

$$T \sin(\phi + \Delta\phi) - T \sin\phi.$$

[1]This amounts to assuming u_x^2 is very small and may be neglected. For, the length of element MM' of the string equals

$$\int_x^{x+\Delta x} \sqrt{1 + u_x^2}\, dx \approx \int_x^{x+\Delta x} dx = \Delta x.$$

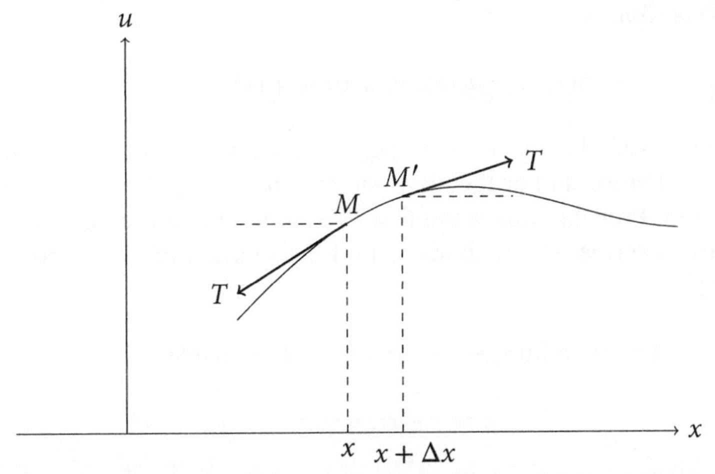

Figure 9.1 The forces acting on element MM' of a vibrating string under uniform tension T

Since we are assuming ϕ is small, we use the approximation $\sin\phi = \tan\phi$ and obtain

$$\begin{aligned}
T\sin(\phi+\Delta\phi) - T\sin\phi &\approx T\tan(\phi+\Delta\phi) - T\tan\phi \\
&= T\left[\frac{\partial u}{\partial x}(x+\Delta x, t) - \frac{\partial u}{\partial x}(x, t)\right] \\
&= T\frac{\partial^2 u}{\partial x^2}(x+\theta\Delta x, t)\,\Delta x,\ 0<\theta<1 \\
&\approx T\frac{\partial^2 u}{\partial x^2}(x, t)\,\Delta x.
\end{aligned}$$

Next, let ρ be the linear density, that is, mass per unit length, of the string. Applying the Newton's second law of motion, to the small element of the string under consideration, we obtain

$$\rho\Delta x\frac{\partial^2 u}{\partial t^2}(x, t) = T\frac{\partial^2 u}{\partial x^2}(x, t)\,\Delta x.$$

Dividing by Δx throughout and putting $c^2 = \frac{T}{\rho}$, results in (9.1).

9.2 CAUCHY PROBLEM ON THE LINE

We begin with the wave equation in one (space) dimension and discuss its solution by two different methods. Consider the homogeneous wave equation

$$\Box_c u \equiv u_{tt} - c^2 u_{xx} = 0,\ t>0,\ x\in\mathbb{R}, \tag{9.2}$$

with the initial conditions

$$u(x,0) = u_0(x),\ u_t(x,0) = u_1(x),\ x \in \mathbb{R}. \tag{9.3}$$

Here the constant c, called the *speed of propagation*, is a positive real number. The operator $\Box_c$ is called the *D'Alembertian* or the *wave operator*, u_0, u_1 are given functions and u is the unknown function. Probably this is the first PDE in the literature derived by D'Alembert in 1747. We now describe two methods to find the solution of (9.2) satisfying the initial conditions (9.3).

Method 1. The first method invokes the *characteristic* variables

$$\xi = x + ct,\ \eta = x - ct.$$

These variables arise out of the real characteristics of the equation (9.2) possesses. Recall the analysis discussed in Chapter 6. It is straightforward to verify that $u_{tt} - c^2 u_{xx} = -4c^2 u_{\xi\eta}$ and so, (9.2) becomes $u_{\xi\eta} = 0$. It is readily verified that the general solution of the latter equation is given by

$$u = F(\xi) + G(\eta) = F(x + ct) + G(x - ct), \tag{9.4}$$

where F, G are arbitrary C^2 functions defined on $\mathbb{R}$. We use the initial conditions (9.3) to determine F and G. We have

$$F(x) + G(x) = u_0(x),$$
$$cF'(x) - cG'(x) = u_1(x).$$

Integrating the second equation and then solving the resulting equations, we obtain

$$F(x) = \frac{1}{2}u_0(x) + \frac{1}{2c}\int_0^x u_1(s)\,ds + k,$$

$$G(x) = \frac{1}{2}u_0(x) - \frac{1}{2c}\int_0^x u_1(s)\,ds - k,$$

where k is a constant of integration. Substituting these expressions into (9.4), we see that the solution u is given by

$$u(x,t) = \frac{1}{2}\left(u_0(x+ct) + u_0(x-ct)\right) + \frac{1}{2c}\int_{x-ct}^{x+ct} u_1(s)\,ds. \tag{9.5}$$

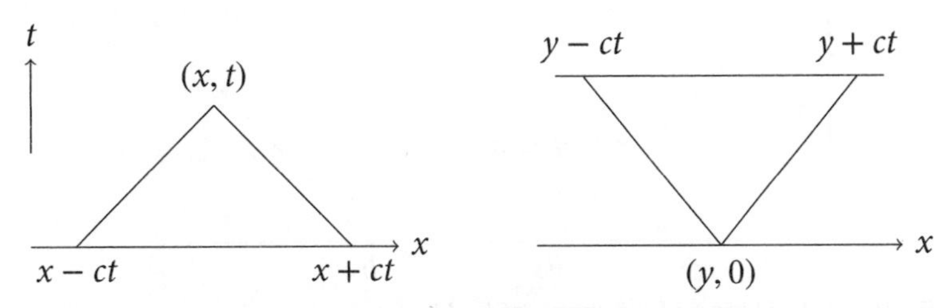

Figure 9.2 Domain of dependence and range of influence

This is called the *D'Alembert's formula*. We have thus proved the following theorem:

Theorem 9.1. Suppose the initial conditions in (9.3) satisfy that $u_0 \in C^2(\mathbb{R})$ and $u_1 \in C^1(\mathbb{R})$. Then, (9.2) has a unique C^2 solution u satisfying the initial conditions in (9.3). Further, u is given by the D'Alembert's formula (9.5). □

We remark that when the initial data u_0 and u_1 are less smooth than stated in Theorem 9.1, we will not get a classical solution of (9.2). In this case we need to search for a *weak solution* of the wave equation. We briefly discuss this concept of solution in Section 9.5.

Domain of Dependence, Range of Influence and Domain of Determinacy: We now make the following observations based on the D'Alembert's formula (9.5). The value of the solution u at (x, t), $t > 0$ depends on the values of the initial data only in the interval $[x - ct, x + ct]$ on the initial line $t = 0$, that is the x-axis. This is referred to as the *domain of dependence* (of the solution) at (x, t). Similarly, a point y on the initial line can *influence* the value of u for some $t > 0$, only in a line segment $\{(x, t) : x \in [y - ct, y + ct]\}$. This is referred to as the *range of influence* of the point $(y, 0)$. These are illustrated in Figure 9.2.

Thus, the initial values in an interval $[a, b]$ on $t = 0$ will influence the values of the solution u in the region

$$\{(x, t) : x \in [a - ct, b + ct],\ t > 0\}.$$

This is the region between the diverging characteristics $x + ct = a$ and $x - ct = b$ drawn from the points $(a, 0)$ and $(b, 0)$, $a < b$, respectively (see Figure 9.3). On the other hand, the converging characteristics emanating from $(a, 0)$ and $(b, 0)$ meet when $t = \frac{1}{2c}(b - a)$. Thus, the values of the solution u in the triangle with vertices at $(a, 0)$, $(b, 0)$ and $\left(\frac{a+b}{2}, \frac{b-a}{2c}\right)$ are determined by the initial values within the interval $[a, b]$ at $t = 0$ (see Figure 9.4). This triangular region is referred to as the *domain of determinacy* of the interval $[a, b]$.

Method 2: This is similar to the method of characteristics. Here, we factorize the wave operator into two first-order operators:

$$\Box_c = (\partial_t - c\partial_x)(\partial_t + c\partial_x).$$

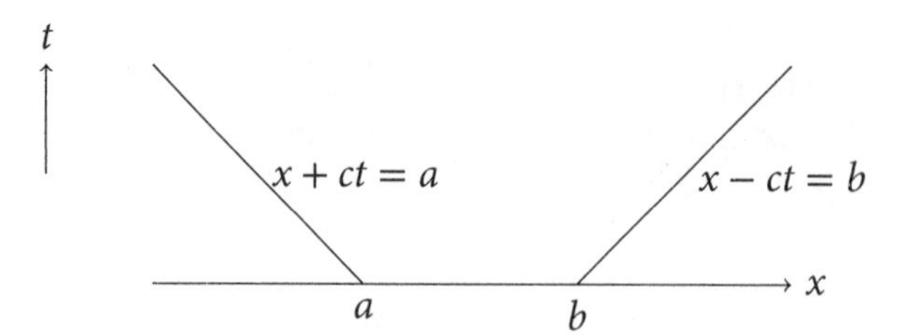

Figure 9.3 Range of influence of $[a, b]$

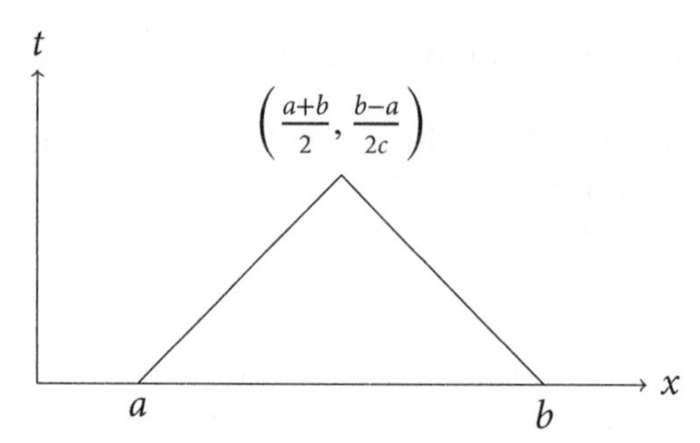

Figure 9.4 Domain of determinacy of $[a, b]$

Let $v = (\partial_t + c\partial_x)u$. Then, solving (9.2) is equivalent to solving the following two first-order equations:

$$(\partial_t + c\partial_x)u = v \text{ and } (\partial_t - c\partial_x)v = 0. \tag{9.6}$$

The first-order homogeneous equation $(\partial_t - c\partial_x)v = 0$ is solved by the method of characteristics and is given by $v(x, t) = V(x + ct)$, for an arbitrary C^1 function V. Obviously, $v(x, 0) = V(x)$, so V comes from the initial condition. Next, we consider the inhomogeneous equation $(\partial_t - c\partial_x)u = v(x, t)$. The following procedure, called *Duhamel's principle*, describes a way to obtain the solution of the inhomogeneous equation by reducing the problem to a homogeneous equation. Fix $s \in [0, t]$ and consider the homogeneous equation $(\partial_t - c\partial_x)w = 0$ in the region $t > s$ with the initial condition $w(x, t) = v(x, s)$ for $t = s$. Let us denote the solution by $w(x, t; s)$ to emphasize the dependence on s. By using the translation invariance of the wave equation, we at once obtain

$$w(x, t; s) = v(x + c(t - s), s), \text{ for } t \geq s.$$

Now let

$$u(x, t) = \tilde{u}(x + ct) + \int_0^t w(x, t; s)\, ds = \tilde{u}(x + ct) + \int_0^t v(x + c(t - s), s)\, ds. \tag{9.7}$$

Here $\tilde{u}$ is an arbitrary C^1 function defined in $\mathbb{R}$. The easy verification that u above satisfies the inhomogeneous equation $(\partial_t - c\partial_x)u = v(x,t)$ is left as an exercise and the initial condition $u(x,0) = \tilde{u}(x)$ for $x \in \mathbb{R}$. Formula (9.7) is called *Duhamel's formula*.

We now use these observations to solve any general kth order equation, in which the operator factors into linear factors. More precisely, consider the initial value problem (IVP):

$$(\partial_t - c_1\partial_x) \cdots (\partial_t - c_k\partial_x)u = 0,$$
$$\partial_t^j u(x,0) = u_{0j}(x),\ j = 0, 1, \ldots, k-1.$$

Here the real constants c_i may or may not be distinct and $k \geq 2$. Since the general case follows from an induction argument, suffices to consider the case $k = 2$. Thus, we consider the following IVP:

$$(\partial_t - c_1\partial_x)(\partial_t - c_2\partial_x)u = 0,\ x \in \mathbb{R}, t > 0 \tag{9.8}$$
$$u(x,0) = u_0(x),\ \partial_t u(x,0) = u_1(x),\ x \in \mathbb{R}. \tag{9.9}$$

Putting $v = (\partial_t - c_2\partial_x)u$, we see that $(\partial_t - c_1\partial_x)v = 0$ and therefore,

$$v(x,t) = u_1(x + c_1 t) - c_2 u_0'(x + c_1 t).$$

Then, since u satisfies the inhomogeneous equation $(\partial_t - c_2\partial_x)u = v$, we obtain, using the Duhamel's formula (9.7), that

$$\begin{aligned} u(x,t) &= u_0(x + c_2 t) + \int_0^t v(x + c_2(t-s), s)\, ds \\ &= u_0(x + c_2 t) + \int_0^t u_1(x + c_2(t-s) + c_1 s)\, ds \\ &\quad - c_2 \int_0^t u_0'(x + c_2(t-s) + c_1 s)\, ds \\ &= u_0(x + c_2 t) + \int_0^t u_1(x + c_2 t + (c_1 - c_2)s)\, ds \\ &\quad - c_2 \int_0^t u_0'(x + c_2 t + (c_1 - c_2)s)\, ds. \end{aligned} \tag{9.10}$$

Looking at the variable of integration, we consider the following two cases:

(1) Case $c_1 \neq c_2$: In this case, by changing the variable of integration, we obtain

$$u(x,t) = \left[\frac{c_1}{c_1 - c_2} u_0(x + c_2 t) - \frac{c_2}{c_1 - c_2} u_0(x + c_1 t)\right] + \frac{1}{c_1 - c_2} \int_{x+c_1 t}^{x+c_2 t} u_1(\xi)\, d\xi. \quad (9.11)$$

Taking $c_2 = -c_1 = c$ in (9.11), we obtain the D'Alembert's formula (9.5).

(2) Case $c_1 = c_2 = c$, say. Then, the integration variable s is absent in both the integrals on the right side of (9.10) and we obtain

$$u(x,t) = u_0(x + ct) - ctu_0'(x + ct) + tu_1(x + ct). \quad (9.12)$$

Looking at the solution given in (9.12), we see that the initial conditions u_0 and u_1 should satisfy that $u_0 \in C^3(\mathbb{R})$ and $u_1 \in C^2(\mathbb{R})$ in order that the solution u is C^2. Thus, there is *loss of regularity* in the problem: more smoothness of the initial data is required in order to get less smooth solution. The operator $(\partial_t - c\partial_x)^2$ is said to be *weakly hyperbolic*.[2] In higher dimensions, such operators do cause difficulties in proving existence and regularity results.

The general case of k speeds $c_1, \ldots, c_k$, distinct or otherwise, follows by an induction argument and the above discussion of the two cases. This is left as an exercise.

First-Order System: The above procedure also extends to a system of first-order equations in one (space) variable. Consider the first-order system

$$u_t + Au_x = 0, \quad (9.13)$$

with initial conditions $u(x,0) = u_0(x)$, $x \in \mathbb{R}$. Here, A is a given real $N \times N$ matrix, u, u_0 are $\mathbb{R}^N$ valued functions and N is a positive integer. The system (9.13) is said to be

(1) *Hyperbolic*,[3] if all the eigenvalues of A are real; in some textbooks the terminology *weakly hyperbolic* is also used.
(2) *Strictly hyperbolic* if the eigenvalues of A are real and distinct.
(3) *Strongly hyperbolic* if the eigenvalues of A are real and A is diagonalizable.
(4) *Symmetric hyperbolic* if the matrix A is symmetric.

By the Jordan decomposition theorem, there is non-singular matrix C such that $C^{-1}AC = \Lambda$. The matrix Λ would be a diagonal matrix if the matrix A is diagonalizable; otherwise, it

[2]Though there is only one real characteristic here, this equation is *not parabolic*.

[3]We will not go into the motivation of this definition, as it is bit complicated.

would be a block diagonal matrix, with *Jordan blocks* on its diagonal. A typical Jordan block is a square matrix of the form

$$\begin{bmatrix} \lambda & 1 & 0 & \cdots & 0 \\ 0 & \lambda & 1 & \cdots & 0 \\ \cdots & \cdots & \cdots & \cdots & \cdots \\ 0 & 0 & 0 & 0 & \lambda. \end{bmatrix}$$

If we put $v = C^{-1}u$, then each component v_i of v satisfies either a homogeneous or inhomogeneous first-order equation, whose solution may be explicitly written down by using the above-described procedure. Note that the eigenvalues of A play the roles of speeds of propagation for different components of u.

9.2.1 Inhomogeneous Equation: Duhamel's Principle

Similar to the first-order inhomogeneous equation considered above, we obtain the solution of second-order inhomogeneous equation by using the Duhamel's principle.[4] Consider the inhomogeneous wave equation

$$u_{tt} - c^2 u_{xx} = f(x,t),\ x \in \mathbb{R}, t > 0, \tag{9.14}$$

with initial conditions (9.3). Because of the linearity of the equation (the superposition principle holds), the required solution is sum of two solutions: one is the solution of the inhomogeneous equation (9.14) with zero initial conditions and another, solution of the homogeneous equation (9.2) satisfying the initial conditions (9.3). Since we already know that the solution to the latter problem is given by the D'Alembert's formula, it suffices to consider the inhomogeneous equation with zero initial conditions. The Duhamel's principle converts this problem to solving a homogeneous equation with a suitable initial data.

Fix $s \geq 0$ and consider the following IVP:

$$\left.\begin{array}{l} v_{tt} - c^2 v_{xx} = 0,\ x \in \mathbb{R}, t > s \\ v(x,s) = 0,\ v_t(x,s) = f(x,s). \end{array}\right\} \tag{9.15}$$

Since the wave operator is translation invariant (change the variable t to $\tau = t-s$), we readily see that the solution of (9.15) is given by the D'Alembert's formula

$$v(x,t;s) = \frac{1}{2c} \int_{x-c(t-s)}^{x+c(t-s)} f(\xi,s)\, d\xi. \tag{9.16}$$

---

[4]This principle applies more generally to any linear equation or system of the form $u_t + Lu = f$, where the differential operator L does not involve the time variable t.

The notation on the left side of (9.16) is to stress the dependence on s. We now claim that the solution u of the inhomogeneous equation (9.14) with zero initial conditions is given by

$$u(x,t) = \int_0^t v(x,t;s)\,ds = \frac{1}{2c}\int_0^t \int_{x-c(t-s)}^{x+c(t-s)} f(\xi,s)\,d\xi\,ds, \tag{9.17}$$

provided that the forcing term f is a C^1 function. Note that the right-most double integral in (9.17) is the integral over the *characteristic triangle* with vertices at (x,t), $(x-ct,0)$ and $(x+ct,0)$.

It is readily seen that $u(x,0)=0$ for all x. By the familiar formula of differentiation under the integral sign, we obtain

$$u_t(x,t) = \frac{1}{2}\int_0^t \big(f(x+c(t-s),s) + f(x-c(t-s),s)\big)\,ds$$

$$u_{tt}(x,t) = f(x,t) + \frac{c}{2}\int_0^t \left(\frac{\partial f}{\partial x}(x+c(t-s),s) - \frac{\partial f}{\partial x}(x-c(t-s),s)\right)ds$$

$$u_{xx}(x,t) = \frac{1}{2c}\int_0^t \left(\frac{\partial f}{\partial x}(x+c(t-s),s) - \frac{\partial f}{\partial x}(x-c(t-s),s)\right)ds.$$

The first equation gives that $u_t(x,0)=0$ for all x and the other two equations show that u satisfies the inhomogeneous equation (9.14). We can now write down the solution of the inhomogeneous equation (9.14), satisfying the initial conditions (9.3):

$$u(x,t) = \frac{1}{2}\big(u_0(x+ct) + u_0(x-ct)\big) + \frac{1}{2c}\int_{x-ct}^{x+ct} u_1(\xi)\,d\xi + \frac{1}{2c}\int_0^t \int_{x-c(t-s)}^{x+c(t-s)} f(\xi,s)\,d\xi ds. \tag{9.18}$$

Using the above representation, it is now straightforward to derive *continuous dependence* of the solution u on the data. For this purpose, assume, in addition, that the functions u_0 and u_1 are bounded functions on $\mathbb{R}$ and that for any fixed $T>0$, the function f has the following finite *norm*:

$$\|f\|_T = \max_{t\in[0,T]} \sup_{x\in\mathbb{R}} |f(x,t)| < \infty.$$

We also introduce the following norms:

$$|u_0|_\infty = \sup_{x\in\mathbb{R}} |u_0(x)|$$

with a similar definition for u_1 and

$$|u(\cdot,t)|_\infty = \sup_{x\in\mathbb{R}} |u(x,t)|, \text{ for } t > 0.$$

We have

Theorem 9.2. Let $T > 0$ be fixed. Then, the solution u given by (9.18) satisfies the following estimate: for any $t \in [0, T]$,

$$|u(\cdot,t)|_\infty \leq |u_0|_\infty + T|u_1|_\infty + \frac{T^2}{2}\|f\|_T. \tag{9.19}$$

□

Proof From (9.18), it follows that

$$|u(x,t)| \leq |u_0|_\infty + t|u_1|_\infty + \|f\|_T \int_0^t (t-s)\, ds,$$

for any $0 < t \leq T$. The estimate (9.19) immediately follows from the above inequality. □

Similar estimates may be obtained for u_x, u_t with appropriate assumptions on the initial conditions. Though the estimate (9.19) is adequate in the present scenario, it is not a useful estimate when we move to equations with variable coefficients or non-linear equations. The more appropriate estimates for the solution of a general hyperbolic equation or system, are the so-called *energy estimates*. Such estimates are quite useful in establishing existence, uniqueness and regularity results for hyperbolic equations or systems. In the analysis, we obtained the estimate using the representation of the solution. However, such a representation in general is not available and we need to derive the estimate(s) of the unknown solution (assuming its existence) using the PDE under consideration and some integration tools. Such estimates are known as *a priori estimates* and play a crucial role in establishing existence and uniqueness of solutions to PDE. Derivation of a priori estimates is the trademark of the modern PDE analysis. In the present case, we obtain one such estimate.

Let u be the solution of the homogeneous equation (9.2) satisfying the initial conditions (9.3). Multiply (9.2) by u_t and integrate the resulting equation[5] over $\mathbb{R}$:

$$\int_{-\infty}^{\infty} u_t u_{tt}\, dx - c^2 \int_{-\infty}^{\infty} u_t u_{xx}\, dx = 0.$$

[5]Do the integration over the interval $(-R, R)$ and let $R \to \infty$.

We can write $u_t u_{tt} = \frac{1}{2}\frac{\partial}{\partial t}(u_t^2)$. Next, integrate by parts in the second integral and assume that the boundary terms vanish. After this, the integrand becomes

$$u_{tx}u_x = \frac{1}{2}\frac{\partial}{\partial t}(u_x^2).$$

We therefore obtain that

$$\frac{1}{2}\frac{d}{dt}\int_{-\infty}^{\infty}\left(u_t^2 + c^2 u_x^2\right)\, dx = 0.$$

If we denote by $E(t)$, called the *total energy* at time t,

$$E(t) = \int_{-\infty}^{\infty}\left(u_t^2(x,t) + c^2 u_x^2(x,t)\right)\, dx,$$

it immediately follows that $E(t) = E(0)$, which is the statement that the total energy is *conserved*. In the situation of the variable coefficients, the total energy may not be conserved, but it is possible to estimate $E(t)$ in terms of the initial energy $E(0)$, which proves to be quite useful.

With the aid of the D'Alembert's solution, we can readily obtain the expression for u_x and u_t. Using these expressions, we invite the reader to come up with appropriate assumptions on the initial conditions u_0 and u_1, so that the computations done in the previous paragraph become valid statements.

9.2.2 Characteristic Parallelogram

We now describe another characterization of the solution of the homogeneous wave equation. Consider a parallelogram in the upper half $x - t$ plane $t > 0$, formed by the characteristics of the wave operator, namely the straight lines $x \pm ct = \text{constant}$. If (x, t) is a point in the upper half plane, then for h, k small, the parallelogram with vertices as shown in Figure 9.5 will be one such parallelogram. If u is a solution of the homogeneous equation (9.2), it is readily seen that

$$u(A) + u(C) = u(B) + u(D), \tag{9.20}$$

using the D'Alembert's formula, where A, B, C and D are the vertices of the parallelogram shown in Figure 9.5. In fact, (9.20) characterizes the solutions of the homogeneous equation.

Theorem 9.3. Let u be any C^2 function satisfying the identity (9.20) for all characteristic parallelograms. Then, u satisfies the homogeneous equation (9.2). □

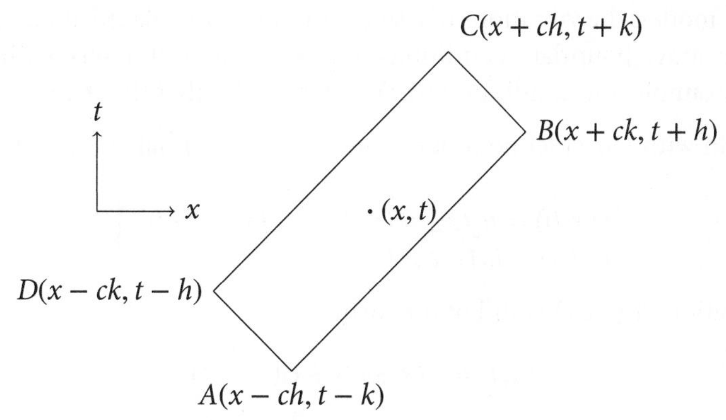

Figure 9.5 A characteristic parallelogram

Proof If we apply Taylor's formula at each of the vertices A, B, C, D and use (9.20), the result follows by letting $h, k \to 0$. □

9.3 CAUCHY PROBLEM IN A QUADRANT (SEMI-INFINITE STRING)

We now consider the wave equation in the first quadrant of the x, t plane:

$$u_{tt} - c^2 u_{xx} = 0, \ x > 0, t > 0. \tag{9.21}$$

In addition to the initial conditions at $t = 0$, we also need to provide *boundary* conditions on the *boundary* $x = 0$ of the quadrant under consideration. The boundary conditions are usually of three types:

(1) *Dirichlet boundary condition*: Here $u(0, t), t > 0$ is prescribed. This is almost equivalent to prescribing the *tangential derivative* u_t along the t-axis.
(2) *Neumann boundary condition*: Here the *normal derivative* u_x is prescribed on the boundary $x = 0$, where

$$u_x(0, t) = \lim_{h \to 0+} \frac{u(h, t) - u(0, t)}{h}, \ t > 0.$$

(3) Since the vectors along the tangent and the normal are linearly independent, the derivative along any other direction can be prescribed by a suitable linear combination of conditions in (1) and (2) above. This type of boundary condition is termed as *mixed* or *Robin boundary condition*. In this case either $\alpha u + \beta u_x$ or $\alpha u_t + \beta u_x$ is prescribed on $x = 0$ for suitable real α and β.

Equation (9.21) models the vibrations in a semi-infinite string placed along the positive real axis $[0, \infty)$. The above boundary conditions imposed at $x = 0$ describe different physical situations. For example, the condition $u(0, t) = 0$ indicates that the string is tied at $x = 0$.

Case 1: We begin with Dirichlet boundary condition. The initial and boundary conditions are of the form

$$\left.\begin{array}{l} u(x, 0) = u_0(x),\ u_t(x, 0) = u_1(x),\ x > 0, \\ u(0, t) = h(t),\ t > 0. \end{array}\right\} \tag{9.22}$$

The general solution of (9.21) is still of the form

$$u(x, t) = F(x + ct) + G(x - ct),$$

for suitable functions F and G. Let (x, t) be a point in the first quadrant satisfying $x > ct$ (see Figure 9.6). Then, since the domain of dependence of (x, t) is the interval $[x - ct, x + ct]$ on the positive x-axis, as $x > ct$, we obtain

$$u(x, t) = \frac{1}{2}\left(u_0(x + ct) + u_0(x - ct)\right) + \frac{1}{2c}\int_{x-ct}^{x+ct} u_1(s)\, ds,\ x > ct \tag{9.23}$$

given by the D'Alembert's formula. Next, consider the case $x < ct$. Using the general solution

$$u(x, t) = F(x + ct) + G(x - ct),$$

we obtain, making use of the boundary condition on $x = 0$,

$$h(t) = F(ct) + G(-ct),\ t > 0.$$

As before, we have

$$F(x) = \frac{1}{2}u_0(x) + \frac{1}{2c}\int_0^x u_1(\xi)\, d\xi + k,$$

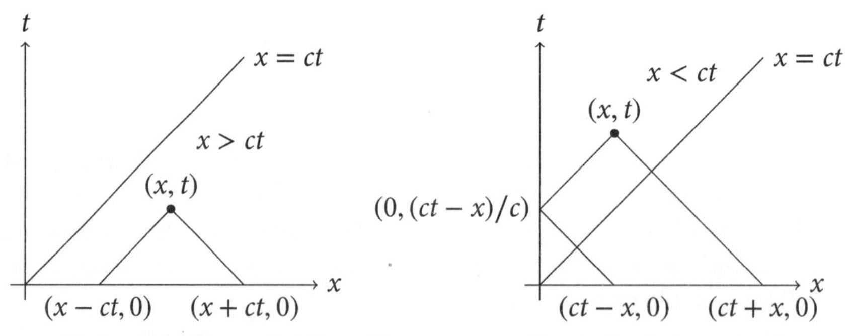

Figure 9.6 Characteristics of the wave equation in the first quadrant

where k is a constant. Therefore,

$$G(-t) = h\left(\frac{t}{c}\right) - \frac{1}{2}u_0(t) - \frac{1}{2c}\int_0^t u_1(\xi)\,d\xi - k,\ t > 0.$$

Therefore, at the point (x, t) with $x < ct$,

$$\begin{aligned} u(x,t) &= F(x+ct) + G(x-ct) \\ &= \left[\frac{1}{2}u_0(x+ct) + \frac{1}{2c}\int_0^{x+ct} u_1(\xi)\,d\xi + k\right] + \\ &\quad + \left[h\left(\frac{ct-x}{c}\right) - \frac{1}{2}u_0(ct-x) - \frac{1}{2c}\int_0^{ct-x} u_1(\xi)\,d\xi - k\right] \\ &= h\left(\frac{ct-x}{c}\right) + \frac{1}{2}\left[u_0(ct+x) - u_0(ct-x)\right] \\ &\quad - \frac{1}{2c}\int_{ct-x}^{ct+x} u_1(\xi)\,d\xi. \end{aligned} \tag{9.24}$$

A schematic diagram showing the characteristics in the two cases $x > ct$ and $x < ct$ are depicted in Figure 9.6.

Remark 9.4. When $h \equiv 0$, the solution can also be obtained by a simple reflection. Define the function $\tilde{u}$ by

$$\tilde{u}(x,t) = \begin{cases} u(x,t), & \text{for } x > 0, \\ -u(-x,t), & \text{for } x < 0, \end{cases}$$

and similar definitions for $\tilde{u}_0$ and $\tilde{u}_1$. Then, it is straightforward to check that $\tilde{u}$ satisfies the wave equation in $\mathbb{R}$ with initial conditions $\tilde{u}_0$ and $\tilde{u}_1$. Once $\tilde{u}$ is determined using D'Alembert's formula, we can obtain u. The details are left as an exercise. □

We now determine the values of the solution $u(x, t)$ on the characteristic line $x = ct$ by requiring it to be a C^2 function in the first quadrant $x > 0, t > 0$. By examining the expressions in (9.23) and (9.24), we see that $u(x, t)$ is continuous across $x = ct$ provided that $u_0(0) = h(0)$. This is called a *compatibility condition*. We leave it as an exercise to show that the following compatibility conditions must be satisfied in order that u is C^2 in the first quadrant:

$$u_0(0) = h(0),\ u_1(0) = h'(0),\ c^2u_0''(0) = h''(0). \tag{9.25}$$

Case 2: We now consider the Neumann boundary problem, namely

$$\begin{aligned}
&u_{tt} - c^2 u_{xx} = 0,\ x > 0, t > 0 \\
&u(x,0) = u_0(x),\ u_t(x,0) = u_1(x),\ x > 0, \\
&u_x(0,t) = h(t),\ t > 0,
\end{aligned}$$

where h is a given smooth function. The procedure for determining the solution essentially remains the same. Again, in the region $x > ct$ (see Figure 9.6) there is no change and the solution is given by the D'Alembert's formula. In the region $x < ct$, the condition

$$h(t) = F'(ct) + G'(-ct),\ t > 0,$$

needs to be satisfied, assuming that the general form of the solution is

$$u(x,t) = F(x + ct) + G(x - ct).$$

As before,

$$F(x) = \frac{1}{2}u_0(x) + \frac{1}{2c}\int_0^x u_1(\xi)\,d\xi + k_1,\ x > 0,$$

with k_1 an arbitrary constant. We have

$$G'(-t) = h\left(\frac{t}{c}\right) - F'(t),\ t > 0.$$

Upon integration and using the expression for F in terms of the initial data, we obtain

$$G(-t) = -\int_0^t h\left(\frac{s}{c}\right)\,ds + \frac{1}{2}u_0(t) + \frac{1}{2c}\int_0^t u_1(\xi)\,d\xi + k_2,\ t > 0,$$

with k_2 an arbitrary constant. Thus, we have in the region $x < ct$,

$$\begin{aligned}
u(x,t) = &-\int_0^{ct-x} h\left(\frac{s}{c}\right)\,ds + \frac{1}{2}\left(u_0(ct+x) + u_0(ct-x)\right) \\
&+ \frac{1}{2c}\int_{ct-x}^{ct+x} u_1(\xi)\,d\xi + k, \qquad (9.26)
\end{aligned}$$

where k is an arbitrary constant. As in the previous case of the Dirichlet boundary condition, it is easily checked that the constant $k = 0$ for the solution u to be continuous on the characteristic line $x = ct$. The compatibility conditions that need to be satisfied by the initial and boundary conditions for u to be C^2 in the first quadrant, including the line $x = ct$, are now given by

$$u_0'(0) = h(0),\ u_1'(0) = h'(0).$$

A somewhat lengthy verification of these conditions is left as an exercise to the reader.

Case 3: The case of mixed boundary conditions is treated in the exercises.

9.4 WAVE EQUATION IN A FINITE INTERVAL

We now consider the wave equation in a finite interval $(0, L)$:

$$u_{tt} - c^2 u_{xx} = 0,\ 0 < x < L, t > 0, \tag{9.27}$$

with initial conditions prescribed in (9.3) and any one of the three boundary conditions prescribed on the boundary lines $x = 0$ and $x = L$. As in the case of the problem in the first quadrant, the initial and boundary conditions need to satisfy certain compatibility conditions in order that the solution u be C^2 in the region $(0, L) \times \{t > 0\}$. The solution will be determined depending on the position of the point $(x, t), t > 0$ as schematically shown in Figure 9.7. The straight line segments that are drawn in this figure are respectively the characteristic lines $x - ct =$ constant originating on the line $x = 0$, and the characteristic lines $x + ct$ =constant originating from the line $x = L$. Referring to Figure 9.7, we see that the points (x, t) in region *I* are not influenced by the boundary conditions at $x = 0$ and $x = L$. Thus, the D'Alembert's formula applies for the solution. In region *II*, the points (x, t) are influenced by the boundary condition on the line $x = 0$. By constructing suitable characteristic parallelogram with one vertex at (x, t), we use the parallelogram property to

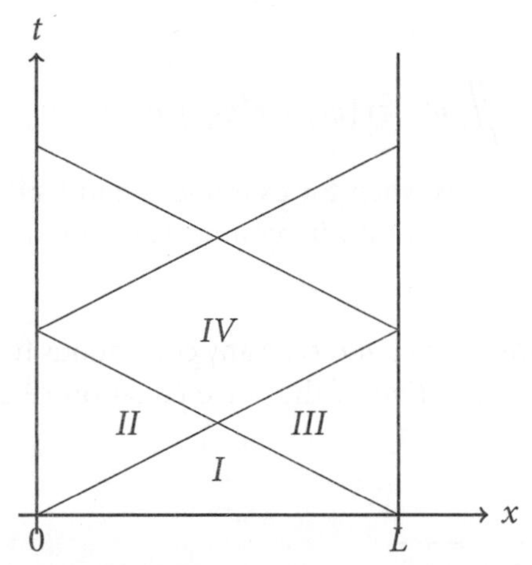

Figure 9.7 Characteristics of the wave equation in a finite interval

obtain the value $u(x, t)$. Similar arguments apply for the region *III*, where the values of the solution will be influenced by the boundary condition on the line $x = L$. We continue with similar arguments to find $u(x, t)$ for any $x \in (0, L)$ and $t > 0$. We ask the reader to write down formulas in different regions and for different boundary conditions.

9.5 NOTION OF A WEAK SOLUTION

When the speed of propagation c is a function of x, it is no longer possible to obtain the solution of the wave equation in explicit form. The same remark also applies when, for example, the inhomogeneous term f is a non-linear function of u also. In such situations, it is in general not possible to prove an existence result in the class of C^2 functions. In the modern development of PDEs, which began sometime in the middle of last century, powerful tools were discovered to tackle the problem of existence and uniqueness of solutions to PDEs. One such tool is the notion of a *weak solution* to a PDE. Roughly speaking, a weak solution of a PDE satisfies an integral relation involving the given partial differential operator, whereas the usual solution, which we call a *strong* or *classical solution* satisfies the given PDE point-wise, at all the points where the solution is defined. Below, we shall explain the concept of a weak solution in the context of the wave equation (9.2).

Suppose u is a strong solution of (9.2). This means that u is a C^2 function that satisfies equation (9.2) for all $x \in \mathbb{R}$ and $t > 0$. Then, by multiplying this equation by a C^2 (or a C^∞ function) ϕ and by integrating, we obtain

$$0 = \int_0^\infty \int_{\mathbb{R}} \phi(x, t) \left(u_{tt} - c^2 u_{xx}\right) \, dxdt.$$

In what follows we omit the limits in the integral signs, for brevity. If we now integrate by parts, then

$$\iint u(x, t) \left(\phi_{tt} - c^2 \phi_{xx}\right) \, dxdt = 0.$$

There will be no boundary terms when for example ϕ and its derivatives vanish outside a bounded set. That is, ϕ is a C^∞ function with compact support and we call all such functions as *test functions*.

Definition 9.5 (Weak Solution). Let $u(x, t)$ be any continuous function[6] defined in $\mathbb{R} \times \mathbb{R}^+$. Then, u is called a *weak solution* of the wave equation (9.2) if the following integral

[6] If Lebesgue integration is used, then suffices to assume that u is locally integrable.

relation

$$\iint u(x,t)\left(\phi_{tt}-c^2\phi_{xx}\right)\,dxdt=0, \tag{9.28}$$

holds for all the test functions ϕ. □

It is possible to include the initial conditions in the definition of a weak solution, but will not do it here to keep the presentation simple. It is readily seen that any strong solution is a weak solution. Conversely, any weak solution that possesses second-order continuous derivatives is a strong solution. This follows by integration by parts and the following simple observation:

If $g(x,t)$ is a continuous function and $\iint g(x,t)\phi(x,t)\,dxdt = 0$ for all test functions ϕ, then $g \equiv 0$.

However, we shall now see that there are many continuous functions, not even possessing first-order derivatives, which are weak solutions of the wave equation. Obviously, these are not strong solutions.

Let $v : \mathbb{R} \to \mathbb{R}$ be any continuous function. Define u by $u(x,t) = v(x+ct)$ (similarly, $v(x-ct)$) for $(x,t) \in \mathbb{R}\times\mathbb{R}$. In order to show that u satisfies (9.28), consider the integral on the left side of (9.28) and make the (non-singular) change of variables: $\xi = x+ct,\ \tau = x-ct$. Then,

$$\iint u(x,t)\left(\phi_{tt}-c^2\phi_{xx}\right)\,dxdt = k\iint v(\xi)\phi_{\xi\tau}\,d\xi d\tau,$$

where k is a constant, resulting from the change of variables. By Fubini's theorem, we can write the second integral as an iterated integral:

$$\int v(\xi)\,d\xi \int \phi_{\xi\tau}\,d\tau.$$

It is easy to check that if ϕ and its derivatives vanish outside a bounded set as a function of x and t, the same holds when we consider it as a function of ξ and τ. The integral with respect to τ vanishes as ϕ and its derivatives vanish outside a bounded set. This shows that u is a weak solution of the wave equation (9.2).

9.6 GENERAL SECOND-ORDER EQUATIONS

In the previous sections, we have discussed the derivation of solutions to the wave equation in different situations. In all these cases, the initial conditions were imposed on the non-characteristic initial curve, namely the line $t = 0$. A natural question that arises is the possibility of prescribing the initial conditions on a general non-characteristic curve prescribed by the equation $t = \phi(x)$. In such a situation, the solution is sought at all points (x,t) with $t > \phi(x)$. Assume that u is a C^2 solution of the inhomogeneous wave equation

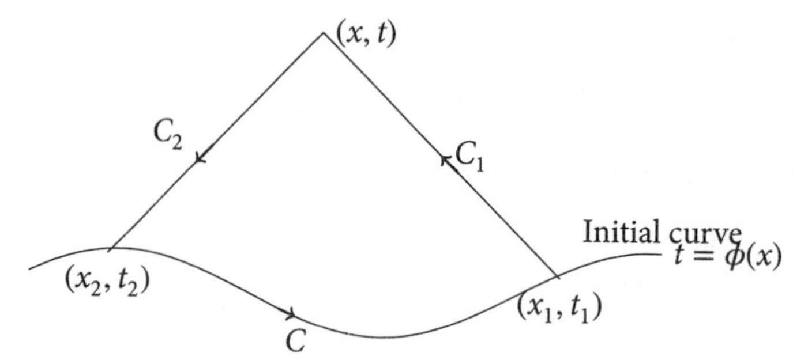

Figure 9.8 A characteristic triangle

(9.14) in the region $t > \phi(x)$ with prescribed conditions on the initial curve $t = \phi(x)$. Consider a point (x, t) with $t > \phi(x)$ and draw the characteristics C_1 and C_2 through it to meet the initial curve at (x_1, t_1) and (x_2, t_2), respectively (see Figure 9.8). Let C be the part of the initial curve between (x_1, t_1) and (x_2, t_2) and consider the domain D bounded by C_1, C_2 and C. Writing $u_{tt} - c^2 u_{xx} = (u_t)_t - (c^2 u_x)_x$ and using the Green's identity in the domain D, we obtain

$$\int_{\widehat{C}} -c^2 u_\xi \, d\tau - u_\tau \, d\xi = \iint_D f(\xi, \tau) \, d\xi d\tau, \tag{9.29}$$

where the integral on the left is a line integral with $\widehat{C}$ denoting the union of the curves C_1, C_2 and C. Since we wish to find a formula for $u(x, t)$, we have used the notations ξ and τ for the running variables.

The curve C_1 is described by the equation $\xi = x + c(t - s)$, $\tau = s$, $s \in [t_1, t]$, with $\xi = x_1$ when $\tau = t_1$. Therefore, we have

$$-\int_{C_1} c^2 u_\xi \, d\tau + u_\tau \, d\xi = \int_{t_1}^{t} \left(c u_\tau(x + c(t - s), s) - c^2 u_\xi(x + c(t - s), s) \right) ds.$$

We observe that the integrand in the integral on the right can be written as

$$c \frac{d}{ds} u(x + c(t - s), s), \; s \in [t_1, t].$$

Therefore,

$$-\int_{C_1} c^2 u_\xi \, d\tau + u_\tau \, d\xi = c \int_{t_1}^{t} \frac{d}{ds} u(x + c(t - s), s) \, ds = c \left(u(x, t) - u(x_1, t_1) \right).$$

Similarly, we have

$$-\int_{C_2} c^2 u_\xi \, d\tau + u_\tau \, d\xi = c\left(u(x,t) - u(x_2,t_2)\right).$$

Therefore, from (9.29), we conclude that

$$u(x,t) = \frac{1}{2}\left(u(x_1,t_1) + u(x_2,t_2)\right) + \frac{1}{2c}\int_C (u_\xi \, d\tau + u_\tau \, d\xi) + \frac{1}{2c}\iint_D f(\xi,\tau)\, d\xi d\tau. \quad (9.30)$$

For the formula (9.30) to be valid, certain restrictions on the initial curve apply. Our basic assumption is that the two characteristics from any point (x,t) intersect the initial curve exactly at two points; this is the *transversal* condition. Another way to see this is the following: as the point (x,t) approaches the initial curve, we want the points (x_1,t_1) and (x_2,t_2) approach the point (x,t). In other words, the characteristic triangle should collapse to a point. It is not difficult to see that if the initial curve $t = \phi(x)$ has the property that $\phi' > 1/c$ at any point, then the above-mentioned property does not hold and so, the formula (9.30) does not describe a solution of the wave equation. This leads to the following definition:

Definition 9.6. A smooth curve $t = \phi(x)$ is said to be *space-like* if $|\phi'(x)| < 1/c$ for all x and *time-like* if $|\phi'(x)| > 1/c$ for all x. □

Thus, space-like curves have the absolute values of their slopes $< 1/c$. For example, when $c = 1$, the functions ϕ defined by

$$\phi(x) = \begin{cases} \exp\left(-\frac{1}{x^2}\right), & \text{for } x \neq 0, \\ 0, & \text{for } x = 0, \end{cases}$$

gives a space-like curve. It is now clear that the formula (9.30) for the solution is valid for space-like curves, but not for time-like curves.

Riemann's Method and Goursat's Problem: We now consider a general second-order linear hyperbolic equation and describe a method, due to Riemann, to obtain the solution for a Cauchy problem with initial conditions prescribed on a general curve. Using the characteristic variables (if necessary), it suffices to consider the equation in the following normal form:

$$L(u) \equiv u_{xy} + au_x + bu_y + cu = f, \quad (9.31)$$

where a, b, c are C^1 functions defined in a domain in x, y plane. Note that the lines $x =$ constant and $y =$ constant are the characteristic curves for (9.31). The Cauchy problem for (9.31) consists in finding a solution u satisfying the prescribed initial condition for u, u_x

and u_y on a non-characteristic curve C in D, at least locally in a neighborhood of the initial curve C.

We remark that when the coefficients a, b and c are constants, it is possible to absorb the terms au_x and bu_y in the term u_{xy}, but not possible to do so for the term cu. A change of variable of the form

$$u(x, y) = v(x, y)\exp(\tilde{a}x + \tilde{b}y),$$

for appropriate constants $\tilde{a}$ and $\tilde{b}$ can be used to absorb the term $au_x + bu_y$.

The idea is to multiply (9.31) by a suitable function v, integrate the resulting equation over a characteristic triangle in D and apply the Green's theorem to find a formula for the solution. We now turn to the details.

Fix a point $P = (x_0, y_0) \in D\backslash C$. We wish to derive, following Riemann, a formula for $u(P)$. Let PAB be the characteristic triangle as shown in Figure 9.9, where PA is part of the characteristic $y = y_0$, BP part of the characteristic $x = x_0$ and AB is part of the initial curve C.

Let

$$L^*v = v_{xy} - av_x - bv_y + (c - a_x - b_y)v,$$

be the *adjoint* of L, operating on C^2 functions v. The form of L^* is derived from the requirement that $vLu - uL^*v$ is in divergence form, for all C^2 functions u and v. We have

$$\begin{aligned} vLu - uL^*v &= v(u_{xy} + au_x + bu_y + cu) \\ &\quad - u(v_{xy} - av_x - bv_y + (c - a_x - b_y)v) \\ &= (vu_x)_y - (uv_y)_x + (auv)_x + (buv)_y. \end{aligned}$$

We also can write

$$(vu_x)_y - (uv_y)_x = (vu_y)_x - (uv_x)_y.$$

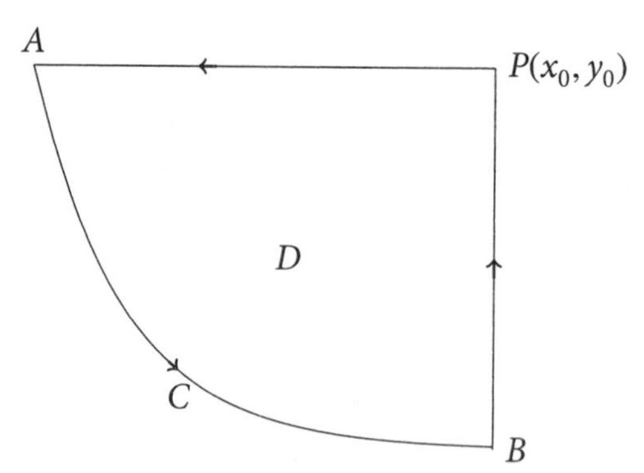

Figure 9.9 A characteristic triangle

Therefore,

$$vLu - uL^*v = \frac{\partial H}{\partial x} + \frac{\partial K}{\partial y} \tag{9.32}$$

with

$$H = \frac{1}{2}\left(vu_y - uv_y\right) + auv \text{ and } K = \frac{1}{2}\left(vu_x - uv_x\right) + buv \tag{9.33}$$

Hence, if D denotes the region bounded by the characteristic triangle PAB, applying the Green's theorem, we obtain that

$$\iint_D (vLu - uL^*v)\,dxdy = \int_{PAB} H\,dy - K\,dx. \tag{9.34}$$

We have, performing an integration by parts,

$$\int_P^A vu_x\,dx = v(A)u(A) - v(P)u(P) - \int_P^A uv_x\,dx.$$

Similarly,

$$\int_B^P vu_y\,dy = v(P)u(P) - v(B)u(B) - \int_B^P uv_y\,dy.$$

Substituting these expressions into (9.34), we obtain after some simplification that

$$\begin{aligned}\iint_D vf\,dxdy &= v(P)u(P) - \frac{1}{2}(v(A)u(A) + v(B)u(B)) \\ &+ \int_B^P u(av - v_y)\,dy + \int_P^A u(v_x - bv)\,dx \\ &+ \int_{AB} \left(\frac{1}{2}(vu_y - uv_y) + auv\right) dy - \left(\frac{1}{2}(vu_x - uv_x) + buv\right) dx \\ &+ \iint_D uL^*v\,dxdy.\end{aligned} \tag{9.35}$$

Here we have used the relation $Lu = f$. We now make the following assumptions on the C^2 function v, that is v solves the adjoint system:

$$\begin{aligned} &L^*v = 0, \text{ in } D, \\ &v_y = av, \text{ on } x = x_0, \\ &v_x = bv, \text{ on } y = y_0, \\ &v(x_0, y_0) = v(P) = 1. \end{aligned} \tag{9.36}$$

Assuming that such a function v exists, which is called the *Riemann function*[7] for the given equation, it follows from (9.35) that

$$\begin{aligned} u(P) = &\frac{1}{2}(v(A)u(A) + v(B)u(B)) \\ &+ \int_{AB} \left(\frac{1}{2}(vu_y - uv_y) + auv\right) dy - \left(\frac{1}{2}(vu_x - uv_x) + buv\right) dx \\ &+ \iint_D vf\,dxdy. \end{aligned} \tag{9.37}$$

Note that every term on the right of (9.37) is known in terms of the initial conditions prescribed on the initial curve and the inhomogeneous term f. Thus, (9.37) gives a formula for the solution u.

Existence of Riemann Function: It remains to prove the existence of the Riemann function satisfying (9.36). This is not a Cauchy problem, as the data is prescribed on characteristics. This is called *Goursat problem*. The equations in (9.36), except the first one, immediately give that

$$v(x_0, y) = \exp\left(\int_{y_0}^{y} b(x_0, s)\,ds\right) \text{ and } v(x, y_0) = \exp\left(\int_{x_0}^{x} a(s, y_0)\,ds\right). \tag{9.38}$$

The existence of a Riemann function will be proved by converting the differential equation in (9.36) into an integral equation. For this, we again use Green's theorem, by choosing $u \equiv 1$ that implies $Lu = c$. Now integrate over the domain Ω enclosed by the rectangle as shown in Figure 9.10. With $u \equiv 1$ and $L^*v = 0$, we have

$$cv = vLu - uL^*v = \frac{\partial}{\partial x}\left(-\frac{1}{2}v_y + av\right) + \frac{\partial}{\partial y}\left(-\frac{1}{2}v_x + bv\right);$$

[7] This is called *Riemann–Green function* in Copson (1975). But Riemann function differs from a Green function as it is not dependent on the initial curve (see John, 1971, 1975).

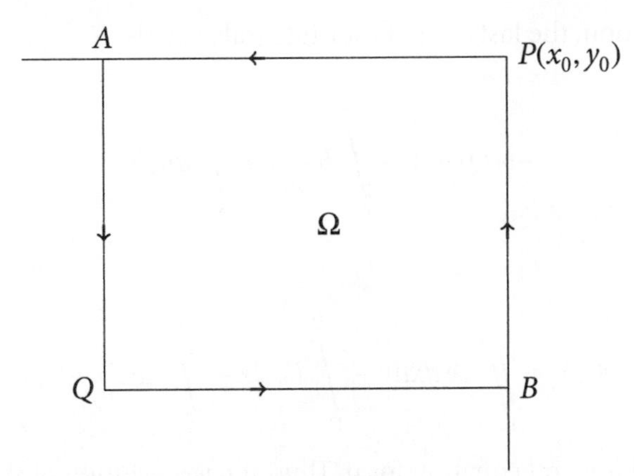

Figure 9.10 Domain for Goursat problem

compare this with (9.32). Therefore, integrating over Ω (see Figure 9.10) and using the Green's theorem, we have

$$\iint_{\Omega} cv\, dxdy = \int_{\partial\Omega} \left(-\frac{1}{2}v_y + av\right) dy - \left(-\frac{1}{2}v_x + bv\right) dx,$$

where $\partial\Omega$ is taken in counter-clockwise direction. Since $\partial\Omega$ is the union of four lines PA, AQ, QB and BP, the line integral on the right-hand side in the above expression is written as the sum of four integrals:

$$\int_P^A \left(\frac{1}{2}v_x - bv\right) dx + \int_Q^B \left(\frac{1}{2}v_x - bv\right) dx$$
$$+ \int_B^P \left(-\frac{1}{2}v_y + av\right) dy + \int_A^Q \left(-\frac{1}{2}v_y + av\right) dy.$$

Using the boundary conditions in (9.36), the first and third integrals can be simplified and we see that the above sum of four integrals simplifies to

$$-\frac{1}{2}\int_P^A v_x\, dx + \int_Q^B \left(\frac{1}{2}v_x - bv\right) dx + \frac{1}{2}\int_B^P v_y\, dy + \int_A^Q \left(-\frac{1}{2}v_y + av\right) dy.$$

Finally, upon integration, the last sum of four integrals equals

$$-v(Q) + 1 - \int_Q^B bv\,dx + \int_A^Q av\,dy.$$

Therefore,

$$v(Q) = \iint_\Omega cv\,dxdy - \int_Q^B bv\,dx + \int_A^Q av\,dy + 1, \tag{9.39}$$

which is the required integral equation for v. Thus, if v is a solution of the Goursat problem (9.36), then v satisfies the integral equation (9.39).

Conversely, it is not difficult to see that if v is a continuous solution of the integral equation (9.39), then v is a solution of the Goursat problem (9.36). We leave this as an exercise. Thus, solving the integral equation (9.39) is equivalent to solving the Goursat problem, which gives the required Riemann function.

We write the integral equation (9.39) as

$$v = Tv + 1,$$

with

$$Tv = \iint_\Omega cv\,dxdy - \int_Q^B bv\,dx + \int_A^Q av\,dy.$$

We think of T as a linear operator acting on continuous functions v defined in a neighborhood of the point $P(x_0, y_0)$. We wish to invoke the classical Banach fixed-point theorem in order to prove the existence and uniqueness of a solution v to the equation $v = Tv + 1$. For this purpose, consider the closed rectangle $\overline{\Omega}$ in which the point $P(x_0, y_0)$ is fixed, the point A is on the line $y = y_0$, the point B is on the line $x = x_0$ and the point $Q(x, y)$ is varying. Let $C(\overline{\Omega})$ be the space of all continuous functions v defined on $\overline{\Omega}$. Then, $C(\overline{\Omega})$ becomes a Banach space with the usual sup norm and T is a linear operator from $C(\overline{\Omega})$ into itself. Since the coefficients a, b, c are assumed to be continuous, we see that

$$|Tv|_\infty \le \frac{1}{2}|v|_\infty,$$

provided that $|x - x_0| + |y - y_0|$ is sufficiently small. Thus, we choose A and B close to the point $P(x_0, y_0)$ so that $|x - x_0| + |y - y_0|$ is sufficiently small for all $(x, y) \in \overline{\Omega}$, which in turn gives the above estimate. Here $|v|_\infty$ denotes the sup norm of v taken over $\overline{\Omega}$. Thus, T is a contraction and therefore the integral equation (9.39) has a unique solution. We have thus proved the following:

Theorem 9.7. For $|x - x_0| + |y - y_0|$ is sufficiently small, the integral equation (9.39) has a unique continuous solution defined on $\overline{\Omega}$. □

9.6.1 An Example

We now consider an example to illustrate the Riemann's method. Consider the following general linear hyperbolic equation in two variables x and t:

$$u_{tt} - a^2 u_{xx} + cu_t + du_x + eu = f$$

where $a > 0$ is a constant and the coefficients c, d, e are constants and f is a given function. It is a simple exercise to see that by a suitable change of variables, the terms cu_t and du_x may be absorbed in the terms involving second-order derivatives.

Example 9.8. Consider the following IVP for the homogeneous equation:

$$\begin{aligned} &u_{tt} - a^2 u_{xx} + \alpha u = 0,\ x \in \mathbb{R},\ t > 0, \\ &u(x, 0) = u_0(x),\ u_t(x, 0) = u_1(x),\ x \in \mathbb{R}. \end{aligned} \tag{9.40}$$

□

Equation (9.40) is referred to as the *telegraph equation*. Here α is a real constant.

We will obtain the solution of the IVP (9.40) by constructing a Riemann function. First some simplifications. If $\alpha > 0$, then by the change of variables $x \mapsto \sqrt{\alpha}x/a$ and $t \mapsto \sqrt{\alpha}t$, reduces equation (9.40) to

$$u_{tt} - u_{xx} + u = 0.$$

If $\alpha < 0$, the equation may be reduced to

$$u_{tt} - u_{xx} - u = 0.$$

Thus, we will assume that $a = 1$ and $\alpha = \pm 1$ in (9.40). First consider the case of positive sign.

Suppose (x_0, t_0), $t_0 > 0$ is an arbitrary point where we wish to find the solution. Suppose u, v are two C^2 functions satisfying the equation $L(u) \equiv u_{tt} - u_{xx} + u = 0$. Using the identity

$$vL(u) - vL(u) = \frac{\partial}{\partial t}(vu_t - uv_t) - \frac{\partial}{\partial x}(vu_x - uv_x)$$

and using the Green's theorem for the domain V bounded by the characteristic triangle Γ with vertices $P(x_0, t_0)$, $Q(x_0 - t_0, 0)$ and $R(x_0 + t_0, 0)$, we obtain

$$\int_\Gamma (vu_t - uv_t)\, dx + \int_\Gamma (vu_x - uv_x)\, dt = 0,$$

where Γ is taken in the counter-clockwise direction. On the characteristic line RP, we have $x = x_0 + t_0 - t$, $0 \le t \le t_0$ and $dx = -dt$. Similarly, on PQ, we have $x = x_0 - t_0 + t$, $t_0 \ge t \ge 0$ and $dx = dt$. Using this in the above line integral and imposing the condition that $v = 1$ on the lines RP and PQ, we obtain, after doing some algebra,

$$u(x_0, t_0) = \frac{1}{2}(u(x_0 + t_0, 0) + u(x_0 - t_0, 0)) + \frac{1}{2}\int_{x_0-t_0}^{x_0+t_0} (vu_t - uv_t)\, dx \tag{9.41}$$

In order to complete the procedure, it remains to find a solution of the Goursat problem $L(v) = 0$ in V, satisfying the condition $v = 1$ on the lines RP and PQ. This is going to be the Riemann function $v(x, t; x_0, t_0)$ of the problem under consideration.

We look for v in the form of a power series:

$$v(x, t) = \sum_{j=0}^{\infty} v_j \gamma^j$$

with $\gamma = (t - t_0)^2 - (x - x_0)^2$; here v_j are functions of x, t to be determined. Note that $\gamma \ge 0$ in V and $= 0$ on the lines RP and PQ. Thus, if we take $v_0 \equiv 1$, then v satisfies the required condition on the lines RP and PQ. Next, the requirement $L(v) = 0$ gives the following recursion relations for v_j:

$$L(v_j) + 4(j + 1)\left[(t - t_0)\frac{\partial v_{j+1}}{\partial t} + (x - x_0)\frac{\partial v_{j+1}}{\partial x}\right] + 4(j + 1)^2 v_{j+1} = 0, \tag{9.42}$$

for $j = 0, 1, \ldots$. The above equations are certainly satisfied if we take all v_j to be constants satisfying the recursion relations

$$v_{j+1} = -\frac{L(v_j)}{4(j + 1)^2} = -\frac{v_j}{4(j + 1)^2}.$$

Since $v_0 = 1$, these relations give that $v_j = \frac{(-1)^j}{4^j (j!)^2}$ for $j = 0, 1, 2, \ldots$. Thus, we have the Riemann function

$$v(x, t; x_0, t_0) = \sum_{j=0}^{\infty} \frac{(-1)^j \gamma^j}{4^j (j!)^2} = J_0(\sqrt{\gamma}), \tag{9.43}$$

where J_0 is the Bessel's function of the first kind of order zero (see, for instance, Abramowitz and Stegun, 1972). Plugging this expression of the Riemann function v into (9.41), we finally obtain the expression for the solution of the IVP (9.40) (with $a = 1$ and $\alpha = 1$):

$$u(x_0, t_0) = \frac{1}{2}(u_0(x_0 + t_0) + u_0(x_0 - t_0)) + \frac{1}{2}\int_{x_0-t_0}^{x_0+t_0} \left(u_1(x)J_0(s) + u_0(x)\frac{\partial J_0}{\partial t_0}(s) \right) dx \quad (9.44)$$

with $s = \sqrt{t_0^2 - (x - x_0)^2}$. If we take $\alpha = -1$, then we only need to change J_0 to I_0 in the representation (9.44), where I_0 is the Bessel's function of the second kind of order zero: $I_0(\sqrt{\gamma}) = \sum_{j=0}^{\infty} \frac{\gamma^j}{4^j(j!)^2}, \gamma \geq 0.$

In the next chapter, we derive the same formula using the solution of the wave equation in two dimensions and Hadamard's method of descent.

9.7 NOTES

In this chapter, we have dealt with one-dimensional wave equation in detail, by considering the Cauchy (initial value) problem, initial-boundary value problems in a quarter plane and a bounded interval. The importance of characteristics, characteristic variables is stressed throughout. The continuous dependence of the solution on the data is shown in a restricted sense. In general, the hyperbolic problems the useful estimates for a solution are the so-called *energy estimates*, which are part and parcel of general existence, uniqueness and other related results. As we see later, the methods discussed in this chapter do not readily extend to equations in higher (space) dimensions; nevertheless, they play an important role in problems with some symmetry, where the number of variables may be brought down to one. We have also considered general second-order linear equations in two variables. A formula for the solution, in a small neighborhood of the initial curve, is obtained using the Riemann function. Of course, we cannot expect the solution to be in a neat form as in D'Alembert's formula, owing to the variable coefficients and lower-order terms.

The material covered here is quite standard. Some references are Prasad and Ravindran (1996), Koshlyakov *et al.* (1964), Rubinstein and Rubinstein (1998), Renardy and Rogers (2004), and John (1971, 1975, 1978).

9.8 EXERCISES

1. Let u be a C^2 solution of the wave equation $u_{tt} - c^2u_{xx} = f(x, t)$ in the upper half plane $x \in \mathbb{R}$, $t > 0$ satisfying the initial conditions $u(x, 0) = u_0(x)$, $u_t(x, 0) = u_1(x)$, $x \in \mathbb{R}$. By integrating the wave equation over the characteristic triangle with vertices (x, t), $(x - ct, 0)$

and $(x+ct, 0)$, and using Green's theorem. (Hint: Write $u_{tt} - c^2 u_{xx} = (u_t)_t - c^2(u_x)_x$), derive the formula for the solution and compare the same with one given in the text.)

This gives one more proof of the uniqueness of the solution to the IVP.

2. If $c_1, \ldots, c_k$ are distinct real numbers, show that the general solution of the one-dimensional equation $\prod_{j=1}^{k}(\partial_t - c_j\partial_x)u = 0$ is given by $u(x,t) = \sum_{j=1}^{k} F_j(x + c_j t)$ where F_j are smooth functions. (Hint: Use an induction argument.)
3. If c is a real number and $k \geq 1$ is an integer, show that the general solution of the one-dimensional equation $(\partial_t - c\partial_x)^k u = 0$ is given by $u(x,t) = \sum_{j=1}^{k} t^{j-1} F_j(x + c_j t)$ where F_j are smooth functions. (Hint: Use an induction argument.)
4. Derive the formula (9.24) using the characteristic parallelogram property.
5. Provide the details of the statement made in Remark 9.4 and the statements made after that regarding initial-boundary value problems in the first quadrant.
6. (Mixed boundary value problem) Consider the wave equation (9.21) in the first quadrant and impose the following mixed boundary condition on the boundary $x = 0$:

$$u_t + \alpha u_x = 0, \; x = 0, t > 0,$$

and the initial conditions at $t = 0$, $x > 0$.

   a. If $\alpha \neq c$, derive a formula for the solution.
   b. If $\alpha = c$, show that a solution in general does not exist, but exists if the initial conditions satisfy some additional conditions. Interpret the boundary condition in this case geometrically.

7. Prove the equivalence statement made in the text regarding the solutions of the integral equation (9.39) and the Goursat problem (9.36).

CHAPTER 10

Wave Equation in Higher Dimensions

10.1 INTRODUCTION

In this chapter, we study the wave equation in higher (space) dimensions and analyze different problems associated with it: the Cauchy problem (initial value problem [IVP]), initial-boundary value problem in half-space, and so on. As noted in the introductory chapter, the wave equation arises in many physical contexts and it is a fundamental equation that has influenced the analysis of solutions of general hyperbolic equations and systems. Unlike the heat equation, the nature of solution of the wave equation depends on being the (space) dimension odd or even, except for one-dimensional case. This is also the reason to take up separately the study of the wave equation in higher dimensions. We also learn that the solution of the wave equation in even dimensions may be obtained from the solution in odd dimensions, by the *method of descent*. We begin with the following Cauchy problem for the homogeneous wave equation in the free space $\mathbb{R}^n$:

$$\Box_c u \equiv u_{tt} - c^2 \Delta u = 0,\ x \in \mathbb{R}^n, t > 0, \tag{10.1}$$

$$u(x,0) = u_0(x),\ u_t(x,0) = u_1(x),\ x \in \mathbb{R}^n. \tag{10.2}$$

Here $n \geq 2$ is an integer, the *(spatial) dimension*, $c > 0$ is a constant, the *speed of propagation* and u_0, u_1 are given smooth functions, the initial values. We describe two methods to find a formula for the solution of (10.1) and (10.2).

The general references for this chapter are Ladyzhenskaya (1985), Rauch (1992), Mitrea (2013), Pinchover and Rubinstein (2005), McOwen (2005), Trèves (2006), Courant and Hilbert (1989), John (1971, 1975, 1978), DiBenedetto (2010), Renardy and Rogers (2004), Prasad and Ravindran (1996), Salsa (2008), Mikhailov (1978), Benzoni-Gavage and Serre (2007), Evans (1998), Kreiss and Lorenz (2004), and Vladimirov (1979, 1984).

10.2 THREE-DIMENSIONAL WAVE EQUATION: METHOD OF SPHERICAL MEANS

To get an idea how this method works, we first consider a special case. Suppose the functions u_0 and u_1 are *radial* functions, that is,

$$u_0(x) = u_0(|x|) \text{ and } u_1(x) = u_1(|x|),$$

where $r^2 \equiv |x|^2 = x_1^2 + \cdots + x_n^2$. Also, extend u_0 and u_1 for $r < 0$ by defining $u_0(-r) = u_0(r)$ and $u_1(-r) = u_1(r)$. In this situation, we can expect a solution u of (10.1) also to be radial in x, that is $u(x,t) = u(r,t)$. For such a function u, we have

$$\Delta u = \frac{\partial^2}{\partial r^2} + \frac{n-1}{r}\frac{\partial}{\partial r}.$$

Therefore, (10.1) reduces to

$$\frac{\partial^2 u}{\partial t^2} - c^2\left(\frac{\partial^2}{\partial r^2} + \frac{n-1}{r}\frac{\partial}{\partial r}\right)u = 0. \tag{10.3}$$

Consider the case $n = 3$. Then, the function $v = ru$ satisfies the one-dimensional equation

$$v_{tt} - c^2 v_{rr} = 0,$$

with initial conditions $v(r,0) = ru_0(r)$ and $v_t(r,0) = ru_1(r)$, as follows from (10.3). Thus, v can be easily determined using the D'Alembert's formula and, so is u.

In general, therefore, our strategy would be to find a suitable function from u, which satisfies (10.3). Then, from the one-dimensional case, treating r as the space variable, we will be able to derive an expression for the solution u, at least in the case $n = 3$. The suitable function we are looking for turns out to be the *spherical mean function* of u, namely the averages of u over the spheres around a given point x. We have already encountered this object in the discussion of MVP for harmonic functions. As we see below, it is easy to obtain u from its spherical mean function.

Spherical Mean Function: Given a C^2 function h defined on $\mathbb{R}^n$, define its *spherical mean function*, denoted by M_h, by

$$M_h(x,r) = \frac{1}{\sigma_n r^{n-1}} \int_{|x-y|=r} h(y)\,dS(y), \tag{10.4}$$

for $x \in \mathbb{R}^n$ and $r > 0$. The integration is over the sphere of radius r, centred at x and $\sigma_n r^{n-1}$ is the surface measure of this sphere with $\sigma_n = 2\pi^{n/2}/\Gamma(n/2)$ denoting the surface measure of the unit sphere in $\mathbb{R}^n$; Γ is the Euler gamma function. By a change of variable, (10.4) can be written as

$$M_h(x,r) = \frac{1}{\sigma_n} \int\limits_{|\xi|=1} h(x+r\xi)\, dS(\xi). \tag{10.5}$$

It is to be noted that we are keeping x fixed and varying r; that is, we consider all the averages around x. The form of (10.5) enables us to define M_h for all real r and it is readily seen that $M_h(x,-r) = M_h(x,r)$, that is M_h is an even function of r. Next we compute $\frac{\partial}{\partial r} M_h(x,r)$. Using (10.5) and (10.4), We have,

$$\begin{aligned}
\frac{\partial}{\partial r} M_h(x,r) &= \frac{1}{\sigma_n} \int\limits_{|\xi|=1} \sum_{i=1}^{n} \frac{\partial h}{\partial x_i}(x+r\xi)\xi_i\, dS(\xi) \\
&= \frac{r}{\sigma_n} \int\limits_{|\xi|<1} \Delta_x h(x+r\xi)\, d\xi, \text{ using divergence theorem,} \\
&= \frac{1}{\sigma_n r^{n-1}} \Delta_x \int\limits_{|y-x|<r} h(y)\, dy, \text{ by the change of variables,} \\
&= \frac{1}{\sigma_n r^{n-1}} \Delta_x \int_0^r d\rho \int\limits_{|y-x|=\rho} h(y)\, dS(y), \text{ by using spherical coordinates,} \\
&= \frac{1}{r^{n-1}} \Delta_x \int_0^r \rho^{n-1} M_h(x,\rho)\, d\rho, \text{ using (10.4).}
\end{aligned}$$

This immediately gives

$$\frac{\partial}{\partial r}\left(r^{n-1} \frac{\partial}{\partial r} M_h(x,r)\right) = r^{n-1} \Delta_x M_h(x,r).$$

The above equation can be written as

$$\left(\frac{\partial^2}{\partial r^2} + \frac{n-1}{r}\frac{\partial}{\partial r}\right) M_h(x,r) = \Delta_x M_h(x,r). \tag{10.6}$$

This is called the *Darboux equation* (The notation Δ_x in the above expressions means the Laplacian taken with respect to the x variables). We also have

$$M_h(x,0) = h(x) \text{ and } \frac{\partial}{\partial r} M_h(x,0) = 0.$$

Case $n = 3$ (Euler–Poisson–Darboux Equation): Though the equation (10.6) is valid for all $n \geq 2$, the analysis is simpler when $n = 3$. In fact, it is now straightforward to obtain an expression for a C^2 solution $u(x, t)$ of (10.1) and (10.2) for general smooth initial data when $n = 3$. The spherical mean function $M_u(x, r, t)$ of u satisfies the equation[1]

$$\frac{\partial^2}{\partial t^2} M_u(x, r, t) = c^2 \left(\frac{\partial^2}{\partial r^2} + \frac{2}{r}\frac{\partial}{\partial r} \right) M_u(x, r, t), \tag{10.7}$$

in view of (10.6) and (10.1). This is a partial differential equation (PDE) in r and t variables and the variable x plays the role of a parameter. Equation (10.7) is called the *Euler–Poisson–Darboux equation.* As seen earlier, the function v defined by $v(x, r, t) = rM_u(x, r, t)$ satisfies the one-dimensional wave equation

$$v_{tt} - c^2 v_{rr} = 0,$$

with initial conditions

$$v(x, r, 0) = rM_u(x, r, 0) = rM_{u_0}(x, r)$$

and

$$\frac{\partial}{\partial t} v(x, r, 0) = r\frac{\partial}{\partial t} M_u(x, r, 0) = rM_{u_1}(x, r).$$

Hence, by D'Alembert's formula, we get

$$v(x, r, t) = \frac{1}{2} \left[(r + ct)M_{u_0}(x, r + ct) + (r - ct)M_{u_0}(x, r - ct) \right] + \frac{1}{2c} \int_{r-ct}^{r+ct} sM_{u_1}(x, s)\, ds.$$

Since $M_{u_0}(x, r)$ and $M_{u_1}(x, r)$ are even functions of r, we obtain

$$M_u(x, r, t) = \frac{1}{2r} \left[(ct + r)M_{u_0}(x, r + ct) - (ct - r)M_{u_0}(x, ct - r) \right] + \frac{1}{2rc} \int_{ct-r}^{ct+r} sM_{u_1}(x, s)\, ds,$$

---

[1] For general n, the function M_u satisfies the Euler–Poisson–Darboux equation

$$\frac{\partial^2}{\partial t^2} M_u(x, r, t) = c^2 \left(\frac{\partial^2}{\partial r^2} + \frac{n-1}{r}\frac{\partial}{\partial r} \right) M_u(x, r, t).$$

for $r \neq 0$. Letting $r \to 0$, the left side tends to $u(x,t)$. The second term on the right tends to $tM_{u_1}(x,ct)$ as $r \to 0$. We convert the limit of the first term on the right as $r \to 0$ to that of a derivative[2] to see that it tends to $\frac{\partial}{\partial t}(tM_{u_0}(x,ct))$. Thus,

$$u(x,t) = tM_{u_1}(x,ct) + \frac{\partial}{\partial t}(tM_{u_0}(x,ct))$$
$$= \frac{1}{4\pi c^2 t} \int_{|y-x|=ct} u_1(y)\,dS(y) + \frac{\partial}{\partial t}\left(\frac{1}{4\pi c^2 t} \int_{|y-x|=ct} u_0(y)\,dS(y)\right). \tag{10.8}$$

The representation (10.8) is known as *Kirchhoff's formula.* By carrying out the t differentiation, we can also write Kirchoff's formula as follows: We have

$$\frac{\partial}{\partial t}\left(tM_{u_0}(x,ct)\right) = M_{u_0}(x,ct) + t\frac{\partial}{\partial t}M_{u_0}(x,ct)$$

and

$$\frac{\partial}{\partial t}M_{u_0}(x,ct) = \frac{1}{4\pi c^2 t^2} \int_{|y-x|=ct} \nabla u_0(y)\cdot(y-x)\,dS(y).$$

Thus, Kirchoff's formula (10.8) is rewritten as

$$u(x,t) = \frac{1}{4\pi c^2 t^2} \int_{|y-x|=ct} \left[tu_1(y) + u_0(y) + \nabla u_0(y)\cdot(y-x)\right]\,dS(y). \tag{10.9}$$

The above formula brings out the essential features of the solution in the case $n = 3$. Thus, any C^2 solution of the Cauchy problem (10.1) and (10.2) is given by (10.8) and hence unique. We state the foregoing discussion as a theorem:

Theorem 10.1. Suppose $u_0 \in C^3(\mathbb{R}^3)$ and $u_1 \in C^2(\mathbb{R}^3)$, then the function u given by Kirchoff's formula (10.8) or (10.9) is the C^2 solution of the Cauchy problem (10.1) and (10.2) for the wave equation in the case $n = 3$. □

Notice that the solution u is *less regular* than the initial data, due to the presence of ∇u_0 in (10.9). This is known as the *focusing effect*, present when $n = 3$ (more generally for $n > 1$). Earlier, we have observed that for $n = 1$, the solution is as smooth as the initial data. In contrast to this point-wise behavior of u, the *energy* of u for any time $t > 0$ is the same as

[2] Let $\phi(t) = ctM_{u_0}(x,ct)$. Then,

$$\frac{d\phi}{dt} = \lim_{r\to 0} \frac{\phi(t+r) - \phi(t-r)}{2r}.$$

the initial energy. The energy of u at time t is defined by

$$E(t) = \frac{1}{2} \int_{\mathbb{R}^3} \left(u_t^2(x,t) + c^2 |\nabla u(x,t)|^2 \right) dx, \tag{10.10}$$

where the integration is over $\mathbb{R}^3$ and ∇u denotes the gradient vector of u with respect to the x variables. The integral in (10.10) is assumed to be finite and represents the sum of kinetic energy and potential energy.

Differentiating (10.10) with respect to t and integrating by parts in the second term of the integral, we see that $\frac{dE}{dt} = 0$, provided that $u(x,t)$ vanishes for sufficiently large $|x|$. Thus,

$$E(t) = E(0) = \frac{1}{2} \int_{\mathbb{R}^3} \left(u_1^2(x,t) + c^2 |\nabla u_0(x)|^2 \right), dx$$

the initial energy.

Remark 10.2. Thus, while studying the existence and uniqueness of solutions to general hyperbolic equations, it is natural to seek solutions with *finite energy*, which is defined by an expression similar to (10.10). □

Huyghens' Principle: We now analyze (10.8) in more detail. At a point (x,t), $t > 0$, the value of $u(x,t)$ depends on the values of the initial data only on the sphere $S(x, ct) \equiv \{y : |y - x| = ct\}$, of radius ct centred at x. This is the *domain of dependence* (see Figure 10.1). Similarly, the values of the initial data at a point y on the initial space $t = 0$ *influence* the value $u(x,t)$, $t > 0$, only if $|x - y| = ct$, that is x lies on the sphere[3] $S(y, ct)$. This is termed as *range of influence* of the point y (see Figure 10.2).

Suppose the initial data u_0 and u_1 have support in the closed ball $B_\rho(0) \equiv \{x : |x| \le \rho\}$ in the space $t = 0$. At a time $t > 0$, u_0 and u_1 can influence the value of $u(x,t)$ if $x \in \cup\{S(y, ct) : y \in B_\rho(0)\}$, which is the union of all the spheres of radius ct, centred at $y \in B_\rho(0)$. More precisely, if $|x - y| = ct$ and $|y| > \rho$, then $u(x,t) = 0$. This will happen if $|x| > \rho + ct$ and $|x - y| = ct$ as

$$|y| = |y - x + x| \ge |x| - |y - x| > (\rho + ct) - ct = \rho.$$

Also, for any x, if we choose t large enough, then any y satisfying $|y - x| = ct$ lies outside $B_\rho(0)$. For,

$$|y| \ge |y - x| - |x| = ct - |x| > \rho,$$

[3]The name *speed of propagation* for c comes from this observation.

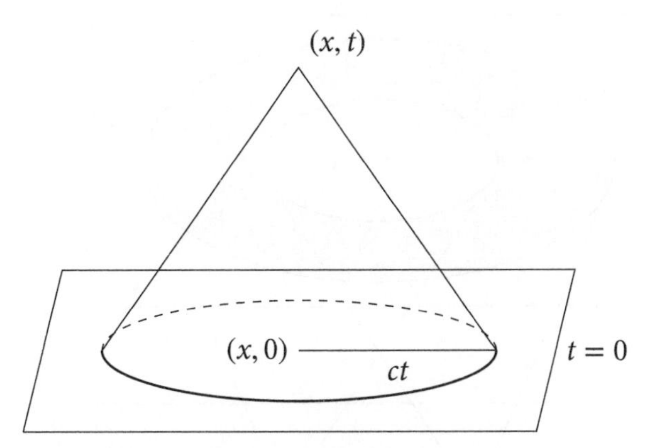

Figure 10.1 Domain of dependence

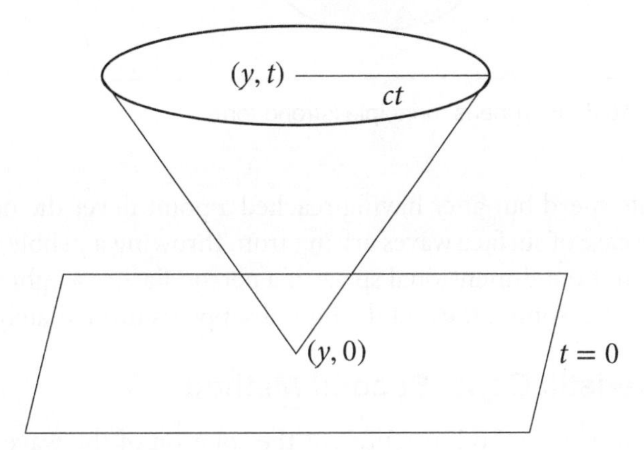

Figure 10.2 Range of influence

provided that $t > (|x| + \rho)/c$. Thus, the support of $u(x, t)$, as a function of x, lies in the spherical shell bounded by the spheres $S(0, ct + \rho)$ and $S(0, ct - \rho)$, provided that t is sufficiently large. Thus, physically, the disturbance originating from a source situated in $B_\rho(0)$ at $t = 0$, will only spread to a shell of thickness 2ρ expanding with velocity c. More precisely, the disturbance occurred in $B_\rho(0)$ at $t = 0$, is felt at some point x at a later time t only if x is situated in the spherical shell bounded by the spheres $S(0, ct + \rho)$ and $S(0, ct - \rho)$. Thus, the disturbance felt at x disappears completely in a finite time. This is termed as the *Huyghens' principle in the strong form* (see Figure 10.3).

This phenomenon is due to the fact that the domain of dependence for the solution $u(x, t)$ is a *surface* in x-space rather than a solid region. Later, when we derive a formula for the solution of the wave equation in arbitrary dimensions, we will see that the Huyghens' principle in the strong form persists in all *odd* dimensions larger or equal to 3. On the other hand, we will see that this principle is *not* true in even dimensions and we have already seen that for $n = 1$ this principle is not true. This means that in even dimensions, the disturbances

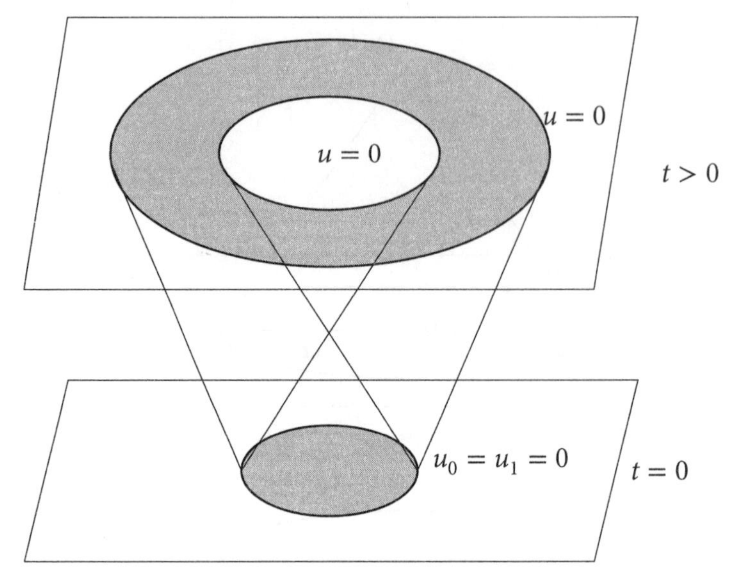

Figure 10.3 Huyghens' principle (strong form)

propagate with finite speed but after having reached a point never die out completely in a *finite* time, as in the case of surface waves arising from throwing a pebble into a water pond. On the other hand in three-dimensional space, if a person flashes a light at an observer, the observer sees the light at some time and the light disappears immediately.

10.2.1 Characteristic Cone: Second Method

We now describe another method to represent the solution of the wave equation in three dimensions. This method is similar to the one used for one-dimensional wave equation, where we integrated the wave equation over a characteristic triangle to obtain the formula for the solution. For the wave equation (10.1), a *characteristic surface* or simply *characteristic* is described as follows:

A smooth surface $\phi(x,t) = 0$ in the $x-t$ space is called a *characteristic surface* (of the wave equation (10.1)) if $\omega(\phi) \equiv \phi_t^2 - c^2|\nabla_x \phi|^2 = 0$ on the surface $\phi(x,t) = 0$; this surface is called a *space-like surface* (respectively, a *time-like surface*) if $\omega(\phi) > 0$ (respectively, $\omega(\phi) < 0$). The surface

$$c^2(t-t_0)^2 - |x-x_0|^2 = 0, \tag{10.11}$$

is a characteristic surface of the wave equation and it is called a *characteristic cone*, with vertex at the point (x_0, t_0). The characteristic cone (10.11) is the boundary of the cones

$$\Gamma^+(x_0,t_0) = \{(x,t) : c(t-t_0) > |x-x_0|\} \text{ and}$$
$$\Gamma^-(x_0,t_0) = \{(x,t) : -c(t-t_0) > |x-x_0|\}.$$

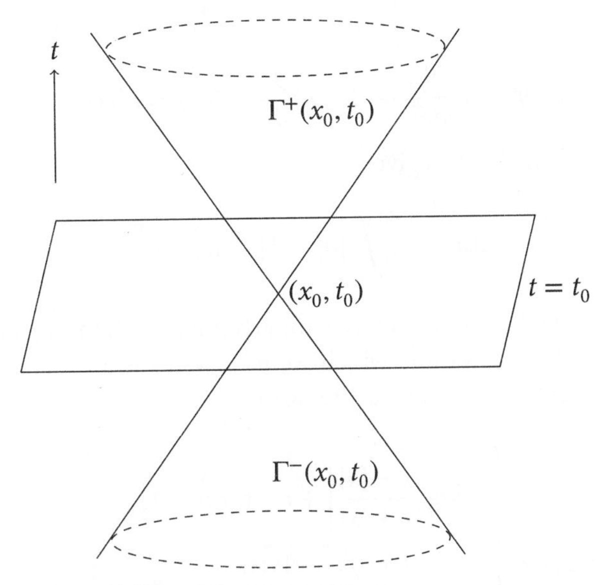

Figure 10.4 Past and future cones

The cones Γ^+ and Γ^- are respectively called the *future* and *past* cones with vertex at (x_0, t_0); (see Figure 10.4). The wave equation has another family of characteristic surfaces, the family of tangent planes to the characteristic cones, namely

$$ct + b \cdot x = C, \tag{10.12}$$

where the vector b and the constant C are arbitrary with $|b| = 1$.

We now proceed to find a formula for the solution of the Cauchy problem for the inhomogeneous wave equation:

$$u_{tt} - \Delta u = f(x, t),\ x \in \mathbb{R}^n, t > 0, \tag{10.13}$$

$$u(x, 0) = u_0(x),\ u_t(x, 0) = u_1(x),\ x \in \mathbb{R}^n. \tag{10.14}$$

Here we have taken the constant $c = 1$, which can always be done by changing the variable t to ct. The functions f, u_0 and u_1 are smooth functions. The procedure below can also be used to derive energy estimates for solutions of general second-order hyperbolic equations (see, for instance, Ladyzhenskaya, 1985).

First we look for some special solution of the homogeneous wave equation (10.13) (that is, $f \equiv 0$) of the form $u(x, t) = v(t/|x|)$, $x \neq 0$. Note that the wave equation (10.1) is invariant under the change of variables: $t \mapsto kt$ and $x \mapsto kx$ for any $k > 0$. Thus, it is but natural to look for a solution in the form of the function v, which would be a constant on the characteristic cone with vertex at $(0, 0)$; see (10.11). Then, v satisfies the ordinary differential equation (ODE)

$$(\eta^2 - 1)\frac{d^2 v}{d\eta^2} + (3 - n)\frac{dv}{d\eta} = 0,\ \eta = t/|x|.$$

The general solution of this ODE is given by

$$v(\eta) = c_1 \int |\eta^2 - 1|^{\frac{n-3}{2}}\, d\eta + c_2,$$

in the intervals $(-\infty, -1), (-1, 1)$ and $(1, \infty)$, for arbitrary constants c_1, c_2. In particular, for $0 < |x| < -t,\ t < 0$, the solution v, which we now write as a function $v(x, t)$ of x and t (though it is a function of $t/|x|$ only) has the form

$$v(x,t) = \begin{cases} c_1 \log\left|\dfrac{t - |x|}{t + |x|}\right| + c_2, & \text{for } n = 1, \\ c_1 \log\dfrac{|t + \sqrt{t^2 - |x|^2}|}{|x|} + c_2, & \text{for } n = 2, \\ c_1 \frac{t}{|x|} + c_2, & \text{for } n = 3. \end{cases}$$

Representation of the Solution when $n = 3$**:** Let u be a C^2 solution of (10.13) and (10.14). Let K be the open cone

$$K = \{(x, t) : x \in \mathbb{R}^3, |x - x_1| < t_1 - t, t_0 < t < t_1\}.$$

Here (x_1, t_1) is a fixed point and $0 \le t_0 < t_1$. Thus, K is part of the past cone with vertex at (x_1, t_1). Denote by Γ the *lateral surface* of K, that is,

$$\Gamma = \{(x, t) : |x - x_1| = t_1 - t, t_0 \le t \le t_1\}$$

and by D its bottom, that is,

$$D = \{(x, t) : |x - x_1| < t_1 - t, t = t_0\}.$$

Assume $f \in C(K \cup D)$. For $(\xi, \tau) \in K$ and $0 < \varepsilon < \tau - t_0$, put

$$K_\varepsilon = \{(x, t) : \varepsilon < |x - \xi| < \tau - t, t_0 < t < \tau\},$$

$$\Gamma_\varepsilon = \{(x, t) : \varepsilon < |x - \xi| = \tau - t, t_0 \le t \le \tau - \varepsilon\},$$

$$D_\varepsilon = \{(x, t) : \varepsilon < |x - \xi| < \tau - t_0, t = t_0\} \text{ and}$$

$$\gamma_\varepsilon = \{(x, t) : |x - \xi| = \varepsilon, t_0 \le t \le \tau - \varepsilon\}.$$

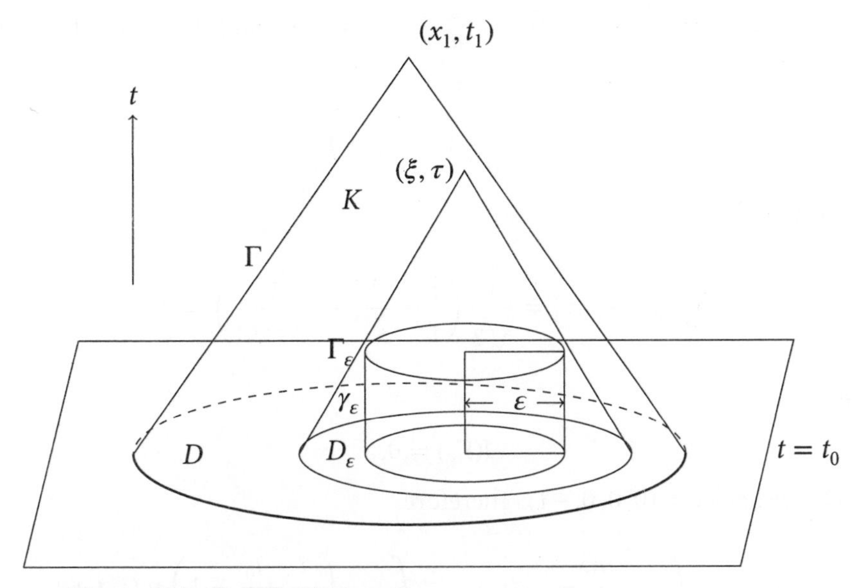

Figure 10.5 Domain of integration

See Figure 10.5. The initial time is t_0 and an expression for $u(\xi, \tau)$ is sought. Consider the function

$$v(x,t) = \frac{t-\tau}{|x-\xi|} + 1.$$

Then, $u, v \in C^2(K_\varepsilon)$ and by Green's identity, we have ($\Box = \Box_1$)

$$v\Box u - u\Box v = -\sum_{i=1}^{3} \partial_i(v\partial_i u - u\partial_i v) + \partial_t(v\partial_t u - u\partial_t v).$$

Note that $\Box v = 0$ in K_ε. Therefore, integrating this identity over K_ε and using the divergence theorem, we obtain that

$$\begin{aligned} \int_{K_\varepsilon} v\Box u \, dxdt &= \int_{\partial K_\varepsilon} \left[-\sum_{i=1}^{3} (v\partial_i u - u\partial_i v)\nu_i + (v\partial_t u - u\partial_t v)\nu_4 \right] dS \\ &= I(\Gamma_\varepsilon) + I(D_\varepsilon) + I(\gamma_\varepsilon), \text{ say.} \end{aligned} \tag{10.15}$$

Here $\nu = (\nu_1, \nu_2, \nu_3, \nu_4)$ is the outward unit normal to the boundary $\partial K_\varepsilon = \Gamma_\varepsilon \cup D_\varepsilon \cup \gamma_\varepsilon$ and dS is the surface measure on this boundary. We now analyze each of these three surface integrals.

On Γ_ε, we have $v = 0$, $\nabla_x v = (t-\tau)\frac{\xi-x}{|\xi-x|^3}$, $v_t = \frac{1}{|x-\xi|}$ and the unit outward normal on Γ_ε is

$$\nu = \frac{1}{\sqrt{2}}\left(\frac{x-\xi}{\tau-t}, 1\right).$$

Consequently, on Γ_ε,

$$\sum_{i=1}^{3} v_{x_i}\nu_i - v_t\nu_4 = \frac{1}{\sqrt{2}}\left(\frac{|x-\xi|^2}{|x-\xi|^3} - \frac{1}{|x-\xi|}\right) = 0.$$

Therefore,

$$I(\Gamma_\varepsilon) = 0. \tag{a}$$

Next, on D_ε, we have $\nu = (0,0,0,-1)$. Therefore,

$$I(D_\varepsilon) = \int\limits_{\varepsilon<|x-\xi|<\tau-t_0} \frac{u(x,t_0)}{|x-\xi|}\,dx + \int\limits_{\varepsilon<|x-\xi|<\tau-t_0} \left(\frac{\tau-t_0}{|x-\xi|} - 1\right) u_t(x,t_0)\,dx.$$

Since $\frac{1}{|x|}$ is locally integrable in $\mathbb{R}^3$ and u is continuous, we obtain that

$$\begin{aligned}\lim_{\varepsilon\to 0} I(D_\varepsilon) = &\int\limits_{|x-\xi|<\tau-t_0} \frac{u(x,t_0)}{|x-\xi|}\,dx \\ &+ \int\limits_{|x-\xi|<\tau-t_0} \left(\frac{\tau-t_0}{|x-\xi|} - 1\right) u_t(x,t_0)\,dx.\end{aligned} \tag{b}$$

Finally, on γ_ε, $\nu = \left(\frac{\xi-x}{|x-\xi|}, 0\right)$. Therefore,

$$\begin{aligned}I(\gamma_\varepsilon) &= -\int_{t_0}^{\tau-\varepsilon} dt \int\limits_{|x-\xi|=\varepsilon} \frac{\partial u}{\partial \nu} v\,dS_x + \int_{t_0}^{\tau-\varepsilon} dt \int\limits_{|x-\xi|=\varepsilon} \frac{\partial v}{\partial \nu} u\,dS_x \\ &= -\int_{t_0}^{\tau-\varepsilon} \left(\frac{t-\tau}{\varepsilon} + 1\right) dt \int\limits_{|x-\xi|=\varepsilon} \frac{\partial u}{\partial \nu}\,dS_x + \int_{t_0}^{\tau-\varepsilon} \frac{t-\tau}{\varepsilon^2}\,dt \int\limits_{|x-\xi|=\varepsilon} u\,dS_x.\end{aligned}$$

On γ_ε, we have $\left|\frac{\partial u}{\partial \nu}\right| \le M$ and therefore, by mean value theorem,

$$|u(x,t) - u(\xi,t)| \le M\varepsilon,$$

where

$$M = \max\{|\nabla u| : |x - \xi| \leq \tau - t, t_0 \leq t \leq \tau\}.$$

Hence,

$$\left| \int_{|x-\xi|=\varepsilon} u \, dS_x \right| \leq 4\pi M \varepsilon^2$$

and

$$\left| \int_{|x-\xi|=\varepsilon} u(x,t) \, dS_x - \int_{|x-\xi|=\varepsilon} u(\xi,t) \, dS_x \right| \leq 4\pi M \varepsilon^3.$$

Therefore,

$$\lim_{\varepsilon \to 0} I(\gamma_\varepsilon) = 4\pi \int_{t_0}^{\tau} (t - \tau) u(\xi, t) \, dt. \tag{c}$$

Substituting the expressions in (a), (b) and (c) into (10.15), we obtain, by letting $\varepsilon \to 0$, that

$$\begin{aligned} 4\pi \int_{t_0}^{\tau} (t - \tau) u(\xi, t) \, dt = &- \int_{|x-\xi|<\tau-t_0} \frac{u(x,t_0)}{|x-\xi|} \, dx \\ &- \int_{|x-\xi|<\tau-t_0} \left(\frac{\tau - t_0}{|x-\xi|} - 1 \right) u_t(x,t_0) \, dx \\ &+ \int_{t_0}^{\tau} dt \int_{|x-\xi|<\tau-t} \left(\frac{t-\tau}{|x-\xi|} + 1 \right) f(x,t) \, dx, \end{aligned}$$

where we have used $\Box u = f$. Differentiating this identity with respect to τ, we get

$$\begin{aligned} 4\pi \int_{t_0}^{\tau} u(\xi, t) \, dt = &(\tau - t_0)^{-1} \int_{|x-\xi|=\tau-t_0} u(x,t_0) \, dS_x \\ &+ \int_{|x-\xi|<\tau-t_0} \left(\frac{u_t(x,t_0)}{|x-\xi|} - 1 \right) u_t(x,t_0) \, dx \\ &+ \int_{t_0}^{\tau} dt \int_{|x-\xi|<\tau-t} \left(\frac{f(x,t)}{|x-\xi|} + 1 \right) dx. \end{aligned}$$

One more differentiation with respect to τ gives

$$u(\xi,\tau) = \frac{\partial}{\partial\tau}\left(\frac{1}{4\pi(\tau-t_0)}\int\limits_{|x-\xi|=\tau-t_0} u(x,t_0)\,dS_x\right) + \frac{1}{4\pi(\tau-t_0)}\int\limits_{|x-\xi|=\tau-t_0} u_t(x,t_0)\,dS_x + \frac{1}{4\pi}\int\limits_{t_0}^{\tau} dt \int\limits_{|x-\xi|=\tau-t} \frac{f(x,t)}{|x-\xi|}\,dS_x.$$

Thus, we arrive at the *Kirchhoff's formula* (replace (ξ,τ) by (x,t)):

$$u(x,t) = \frac{1}{4\pi(t-t_0)}\int\limits_{|x-\xi|=t-t_0} u_t(\xi,t_0)\,dS_\xi + \frac{\partial}{\partial t}\left(\frac{1}{4\pi(t-t_0)}\int\limits_{|x-\xi|=t-t_0} u(\xi,t_0)\,dS_\xi\right) + \frac{1}{4\pi}\int\limits_{|x-\xi|<t-t_0} \frac{f(\xi, t-t_0-|x-\xi|)}{|x-\xi|}\,d\xi. \tag{10.16}$$

This is precisely (10.8) if we take $t_0 = 0, f \equiv 0$ and $c = 1$.

10.3 TWO-DIMENSIONAL WAVE EQUATION: METHOD OF DESCENT

We are now going to discuss the *method of descent*, due to Hadamard, which enables us to obtain the solution of two-dimensional wave equation using that of three-dimensional equation (*descending* from $n = 3$ to $n = 2$). In fact, the procedure can easily be extended to n-dimensional situation.

Consider the Cauchy problem for the homogeneous two-dimensional wave equation and let $u \in C^2(\mathbb{R}^2 \times [0,\infty))$ be the solution of

$$u_{tt} - \Delta u = 0,\ x \in \mathbb{R}^2,\ t > 0, \tag{10.17}$$

$$u(x,0) = u_0(x),\ u_t(x,0) = u_1(x),\ x \in \mathbb{R}^2. \tag{10.18}$$

Define $\tilde{u}(\tilde{x}, t) = u(x, t)$, where $\tilde{x} = (x_1, x_2, x_3) \in \mathbb{R}^3$ and $x = (x_1, x_2) \in \mathbb{R}^2$ and similarly for $\tilde{u}_0$ and $\tilde{u}_1$. Then, $\tilde{u}$ solves the Cauchy problem for the three-dimensional wave equation:

$$\tilde{u}_{tt} - \Delta\tilde{u} = 0, \; \tilde{x} \in \mathbb{R}^3, \; t > 0,$$
$$\tilde{u}(\tilde{x}, 0) = \tilde{u}_0(\tilde{x}), \; \tilde{u}_t(\tilde{x}, 0) = \tilde{u}_1(\tilde{x}), \; \tilde{x} \in \mathbb{R}^3. \tag{10.19}$$

Therefore, from (10.8) with $c = 1$, we obtain that

$$u(x, t) = \tilde{u}(x, 0, t) = \frac{1}{4\pi t} \int_{|\tilde{y}-\tilde{x}|=t} \tilde{u}_1(\tilde{y})\, dS(\tilde{y}) + \frac{\partial}{\partial t}\left(\frac{1}{4\pi t} \int_{|\tilde{y}-\tilde{x}|=t} \tilde{u}_0(\tilde{y})\, dS(\tilde{y}). \right) \tag{10.20}$$

Note that in the above expression, the variable $\tilde{y} = (y_1, y_2, y_3)$ and $\tilde{x} = (x_1, x_2, 0)$. Furthermore, $dS_{\tilde{y}}$ is the surface measure of the three-dimensional sphere $|\tilde{y} - \tilde{x}| = t$. Therefore,

$$|\tilde{y} - \tilde{x}|^2 = (y_1 - x_1)^2 + (y_2 - x_2)^2 + y_3^2$$

and so[4] $dS(\tilde{y}) = \frac{t}{|y_3|} dy_1 dy_2$. Since the regions $\{y_3 > 0\}$ and $\{y_3 < 0\}$ make the same contribution to the integral in (10.20), we get (discard the 0 in the arguments)

$$u(x_1, x_2, t) = \frac{1}{2\pi} \iint_{B_t(x)} \frac{u_1(y_1, y_2)}{\sqrt{t^2 - r^2}}\, dy_1 dy_2 + \frac{\partial}{\partial t}\left(\frac{1}{2\pi} \iint_{B_t(x)} \frac{u_0(y_1, y_2)}{\sqrt{t^2 - r^2}}\, dy_1 dy_2 \right), \tag{10.21}$$

where $B_t(x)$ denotes the two-dimensional open ball centred at $x = (x_1, x_2)$ and radius t: $B_t(x) = \{y = (y_1, y_2) : (x_1 - y_1)^2 + (x_2 - y_2)^2 < t^2\}$ and $r = |y - x|$.

Looking at the expressions in (10.21), we see that the domain of dependence of the solution u at (x, t), $t > 0$ in two-dimensional case is the *disk* $B(x, t)$ and not the *circle* $S(x, t)$. This is one of the main differences between the cases $n = 3$ and $n = 2$. In fact, this difference persists between any odd and even dimensions. We say that the Huyghens' principle holds in the weak form for $n = 2$ (more generally for any even dimension). See the observations made at the end of Section 10.2.

[4]Writing the surface $|\tilde{y} - \tilde{x}| = t$ as $y_3^2 = t^2 - (x_1 - y_1)^2 - (x_2 - y_2)^2$, we have $dS(\tilde{y}) = \left(1 + \left(\frac{\partial y_3}{\partial y_1}\right)^2 + \left(\frac{\partial y_3}{\partial y_2}\right)^2\right)^{1/2}$.

10.3.1 Telegraph Equation

Consider the one-dimensional wave equation with lower-order terms present:

$$w_{tt} - w_{xx} + \alpha w_t + \beta w_x + \gamma w = 0, \; x \in \mathbb{R}, t > 0. \tag{10.22}$$

Here the coefficients α, β and γ are real constants. It is always possible to absorb the terms containing w_t and w_x by a simple change of variable: change $w(x,t)$ to $w(x,t)\exp\left(\frac{\alpha}{2}t - \frac{\beta}{2}x\right)$. However, it is *not* possible to absorb the term containing w. This is the case we want to analyze now. The *telegraph equation* is given by

$$w_{tt} - w_{xx} - \lambda^2 w = 0, \; x \in \mathbb{R}, \; t > 0. \tag{10.23}$$

Here the constant $\lambda > 0$. We impose the initial conditions

$$w(x,0) = 0, \; w_t(x,0) = \psi(x), \; x \in \mathbb{R}. \tag{10.24}$$

Put $x_1 = x$ and consider the function $u(x_1, x_2, t)$ defined by

$$u(x_1, x_2, t) = w(x_1, t)\cos(\lambda x_2).$$

Using (10.23), we see that u satisfies the two-dimensional wave equation

$$u_{tt} - u_{x_1x_1} - u_{x_2x_2} = 0,$$

with initial conditions

$$u(x_1, x_2, 0) = 0, \; u_t(x_1, x_2, 0) = \psi(x_1)\cos(\lambda x_2).$$

Using (10.21) for $u(x_1, x_2, t)$ and taking $x_2 = 0$ (*descending* from $n = 2$ to $n = 1$), we get

$$w(x,t) = w(x_1,t) = u(x_1,0,t) = \frac{1}{2\pi}\iint\limits_{B_t(x)} \frac{\cos(\lambda y_2)\psi(y_1)}{\sqrt{t^2 - r^2}}\,dy_1 dy_2,$$

where $B_t(x)$ is the two-dimensional open ball centred at $(x_1, 0)$ and radius t: $B_t(x) = \{y = (y_1, y_2) : (x_1 - y_1)^2 + y_2^2 < t^2\}$. Performing the integration with respect to y_2 first, we obtain

$$w(x,t) = w(x_1,t) = \frac{1}{2}\int_{x_1-t}^{x_1+t} J_0(\lambda s)\psi(y_1)\,dy_1, \tag{10.25}$$

where $s^2 = t^2 - (x_1 - y_1)^2$ and J_0 is the Bessel's function of the first kind of order 0 and is given by (see, for example, Abramowitz and Stegun, 1972)

$$J_0(z) = \frac{2}{\pi} \int_0^{\pi/2} \cos(z \sin \theta)\, d\theta.$$

We state the foregoing in the following theorem:

Theorem 10.3. Any C^2 solution of the one-dimensional Cauchy problem (10.23) satisfying the initial conditions (10.24) has the representation (10.25). □

From Theorem 10.3, it is not difficult to write down a formula for the solution of the inhomogeneous equation satisfying more general initial conditions:

$$w_{tt} - w_{xx} - \lambda^2 w = f(x,t),\ x \in \mathbb{R},\ t > 0. \tag{10.26}$$

with the initial conditions

$$w(x,0) = \phi(x),\ w_t(x,0) = \psi(x),\ x \in \mathbb{R}. \tag{10.27}$$

This is left as an exercise.

Remark 10.4. In (10.23), if we replace the term $-\lambda^2 w$ by $\lambda^2 w$, we then use the hyperbolic cosine function in place of the cosine function. This change will then produce, in place J_0, the function I_0, the Bessel's function of the second kind, of order 0, in the solution (see Chapter 9). □

Remark 10.5. If we add $\delta(x)$, the Dirac delta function at the origin, to the right side of (10.23), we can write down the solution to this inhomogeneous problem with zero initial data (that is $\psi = 0$), as

$$w(x,t) = \int_0^{(t-|x|)_+} J_0(\lambda \xi)\, d\xi,$$

where $\xi^2 = (t-s)^2 - x^2$ and $a_+ = \max\{a, 0\}$ for $a \in \mathbb{R}$. □

10.4 WAVE EQUATION FOR GENERAL n

We now describe two methods to obtain the representation for the solution of wave equation in arbitrary dimensions $n \geq 3$. The first method exploits the Euler–Poisson–Darboux equation satisfied by the spherical mean function of the solution, which we have already utilized in Section 10.2. More precisely, we consider only *odd* $n \geq 3$; for even n, we can then use the method of descent.

In the second method, following Courant and Hilbert (1989), the solution formula is obtained by an inversion formula in an Abel type integral equation. This method is applicable to all n, even or odd, but the nature of solutions will differ from odd n to even n. In what follows, repeated reference will be made to the Darboux equation (10.6) and Euler–Poisson–Darboux equation (10.7) (with $2/r$ replaced by $(n-1)/r$); u is a C^2 solution of the wave equation (10.1) and M_u is its spherical mean function.

10.4.1 Solution Formula via Euler–Poisson–D'arboux Equation

Motivated by the Euler–Poisson–Darboux equation, we look for a *linear* differential operator L such that the following commutation relation is satisfied:

$$[\partial_r^2, L] \equiv \partial_r^2 L - L\partial_r^2 = (n-1)L\left(\frac{1}{r}\partial_r\right), \tag{10.28}$$

where $\partial_r = \frac{\partial}{\partial r}$. Define the function $N(x, r, t)$ by

$$N(x, r, t) = LM_u(x, r, t).$$

Then, using (10.28), we get

$$\begin{aligned}
\partial_r^2 N(x, r, t) =& L\left(\partial_r^2 + \frac{n-1}{r}\partial_r\right) M_u(x, r, t) \\
=& L\left(\frac{1}{c^2}\frac{\partial^2}{\partial t^2} M_u(x, r, t)\right), \text{ using (10.7)} \\
=& \frac{1}{c^2}\frac{\partial^2}{\partial t^2} LM_u(x, r, t) \\
=& \frac{1}{c^2}\frac{\partial^2}{\partial t^2} N(x, r, t).
\end{aligned}$$

Thus, N satisfies the one-dimensional wave equation

$$N_{tt} - c^2 N_{rr} = 0$$

in r and t variables and the variable x playing the role of a parameter. The initial conditions at $t = 0$ for u give initial conditions for N via M_u. Once we obtain N using the D'Alembert's formula, the formula for u will be obtained using

$$u(x, t) = \lim_{r \to 0} M_u(x, r, t).$$

Thus, we are led to find a differential operator L so that the commutation relation (10.28) is satisfied.

Let m be a non-negative integer and define $\tilde{D}_r = r^{-1}\partial_r$, $r \in \mathbb{R}$, $r \neq 0$. Define the operator L_m by

$$L_m = \tilde{D}_r^m r^{2m+1}, \tag{10.29}$$

that is, $L_m v = \tilde{D}_r^m(r^{2m+1}v)$ for smooth function v of r.

Lemma 10.6. For a smooth function v defined on $\mathbb{R}$, the following relations hold:

$$(1) \quad [\partial_r^2, L_m]v = 2(m+1)L_m\tilde{D}_r v \quad (2) \quad L_m v(r) = \sum_{k=0}^{m} c_{m,k} r^{k+1}\partial_r^k v(r),$$

where the constants $c_{m,k}$ do not depend on v and

$$c_{m,0} = 1 \cdot 3 \cdots (2m+1) = \prod_{j=0}^{m}(2j+1).$$

□

Proof The statement in (1) will be proved by induction. For $m = 0$, a direct computation shows that the commutation relation is true. Assume that (1) is true for some m.

Observe that $\tilde{D}_r$ is a *derivation*, that is,

$$\tilde{D}_r(uv) = u\tilde{D}_r v + v\tilde{D}_r u.$$

Consider

$$\begin{aligned} L_{m+1}v &= \tilde{D}_r^{m+1}(r^{2m+3}v) \\ &= \tilde{D}_r^m\tilde{D}_r(r^{2m+3}v) \\ &= \tilde{D}_r^m[(2m+3)r^{2m+1}v + r^{2m+3}\tilde{D}_r v] \\ &= (2m+3)L_m v + L_m(r\partial_r v). \end{aligned} \tag{10.30}$$

Therefore,

$$\begin{aligned} \partial_r^2 L_{m+1}v &= (2m+3)\partial_r^2 L_m v + \partial_r^2 L_m(r\partial_r v) \\ &= (2m+3)\left[L_m\partial_r^2 v + 2(m+1)L_m(\tilde{D}_r v)\right] + L_m\left(r\partial_r^3 v + 2\partial_r^2 v\right) \\ &\quad + 2(m+1)\left[L_m(\partial_r^2 v) + L_m(\tilde{D}_r v)\right]. \end{aligned} \tag{10.31}$$

The induction hypothesis is used in the last line. On the other hand, using (10.30), we get

$$L_{m+1}\partial_r^2 v = (2m+3)L_m(\partial_r^2 v) + L_m(r\partial_r^3 v). \tag{10.32}$$

Next, using (10.31) and (10.32), we get

$$\begin{aligned}
[\partial_r^2, L_{m+1}]v &= \partial_r^2 L_{m+1}v - L_{m+1}\partial_r^2 v \\
&= 2(m+2)L_m\partial_r^2 v + 4(m+1)(m+2)L_m\tilde{D}_r v \\
&= 2(m+2)L_m(\partial_r^2 v + 2(m+1)\tilde{D}_r v).
\end{aligned}$$

But, using again (10.30), we get

$$\begin{aligned}
L_{m+1}\tilde{D}_r v &= (2m+3)L_m\tilde{D}_r v + L_m(r\partial_r\tilde{D}_r v) \\
&= L_m(\partial_r^2 v + 2(m+1)\tilde{D}_r v).
\end{aligned}$$

Therefore, (1) holds true with m replaced by $m+1$ and the induction is complete.
To prove (2), we use Leibnitz's rule. We have

$$L_m v = \tilde{D}_r^m(r^{2m+1}v) = \sum_{k=0}^{m}\binom{m}{k}\tilde{D}_r^{m-k}(r^{2m+1})\tilde{D}_r^k v.$$

Now

$$\tilde{D}_r^{m-k}(r^{2m+1}) = (2m+1)\cdot(2m-1)\cdots(2k+3)r^{2k+1}$$

and

$$\tilde{D}_r^k v = \sum_{j=0}^{k-1}\frac{a_{k,j}}{r^{k+j}}\partial_r^{k-j}v,$$

for some constants $a_{k,j}$ which do not depend on v. Therefore, we get

$$L_m v = \sum_{k=0}^{m} c_{m,k}r^{k+1}\tilde{D}_r^k v,$$

where the constants $c_{m,k}$ do not depend on v. To determine $c_{m,0}$, observe that

$$\tilde{D}_r^m(r^{2m+1}) = c_{m,0}r.$$

This immediately gives

$$c_{m,0} = 1\cdot 3\cdots(2m+1)$$

as required. This completes the proof of the lemma. □

Some examples are:

- For $n = 3$, $m = 0$ and $L_0 v = rv$. This was seen in Section 10.2.
- For $n = 5$, $m = 1$ and $L_1 v = r^2 \partial_r v + 3rv$. This has been mentioned in John (1978) as an exercise on page 109.
- For $n = 7$, $m = 2$ and $L_2 v = r^3 \partial r^2 v + 9r^2 \partial_r v + 15rv$.

With this preparation, we now proceed to obtain a formula for the solution of the wave equation in odd (space) dimensions, using the Euler–Poisson–Darboux equation. Recall the Cauchy problem stated at the beginning of this chapter:

$$\begin{aligned} \Box_c u \equiv u_{tt} - c^2 \Delta u = 0,\ x \in \mathbb{R}^n,\ t > 0, \\ u(x,0) = u_0(x),\ u_t(x,0) = u_1(x),\ x \in \mathbb{R}^n. \end{aligned} \tag{10.33}$$

Let $n \geq 3$ be odd and u be a C^2 solution of (10.33). Let $m = \frac{n-3}{2}$ and consider the operator L_m introduced in Lemma 10.6. Put

$$N(x,r,t) = L_m M_u(x,r,t).$$

Lemma 10.6 (1) and the discussion preceding it immediately yield the following theorem:

Theorem 10.7. The function N satisfies the one-dimensional wave equation

$$N_{tt} - c^2 N_{rr} = 0$$

along with the initial conditions

$$N(x,r,0) = L_m M_{u_0}(x,r),\ N_t(x,r,0) = L_m M_{u_1}(x,r).$$

□

By D'Alembert's formula, we have

$$\begin{aligned} N(x,r,t) = {} & \frac{1}{2}\left(L_m M_{u_0}(x,r+ct) + L_m M_{u_0}(x,r-ct)\right) + \\ & + \frac{1}{2c}\int_{r-ct}^{r+ct} L_m M_{u_1}(x,s)\,ds \end{aligned}$$

Next, from Lemma 10.6 (2), it follows that

$$\lim_{r\to 0} \frac{N(x,r,t)}{c_{m,0} r} = \lim_{r\to 0} M_u(x,r,t) = u(x,t).$$

Therefore, we get (see the explanation preceding and leading to (10.8))

$$u(x,t) = \lim_{r\to 0} \frac{N(x,r,t)}{c_{m,0} r}$$
$$= \frac{1}{c_{m,0}} \frac{\partial}{\partial t} \left(L_m M_{u_0}(x,ct) \right) + \frac{1}{c_{m,0}} L_m M_{u_1}(x,ct).$$

Of course, for u to be C^2, the initial functions u_0 and u_1 need to be sufficiently smooth. Using the definition of L_m, we rewrite the above formula as (remember $n \geq 3$ is odd and $m = (n-3)/2$):

$$u(x,t) = \frac{1}{c_n} \left[\left(\frac{\partial}{\partial t}\right) \left(\frac{1}{t}\frac{\partial}{\partial t}\right)^{\frac{n-3}{2}} \left(t^{n-2} M_{u_0}(x,ct)\right) + \right.$$
$$\left. + \left(\frac{1}{t}\frac{\partial}{\partial t}\right)^{\frac{n-3}{2}} \left(t^{n-2} M_{u_1}(x,ct)\right) \right], \tag{10.34}$$

where $c_n = c_{m,0} = 1 \cdot 3 \cdots (n-2)$. Of course, the spherical means M_{u_0} and M_{u_1} may be written in terms of surface integrals of the functions u_0 and u_1 respectively. Since $c_3 = 1$, we can easily verify that the formula given in (10.34) coincides with one given in (10.8). We now summarize the foregoing in the form of a theorem:

Theorem 10.8. Assume that $n \geq 3$ is odd, the initial values $u_0 \in C^{\frac{n+3}{2}}(\mathbb{R}^n)$ and $u_1 \in C^{\frac{n+1}{2}}(\mathbb{R}^n)$. Define u by (10.34). Then, u is a C^2 solution of the Cauchy problem (10.33). □

For the sake of completeness, we also state the result for n even, which is derived using the method of descent, from the known result for n odd. For $n \geq 2$ even, define u by

$$u(x,t) = \frac{c_{n+1}}{2\Gamma(n/2)} \left[\left(\frac{\partial}{\partial t}\right) \left(\frac{1}{2t}\frac{\partial}{\partial t}\right)^{\frac{n-2}{2}} \int_0^t \frac{r}{\sqrt{t^2-r^2}} \left(r^{n-2} M_{u_0}(x,cr)\right)\, dr + \right.$$
$$\left. + \left(\frac{1}{t}\frac{\partial}{\partial t}\right)^{\frac{n-2}{2}} \int_0^t \frac{r}{\sqrt{t^2-r^2}} \left(t^{n-2} M_{u_1}(x,cr)\right) dr \right], \tag{10.35}$$

where Γ is the Euler's gamma function.

Theorem 10.9. Assume that $n \geq 2$ is even and $u_0 \in C^{\frac{n+4}{2}}(\mathbb{R}^n)$ and $u_1 \in C^{\frac{n+2}{2}}(\mathbb{R}^n)$. Then, u, defined in (10.34), is a C^2 solution of the Cauchy problem (10.33). □

In the above theorem, the initial conditions at $t = 0$ are satisfied in the sense of limiting values: For any $x_0 \in \mathbb{R}^n$,

$$\lim_{(x,t)\to(x_0,0+)} u(x,t) = u_0(x_0)$$

and

$$\lim_{(x,t)\to(x_0,0+)} u_t(x,t) = u_1(x_0).$$

A somewhat lengthy verification is left as an exercise. We also invite the reader to compare the above formulas with the ones that will be derived in the next subsection.

Remark 10.10. (*Loss of Regularity*) When $n \geq 3$ and odd, we need $u_0 \in C^{\frac{n+3}{2}}(\mathbb{R}^n)$ in order for u to be a C^2 solution, thus losing a regularity of order $\frac{n-1}{2}$. Whereas, for $n \geq 2$ and even, we need $u_0 \in C^{\frac{n+4}{2}}(\mathbb{R}^n)$ in order for u to be a C^2 solution, losing a regularity of order $\frac{n}{2}$. Thus, there is more loss in the order of regularity in even dimensions. However, the case of $n = 1$ is different, where the solution is as smooth as the initial data. □

10.4.2 An Inversion Method

Following Courant and Hilbert (1989), we now describe a method of obtaining a representation for the solution of the wave equation in any dimension $n \geq 3$.

We begin with a discussion of the spherical mean function of certain specific function. Recall that for an arbitrary function $h \in C^2(\mathbb{R}^n)$, its spherical mean function $v(x,r) = M_h(x,r)$ is defined by

$$v(x,r) = \frac{1}{\sigma_n} \int_{|\xi|=1} h(x + r\xi)\, dS(\xi), \; r \in \mathbb{R}, \tag{10.36}$$

satisfies the Darboux equation

$$v_{rr} + \frac{n-1}{r} v_r = \Delta_x v, \tag{10.37}$$

and the initial conditions $v(x,0) = h(x)$ and $v_r(x,0) = 0$ (v is an even function of r). Suppose now that h is a function of only one variable, say x_1: $h = h(x_1)$. Then, integrating with respect to the variables $x_2, \ldots, x_n$ in the integral in (10.36), it may be written as (replacing x_1 by x)

$$v(x,r) = \frac{\sigma_{n-1}}{\sigma_n} \int_{-1}^{1} h(x + r\mu) \left(1 - \mu^2\right)^{\frac{n-3}{2}} d\mu, \; x, r \in \mathbb{R}, \tag{10.38}$$

and satisfies the equation

$$v_{rr} + \frac{n-1}{r} v_r = v_{xx}. \tag{10.39}$$

The integral in (10.36) transforms into the integral in (10.38) by writing $\xi = (\xi_1, \xi')$ and performing the surface integral first with ξ' variables. Now differentiate (10.38) twice with respect to x to obtain

$$v_{xx} = \frac{\sigma_{n-1}}{\sigma_n} \int_{-1}^{1} h''(x + r\mu) \left(1 - \mu^2\right)^{\frac{n-3}{2}} \, d\mu,$$

which may be used to replace the right-hand side in (10.39). Since x plays only the role of a parameter, the above analysis may be summarized as follows:

If a function v defined on $\mathbb{R}$ and a function $h \in C^2(\mathbb{R})$ are connected by the relation

$$v(r) = \int_{-1}^{1} h(r\mu) \left(1 - \mu^2\right)^{\frac{n-3}{2}} \, d\mu,$$

then,

$$v_{rr} + \frac{n-1}{r} v_r = \int_{-1}^{1} h''(r\mu) \left(1 - \mu^2\right)^{\frac{n-3}{2}} \, d\mu.$$

We now extend the above idea to study the n-dimensional wave equation. In this regard, we may now consider h to depend on $x \in \mathbb{R}^n$, which will be treated as a parameter, that is, we consider $h = h(x, r\mu)$. Accordingly, we write $v = v(x, r)$. To see the connection of this analysis to the solution of the wave equation, let u to be a C^2 solution of the wave equation

$$u_{tt} = \Delta u, \; x \in \mathbb{R}^n, \; t > 0,$$

satisfying the initial conditions

$$u(x, 0) = \psi(x), \; u_t(x, 0) = 0, \; x \in \mathbb{R}^n.$$

The function u may then be extended to $t < 0$ by setting $u(x, t) = u(x, -t)$. Now consider the function (temporarily we use r in place of t)

$$v(x, r) = \frac{\sigma_{n-1}}{\sigma_n} \int_{-1}^{1} u(x, r\mu) \left(1 - \mu^2\right)^{\frac{n-3}{2}} \, d\mu$$

corresponding to the function $u(x, r\mu)$. Then,

$$v(x, 0) = u(x, 0) = \psi(x),\ v_r(x, 0) = 0.$$

Also,

$$\Delta_x v = v_{rr} + \frac{n-1}{r} v_r.$$

The previous analysis shows that this equation is satisfied by $v(x, r) = M_\psi(x, r)$ and therefore, the solution of the wave equation u must satisfy

$$\frac{2\sigma_{n-1}}{\sigma_n} \int_0^1 u(x, r\mu) \left(1 - \mu^2\right)^{\frac{n-3}{2}} d\mu = M_\psi(x, r), \tag{10.40}$$

where we have used that u is an even function of $t \in \mathbb{R}$. Conversely, if u satisfies the above relation, then we obtain a solution of the wave equation, even in t, in a unique manner. If we can somehow *invert* the relation in (10.40), then we would have obtained u in terms of ψ. We now proceed to do this.

Lemma 10.11. Let $n \geq 3$. Suppose $v(r)$ and $\phi(r)$ are continuous functions defined on $\mathbb{R}$, $\phi(-r) = \phi(r)$ and are connected by the relation

$$v(r) = \int_0^1 \phi(r\mu) \left(1 - \mu^2\right)^{\frac{n-3}{2}} d\mu.$$

Then,

- For $n = 2k + 1$, $k \geq 1$, we have

$$\phi(r) = \frac{2r}{(k-1)!} \left(\frac{d}{dr^2}\right)^k \left(r^{2k-1} v(r)\right).$$

- For $n = 2k$, $k \geq 2$, we have

$$\phi(r) = \frac{2r}{\sqrt{\pi}\,\Gamma\left(\frac{2k-1}{2}\right)} \left(\frac{d}{dr^2}\right)^k \int_0^r \frac{\rho}{\sqrt{r^2 - \rho^2}} \left(\rho^{2k-2} v(\rho)\right) d\rho.$$

□

Here the notation $\frac{d}{dr^2}$ denotes the derivation with respect to the variable r^2; of course, $\frac{d}{dr^2} = \frac{1}{2r}\frac{d}{dr}$.

Proof Make the change of variables: $r = \sqrt{s}$, $r\mu = \sqrt{\sigma}$. Then, the given relation in the lemma is rewritten as

$$2v(\sqrt{s})s^{\frac{n-2}{2}} = \int_0^s \frac{\phi\left(\sqrt{\sigma}\right)}{\sqrt{\sigma}}(s-\sigma)^{\frac{n-3}{2}}\,d\sigma.$$

Set $w(s) = 2v\left(\sqrt{s}\right)s^{\frac{n-2}{2}}$ and $\chi(\sigma) = \frac{\phi\left(\sqrt{\sigma}\right)}{\sqrt{\sigma}}$. Thus,

$$w(s) = \int_0^s \chi(\sigma)\,(s-\sigma)^{\frac{n-3}{2}}\,d\sigma. \tag{10.41}$$

If $n = 2k+1$, then by differentiating k times the relation (10.41), we obtain

$$(k-1)!\,\chi(s) = 2\left(\frac{d}{ds}\right)^k w(s),$$

or in terms of r, v and ϕ,

$$\phi(r) = \frac{2}{(k-1)!}r\left(\frac{d}{dr^2}\right)^k\left(r^{2k-1}v(r)\right).$$

If $n = 2k$, we write $\frac{n-3}{2} = k-1-\frac{1}{2}$ and differentiate $k-1$ times the relation (10.41) to obtain

$$\left(\frac{d}{ds}\right)^{k-1} w(s) = \frac{n-3}{2}\cdot\frac{n-5}{2}\cdots\frac{1}{2}\int_0^s \frac{\chi(\sigma)}{\sqrt{s-\sigma}}\,d\sigma.$$

Therefore,[5]

$$\frac{n-3}{2}\cdot\frac{n-5}{2}\cdots\frac{1}{2}\chi(s) = \frac{1}{\pi}\frac{d}{ds}\int_0^s w^{(k-1)}(\sigma)(s-\sigma)^{-1/2}\,d\sigma.$$

[5] *Liouville transformation*: Let $0 < \alpha < 1$ and the functions f and g defined on $[0,\infty)$ are connected by the relation

$$g(t) = \frac{1}{\Gamma(\alpha)}\int_0^t f(s)(t-s)^{-\alpha}\,ds,$$

then

$$f(t) = \frac{d}{dt}\left(\frac{1}{\Gamma(1-\alpha)}\int_0^t g(s)(t-s)^{1-\alpha}\,ds\right).$$

Using the fact that $\left(\frac{d}{ds}\right)^j w(0) = 0$ for $j = 1, 2, \ldots, \frac{n-2}{2}$ and some simple manipulation, the above relation can be written as

$$\chi(s) = \frac{1}{\sqrt{\pi}\,\Gamma\left(\frac{2k-1}{2}\right)} \left(\frac{d}{ds}\right)^k \int_0^s \frac{w(\sigma)}{\sqrt{s-\sigma}}\, d\sigma,$$

or, in terms of the original variables

$$\phi(r) = \frac{2r}{\sqrt{\pi}\,\Gamma\left(\frac{2k-1}{2}\right)} \left(\frac{d}{dr^2}\right)^k \int_0^r \frac{\rho}{\sqrt{r^2-\rho^2}} \left(\rho^{2k-2} v(\rho)\right) d\rho.$$

This completes the proof of the lemma. □

Returning to the discussion on the solution u of the wave equation

$$u_{tt} - \Delta u = 0$$

satisfying the initial conditions $u(x,0) = \psi(x)$ and $u_t(x,0) = 0$ for $x \in \mathbb{R}^n$, we take $v(r) = M_\psi(x,r)$ and $\phi(r) = \frac{2\sigma_{n-1}}{\sigma_n} u(x,r)$ (r and t are interchangeably used) in the Lemma 10.11. Then, using (10.40), we obtain the following:

- For $n = 2k,\ k \geq 2$,

$$u(x,t) = \frac{t}{\Gamma(k)} \left(\frac{\partial}{\partial t^2}\right)^k \int_0^t \frac{r}{\sqrt{t^2-r^2}} \left(r^{2k-2} M_\psi(x,r)\right) dr$$

- For $n = 2k+1,\ k \geq 1$,

$$u(x,t) = \frac{\sqrt{\pi}\, t}{\Gamma(k+\frac{1}{2})} \left(\frac{\partial}{\partial t^2}\right)^{(n-1)/2} \left(t^{n-2} M_\psi(x,t)\right).$$

Representation Formula for the Solution u**:** We are now in a position to write down the solution formula for the general IVP for the wave equation in arbitrary dimensions. For this, we introduce the following notation: For sufficiently smooth function h defined on $\mathbb{R}^n$, set

$$Q_h(x,t) = \frac{1}{2\Gamma(k)} \left(\frac{1}{2t}\frac{\partial}{\partial t}\right)^{k-1} \int_0^t \frac{r}{\sqrt{t^2-r^2}} \left(r^{2k-2} M_h(x,r)\right) dr$$

for $n = 2k,\ k \geq 2$ and

$$Q_h(x,t) = \frac{\sqrt{\pi}}{2\Gamma(k+\frac{1}{2})}\left(\frac{1}{2t}\frac{\partial}{\partial t}\right)^{k-1}\left(t^{2k-1}M_h(x,t)\right)$$

for $n = 2k+1,\ k \geq 1$. We have

Theorem 10.12. Consider the IVP for the homogeneous wave equation

$$\begin{aligned} &u_{tt} - \Delta u = 0,\ x \in \mathbb{R}^n,\ t > 0 \\ &u(x,0) = \psi(x),\ u_t(x,0) = \phi(x),\ x \in \mathbb{R}^n. \end{aligned}$$

Then, the solution u has the following representation:

$$u(x,t) = \frac{\partial}{\partial t}Q_\psi(x,t) + Q_\phi(x,t).$$

Conversely, suppose $\psi \in C^{\frac{n+3}{2}}(\mathbb{R}^n)$ and $\phi \in C^{\frac{n+1}{2}}(\mathbb{R}^n)$, for $n \geq 3$, odd; and $\psi \in C^{\frac{n+4}{2}}(\mathbb{R}^n)$ and $\phi \in C^{\frac{n+2}{2}}(\mathbb{R}^n)$, for $n \geq 3$, even. Then the function u as given above is a C^2 solution of the wave equation satisfying the initial conditions stated in the theorem. □

A somewhat lengthy verification of the converse statement is left as an exercise.

10.5 MIXED OR INITIAL BOUNDARY VALUE PROBLEM

We now consider the wave equation in a bounded or unbounded region Ω in $\mathbb{R}^n$:

$$\Box_c u = u_{tt} - c^2\Delta u = f(x,t),\ x \in \Omega,\ t > 0, \tag{10.42}$$

with the prescribed initial and boundary conditions:

$$\left.\begin{aligned} &u(x,0) = u_0(x),\ u_t(x,0) = u_1(x),\ x \in \Omega \\ &u(x,t) = 0,\ x \in \partial\Omega,\ t > 0. \end{aligned}\right\} \tag{10.43}$$

The zero Dirichlet condition may be replaced by a non-zero Dirichlet condition. Neumann or mixed boundary conditions may also be prescribed. More generally, the boundary $\partial\Omega$ may be divided into two parts and different kinds of boundary conditions may be prescribed on these two parts of the boundary.

Solutions with Finite Energy: For a solution u of (10.42), the associated *energy* is defined by

$$E(t) = \frac{1}{2} \int_{\Omega} \left(u_t(x,t)^2 + c^2 |\nabla_x u(x,t)|^2 \right) dx \tag{10.44}$$

and we consider only solutions with *finite energy*, that is $E(t) < \infty$ for all $t \geq 0$. The solutions with finite energy are also physically meaningful. Using divergence theorem and (10.42), we have,

$$\begin{aligned}
\frac{dE}{dt} &= \int_{\Omega} \left(u_t u_{tt} + c^2 \sum_{i=1}^{n} u_{x_i} u_{x_i t} \right) dx \\
&= \int_{\Omega} u_t \left(c^2 \Delta u + f \right) dx - c^2 \int_{\Omega} \Delta u u_t \, dx + c^2 \int_{\partial\Omega} u_t \frac{\partial u}{\partial \nu} \, d\sigma \\
&= \int_{\Omega} f u_t \, dx + c^2 \int_{\partial\Omega} u_t \frac{\partial u}{\partial \nu} \, d\sigma.
\end{aligned}$$

Here ν denotes the outward unit normal to the boundary $\partial\Omega$. If, initially, u, u_t vanish for all $x \in \Omega$ and $t = 0$, and $u = 0$ or $\frac{\partial u}{\partial \nu} = 0$ for all $x \in \partial\Omega$, $t \geq 0$, then, with $f \equiv 0$, we see that $\frac{dE}{dt} = 0$ or E is a constant. Since $E(0) = 0$, it follows that $E(t) = 0$ for all $t \geq 0$. But then u_t and all u_{x_i} are zero, so that u is a constant. Since $u = 0$ for $t = 0$, it follows that $u \equiv 0$. Thus, under suitable regularity assumption, we have obtained the following uniqueness result for solutions with finite energy:

Theorem 10.13 (Uniqueness). Assume that Ω is a region in $\mathbb{R}^n$ with smooth boundary $\partial\Omega$. Then, any solution of (10.42) with finite energy and satisfying the initial-boundary conditions (10.43), is unique. □

The above uniqueness result easily extends to equations with Δ replaced by a second-order uniformly elliptic operators. Proving the existence of solutions with finite energy, however, is not at all simple and demands heavy machinery from functional analysis, such as Sobolev spaces and operator theory. These topics are not part of the present book. This of course compels us to make mere statements regarding the existence of solutions with finite energy, without much explanation and/or proofs. The interested reader can certainly look into more advanced books or research articles for understanding the deeper analysis involved. Here we mention a book by Ladyzhenskaya (1985) and an article by Wilcox (1962).

Galerkin Method: There are different methods found in the literature to establish the existence of a solution with finite energy. Below we describe the *Galerkin method* to obtain such solutions for (10.42). This method requires the knowledge of existence of a sequence of real numbers $\{\lambda_n\}$, the *eigenvalues*, and the associated sequence of functions $\{\phi_n\}$, the *eigenfunctions*, satisfying

$$\Delta\phi_n + \lambda_n\phi_n = 0, \tag{10.45}$$

for $n = 1, 2, \ldots$. Further, each ϕ_n satisfies the appropriate boundary condition. We remark that to prove the existence of λ_n and the associated ϕ_n for a general domain with smooth boundary, again requires tools from functional analysis. Such problems are categorized as spectral problems for uniformly elliptic operators. The existence of eigenvalues and the associated eigenfunctions are obtained using the spectral theory for compact operators in suitable Hilbert spaces.

We remark that under fairly general conditions on the regularity of the domain and for a general uniformly elliptic operator, in place of Δ, it is possible to prove the existence of eigenvalues and the associated eigenfunctions, for Dirichlet and Neumann boundary conditions. In case of Δ, for example, and domains with specific geometry – a square, a ball – it is possible to explicitly determine the eigenvalues and the associated eigenfunctions. Further, it may be shown that the eigenfunctions are *complete*. This means that the closure of the linear span of the set of eigenfunctions is the space of all square-integrable functions defined on Ω. This enables us to *expand* any square-integrable function defined in Ω in terms of the eigenfunctions.

We assume that the eigenfunctions ϕ_n also satisfy the zero Dirichlet condition and are *orthonormal*:

$$\int_\Omega \phi_k(x)\phi_l(x)\,dx = \delta_{kl},$$

where δ_{kl} is the Kronecker delta function. The Galerkin method then seeks a solution u of (10.42) in the form of an infinite series

$$u(x,t) = \sum_{k=1}^{\infty} a_k(t)\phi_k(x). \tag{10.46}$$

Formally, plugging the series into the equation in (10.42), we see that the unknown coefficient a_k satisfies the ODE:

$$a_k''(t) + c^2\lambda_k a_k(t) = 0, \tag{10.47}$$

for $k = 1, 2, \ldots$. Using the initial conditions in (10.43) (the boundary condition is automatically satisfied because of the choice of ϕ_k), we obtain the initial conditions for a_k, as the Fourier coefficients of u_0 and u_1:

$$a_k(0) = \int_\Omega u_0(x)\phi_k(x)\,dx, \ a_k'(0) = \int_\Omega u_1(x)\phi_k(x)\,dx. \tag{10.48}$$

Thus solving (10.47) with initial conditions given in (10.48), we get

$$a_k(t) = \int_\Omega \left[u_0(x)\cos(c\lambda_k t) + \frac{u_1(x)\sin(c\lambda_k t)}{c\lambda_k} \right] \phi_k(x)\, dx. \tag{10.49}$$

We still have the task of proving the convergence of the series in (10.46) in some appropriate space, but we will not go into details. See, for example, the paper by Wilcox (1962) and the book by Ladyzhenskaya (1985).

We remark that this procedure also works for parabolic operators $\partial_t - L$, where L is a uniformly elliptic operator with coefficients depending only on x.

10.6 GENERAL HYPERBOLIC EQUATIONS AND SYSTEMS

In this section, we merely introduce the notion of hyperbolicity for general linear equations of any order and for systems of first-order equations. We will not venture into any question regarding the existence and/or uniqueness of solutions, as these topics require advanced tools from functional analysis.

An mth-order linear partial differential operator is given by

$$A = \sum_{j+|\alpha|\le m} a_{\alpha j}(x,t) D_x^\alpha D_t^j, \; x \in \mathbb{R}^n, t \in \mathbb{R},$$

where the coefficients $a_{\alpha j}$ are all real.

Definition 10.14 (Hyperbolicity). The operator A is said to be *hyperbolic* at the point (x, t) if $a_{0m}(x, t) \neq 0$ (that is, the direction of the t-axis is *non-characteristic* at (x, t)) and for any vector $\xi \in \mathbb{R}^n$, all the roots λ of the equation

$$A_m(x,t,\xi,\lambda) \equiv \sum_{j+|\alpha|=m} a_{\alpha j}(x,t)\xi^\alpha \lambda^j = 0$$

are *real*. □

The polynomial A_m is called the *principal symbol* of A.

Definition 10.15 (Strict Hyperbolicity). The operator A is said to be *strictly hyperbolic* if the roots of the polynomial A_m are real and distinct for all non-zero vectors ξ. This case is also referred to as the case of *simple characteristics*. □

In case of systems, the coefficients $a_{\alpha j}$ will be real square matrices. For systems, A is called *hyperbolic* at (x, t) if $a_{0m}(x, t)$ is non-singular and the roots λ of the characteristic equation

$$\det \sum_{j+|\alpha|=m} a_{\alpha j}(x,t)\xi^{\alpha}\lambda^{j} = 0$$

are all real for any $\xi \in \mathbb{R}^n$. If, in addition, these roots are distinct for all non-zero ξ, then A is said to *strictly hyperbolic* at (x, t).

The simplest hyperbolic operators of first order are

$$\frac{\partial}{\partial t} + \sum_{j=1}^{n} a_j(x,t)\frac{\partial}{\partial x_j} + b(x,t)$$

with real a_j and b. The second order-operator of the form

$$\frac{\partial^2}{\partial t^2} - \sum_{i,j=1}^{n} a_{ij}(x,t)\frac{\partial^2}{\partial x_i x_j} + \sum_{j=1}^{n} b_j(x,t)\frac{\partial}{\partial x_j} + c(x,t)$$

with real coefficients is hyperbolic if the matrix $[a_{ij}]$ is symmetric and positive definite. In particular, the D'Alembertian $\Box_c$ is strictly hyperbolic.

Of particular interest are the first-order systems, since many examples of practical interest belong to this class. Consider a first-order system of equations

$$u_t + \sum_{j=1}^{n} A_j(x,t)u_{x_j} + B(x,t)u = f(x,t), \tag{10.50}$$

or, more generally

$$A_0(x,t)u_t + \sum_{j=1}^{n} A_j(x,t)u_{x_j} + B(x,t)u = f(x,t). \tag{10.51}$$

Here $x \in \mathbb{R}^n$, $t \in \mathbb{R}$; u, f are real N-vectors and A_0, A_j and B are real square matrices of order N. We rephrase earlier definitions in the context of (10.50) and (10.51).

Definition 10.16. The system (10.50) is said to be *hyperbolic* if the matrix $\sum_{j=1}^{n} \xi_j A_j$ has only real eigenvalues for all real $\xi \in \mathbb{R}^n$. If, in addition, these eigenvalues are also distinct for all non-zero ξ, then (10.50) is said to be *strictly hyperbolic*. If the matrix $\sum_{j=1}^{n} \xi_j A_j$ has only real eigenvalues and is diagonalizable for all ξ, then (10.50) is said to be *strongly hyperbolic*. It is *symmetric hyperbolic* if all the matrices A_j are symmetric. □

We now make similar definitions with regard to the system (10.51). This system is *hyperbolic* if the matrix A_0 is non-singular and all the roots λ of the characteristic equation

$$\det\left(\lambda A_0 + \sum_{j=1}^{n} \xi_j A_j\right) = 0$$

are real, for all ξ. The definitions of strict hyperbolicity and strong hyperbolicity are similar. This system is called *symmetric hyperbolic* if the matrices A_0 and A_j are all symmetric and A_0 is positive definite.

Remark 10.17. Suppose the matrices in (10.50) are all constant matrices. For the system to be strictly hyperbolic a certain relation should be satisfied between n and N. This was discovered by Lax (1982) and Friedland et al. (1984), in the early 1980s. For example, when $n = 3$, the case is of physical interest, the smallest N for which the system may be a strictly hyperbolic is 7. However, for 2×2 systems, that is $n = 2$, there are strictly hyperbolic systems for arbitrary N. A simple example is the following: □

Take A_1 to be a diagonal matrix with distinct real numbers on the diagonal and A_2 to be a *tridiagonal* symmetric matrix with all non-zero sub-diagonal elements. Then, it is not difficult to see that the matrix $\xi_1 A_1 + \xi_2 A_2$ has real and distinct eigenvalues for all non-zero real ξ. This easy verification is left as an exercise. □

10.7 NOTES: QUASILINEAR EQUATIONS

We have seen that the solution of a first-order non-linear equation such as Burgers' equation, ceases to be smooth after a finite time, in spite of the initial data being very smooth. A similar question may be posed for a quasilinear hyperbolic equation of the form

$$u_{tt} - \Delta u = \sum_{i,j=1}^{n} a_{ij}(D_{t,x}u) u_{x_i x_j},$$

which is a perturbation of the n-dimensional wave equation. Here $D_{t,x}u$ denotes $(u_t, u_{x_1}, \ldots, u_{x_n})$ and $a_{ij} = a_{ji}$ are smooth functions satisfying $a_{ij}(0) = 0$ for all i, j and

$$\sum_{i,j=1}^{n} |a_{ij}(D_{t,x}u)| \leq 1/2, \text{ for } |D_{t,x}u| \text{ small.}$$

We impose the initial conditions

$$u(x, 0) = \varepsilon f(x), \; u_t(x, 0) = \varepsilon g(x),$$

where f, g are smooth functions with compact support; thus ε measures the strength of the initial data. It is necessary to restrict to small ε if we expect a global solution, that is a solution which exists for all $t > 0$. The following example, suggested by L. Nirenberg, illustrates this.

Consider the semi-linear equation

$$u_{tt} - \Delta u = u_t^2 - |\nabla_x u|^2$$

in $\mathbb{R}^3$ with initial data

$$u(x,0) = 0, \ u_t(x,0) = \varepsilon g(x), \ x \in \mathbb{R}^3,$$

where g is a smooth function of compact support. The substitution $v = e^{-u}$ shows that v satisfies the homogeneous wave equation $v_{tt} - \Delta v = 0$ with initial data $v(x,0) = 1$ and $v_t(x,0) = -\varepsilon g(x)$. Therefore,

$$v(x,t) = 1 - \frac{\varepsilon}{4\pi t} \int_{|y-x|=t} g(y)\, dS(y), \ t > 0.$$

Since then $u = \log(1/v)$, the solution u is defined for those t that makes $v > 0$. Suppose g has compact support and bounded below by some $\delta > 0$. From the explicit expression for v, it is clear that $v \leq 1 - \varepsilon\delta t$. Therefore, v remains positive as long as $t = O(1/\varepsilon)$. This example also suggests that in general we cannot expect global existence of solutions.

Returning to the second-order quasilinear equation, when we try to establish the existence of a solution by the method of iterations, there is an immediate difficulty of the iterates not falling into the same function class. Klainerman (1980) overcomes this difficulty by adapting a Nash–Moser–Hörmander type scheme. With this complicated and highly technical procedure of obtaining estimates, Klainerman was able to prove the global existence of a solution for small ε, provided that $n \geq 6$. This left open, in particular, the physically relevant case of $n = 3$.

Using the explicit representation of the solution of the homogeneous wave equation described above, von Wahl (1971) derived $L^p - L^q$ estimates for the solution, for suitable positive exponents p, q. Using the short time existence of the solution to the quasilinear equation (which in itself is non-trivial), the $L^p - L^q$ estimates were used to obtain *a priori* estimates for the solution, so that the method of continuation of existence of solution may be extended to larger time intervals. This programme was successful and is reported in Klainerman and Ponce (1983), again requiring $n \geq 6$.

Fritz John, considering the case of $n = 3$, introduced some new ideas by obtaining estimates on the solution of the homogeneous/ inhomogeneous wave equation, using the generators of the Poincarè group and Lorentz group, instead of just using the usual first-order differential operators, being the generators of the translation group. Using these hard estimates, John (1983) was able to obtain *almost global existence* of solutions to quasilinear equation for small ε. The terminology of *almost global existence* introduced here is to indicate that the time of existence is exponentially large, depending on ε.

Using the ideas in John (1983) and John and Klainerman (1983), Klainerman was successful in deriving $L^2 - L^\infty$ for a general class of functions including the solution of the homogeneous wave equation. These estimates resemble the usual Sobolev estimates, but also involve the generators of the Poincarè and Lorentz groups. Using these estimates, Klainerman succeeded in settling the question of global existence of solutions to quasilinear equation for all dimensions (see Klainerman, 1985). In particular, it was shown that there is global existence for $n \geq 4$ and almost global existence for $n = 3$.

It is perhaps oversimplification of the long and hard work that went into the question of estimating the time of existence of solutions to quasilinear equations, to state that this question boils down to estimating the time of existence T by the finiteness of the integral

$$\varepsilon \int_0^t (1+s)^{-\frac{n-1}{2}}\, ds$$

for all $t < T$. We immediately see that $T = \infty$ if $n \geq 4$, indicating global existence; $T = \exp(O(1/\varepsilon))$ if $n = 3$, indicating almost global existence. Also, $T = O(1/\varepsilon^2)$ for $n = 2$ and $T = O(1/\varepsilon)$ for $n = 1$.

Some of these results can also be found in the Lecture Notes by Hörmander (1988, 1997). This reference also contains an example for the case $n = 3$ exhibiting almost global existence. In this example, by a substitution, the problem is reduced to a Burgers' type equation. Using the explicit information of the time of existence of smooth solutions to the Burgers' equation, almost global existence is established.

10.8 EXERCISES

1. Let $c_1, \ldots, c_k$ be positive and distinct real numbers. Show that the solution of the equation

$$\prod_{j=1}^{k} \Box_{c_j} u = \prod_{j=1}^{k} \left(\partial_t^2 - c_j^2 \Delta\right) u = 0$$

can be written as

$$u(x,t) = \sum_{j=1}^{k} u_j(x,t),$$

where u_j satisfies the equation $\Box_{c_j} u = 0$.

(Hint: Let $n = 3$ and $k = 2$. Then, using Darboux equation (10.6), verify that the function $v = rM_u$ satisfies the one-dimensional equation

$$\left(\partial_t^2 - c_1^2\partial_r^2\right)\left(\partial_t^2 - c_2^2\partial_r^2\right) v = 0.$$

As $c_1 \neq c_2$, the general solution of the above equation is given by

$$v(x, r, t) = F_1(r - c_1t) + F_2(r + c_1t) + F_3(r - c_2t) + F_4(r + c_2t)$$

for smooth functions F_i.)

2. Let $n = 3$ and consider the equation $\Box_c^2 u = 0$, where $c > 0$. Taking sufficiently smooth initial data $\partial_t^j u$ for $j = 0, 1, 2, 3$ at $t = 0$, write down the solution explicitly.
(Hint: As in the previous exercise, the function v now satisfies the equation

$$\left(\partial_t^2 - c^2\partial_r^2\right)^2 v = 0.$$

The general solution of the above equation is given by

$$v(x, r, t) = F_1(r - ct) + tF_2(r - ct) + F_3(r + ct) + tF_4(r + ct)$$

for smooth functions F_i.)

3. From the formula for the solution of the homogeneous wave equation for general n given in the text, derive the formula for the solution of the inhomogeneous wave equation using Duhamel's principle.
4. Consider the IVP for the wave equation

$$u_{tt} - \Delta u = 0,\ x \in \mathbb{R}^n,\ t > 0$$
$$u(x, 0) = \phi(x),\ u_t(x, 0) = \psi(x),\ x \in \mathbb{R}^n.$$

a. Let $V(\phi)$ denote the solution of the above IVP for $\psi = 0$. Verify that the solution of the IVP is given by

$$u(x, t) = V(\phi)(x, t) + \int_0^t V(\psi)(x, s)\, ds.$$

b. Let $U(\psi)$ denote the solution of the IVP with $\phi = 0$. Verify that the solution of the IVP is given by

$$u(x, t) = \frac{\partial}{\partial t} U(\phi)(x, t) + U(\psi)(x, t).$$

5. Let L be linear partial differential operator defined by

$$L = \sum_{k=0}^{m} a_k r^{k+1} \partial_r^k,$$

where m is a non-negative integer and $a_0, \dots, a_m$ are real numbers with $a_m = 1$. Assume that L satisfies the following commutator relation:

$$[\partial_r^2, L] \equiv \partial_r^2 L - L\partial_r^2 = (n-1)L\left(\frac{1}{r}\partial_r\right),$$

where $n \geq 3$ is an integer.

a. Show that the integer n is necessarily odd and $m = (n-3)/2$.
b. Write down the recursive relations to determine the coefficients a_k.
c. Show that the operator L can be written as

$$L = \frac{1}{r^m}\partial_r^m r^{2m+1}, \ r > 0.$$

6. Write down the formula for the solution of the inhomogeneous problem (10.26) and (10.27).
7. Prove Theorem 10.12.
8. If A is a tridiagonal matrix with *all* non-zero sub-diagonal elements, show that the geometric multiplicity of any eigenvalue of A is 1. Use this result to establish the statement made in the second paragraph of Remark 10.17. (Hint: Consider the minor of the element in the first row and last column.)

CHAPTER 11

Cauchy–Kovalevsky Theorem and Its Generalization

11.1 INTRODUCTION

In this chapter, we consider equations with analytic coefficients and discuss the existence and uniqueness of their solutions. Historically, the Cauchy–Kovalevsky Theorem (CKT) is one of the first results in the theory of partial differential equations (PDE) that addressed the question of existence and uniqueness of solutions. Its proof introduced the concepts of estimates that are at the heart of the modern PDE techniques. In fact, these estimates are known as *a priori estimates* for the solution and its derivatives, derived before establishing the existence of a solution. More precisely, assuming the existence of a solution, such estimates are derived. Thus, *a priori* estimates are necessary conditions for the existence of a solution. The strategy is to use *a priori* estimates to define a suitable class of functions in which a solution is sought. The rapid development of modern functional analysis provided impetus to the study of PDE and in the current scenario, the study of PDE may be termed as advanced or applied functional analysis.

We first discuss the CKT for linear equations and then its generalization to a system of linear equations. Many of the books on the subject deal with this classical theorem. We cite here John (1978), Hörmander (1976), Trèves (2006), and Folland (1995), among others. We follow the procedure in Hörmander (1976) very closely for linear equations and Trèves (2006), Caflisch (1990), for linear systems.

We begin with a discussion of *analytic functions* $u(z)$ of n complex variables $z = (z_1, \ldots, z_n) \in \mathbb{C}^n$. We use the following notations throughout this chapter: Let $z_0 \in \mathbb{C}^n$ and let u be a complex-valued function defined in a neighborhood of z_0. The function u is said to be *analytic* at z_0 if u has the power series representation

$$u(z) = \sum_{\alpha} a_\alpha (z - z_0)^\alpha, \tag{11.1}$$

where the power series converges absolutely in a neighborhood of z_0. Here $\alpha = (\alpha_1, \ldots, \alpha_n)$ denotes a multi-index with α_i non-negative integers and $z^\alpha = z_1^{\alpha_1} \cdots z_n^{\alpha_n}$ for $z \in \mathbb{C}^n$. It follows immediately from (11.1) that u is infinitely differentiable in a neighborhood of z_0 and

$$a_\alpha = \frac{1}{\alpha!} D^\alpha u(z_0). \tag{11.2}$$

Here D^α denotes the differential operator $D^\alpha = D_1^{\alpha_1} \cdots D_n^{\alpha_n}$ of order $|\alpha| = \alpha_1 + \cdots + \alpha_n$ and $D_j = \frac{\partial}{\partial z_j}$, $j = 1, 2, \dots, n$. In particular, for each $j = 1, 2, \dots, n$, the function of one complex variable

$$\zeta \mapsto u(z_1, \dots, z_{j-1}, \zeta, z_{j+1}, \dots, z_n),$$

where z_j's are held fixed, is analytic in a neighborhood of z_{0j}. It is not difficult to see that the converse statement is also true. We denote by Ω the *unit polydisc* in $\mathbb{C}^n$:

$$\Omega = \{z \in \mathbb{C}^n : |z_j| < 1, j = 1, 2, \dots, n\}.$$

11.2 CAUCHY–KOVALEVSKY THEOREM

We study here the Cauchy or initial value problem (IVP) for the mth-order linear PDE with complex coefficients:

$$\sum_{|\alpha| \le m} a_\alpha(z) D^\alpha u(z) = f(z),$$

with initial conditions

$$D_j^k(u - \varphi) = 0, \text{ for } z_j = 0 \text{ if } 0 \le k < m, j = 1, 2, \dots, n,$$

where the coefficients $a_\alpha(z)$, the function $f(z)$ and the data $\varphi(z)$ are all analytic in a neighborhood of the origin and m is a positive integer. We begin with a version of CKT with smallness assumption on the coefficients $a_\alpha(z)$.

Theorem 11.1. For $\beta > 0$, consider a differential equation

$$D^\beta u = \sum_{|\alpha| \le |\beta|} a_\alpha D^\alpha u + f, \tag{11.3}$$

along with the initial conditions

$$D_j^k(u - \varphi) = 0, \text{ for } z_j = 0 \text{ if } 0 \le k < \beta_j, j = 1, 2, \dots, n, \tag{11.4}$$

where the coefficients $a_\alpha(z)$, the function f and φ are given analytic functions in a neighborhood of the origin 0 in $\mathbb{C}^n$. If $\sum_{|\alpha| = |\beta|} |a_\alpha(0)|$ is sufficiently small, then the

Cauchy problem (11.3) and (11.4) has a unique analytic solution in a neighborhood of 0. □

The smallness condition stated in the theorem depends only on $|\beta|$. The theorem will be proved by the method of iterations. As a preparation, we first prove some basic results.

Lemma 11.2. Suppose g is an analytic function in the polydisc Ω. Then, the equation $D^\beta u = g$, $\beta > 0$ has a unique solution, analytic in Ω and satisfying the initial conditions

$$D_j^k u(z) = 0, \text{ when } z_j = 0,\ z \in \Omega,\ 0 \le k < \beta_j,\ j = 1, 2, \dots, n.$$

□

Proof If $g(z) = \sum_\alpha g_\alpha z^\alpha$, $z \in \Omega$, where the power series converges absolutely in Ω, then the required unique solution is given by the power series representation

$$u(z) = \sum_\alpha g_\alpha \frac{\alpha!}{(\alpha+\beta)!} z^{\alpha+\beta}.$$

□

Lemma 11.3. Let $v(\zeta)$ be an analytic function of one complex variable in $\{|\zeta| < 1\} \subset \mathbb{C}$ such that $v(0) = 0$ and $|v'(\zeta)| \le C(1-|\zeta|)^{-a-1}$ for $|\zeta| < 1$ and some positive constants C and a. Then,

$$|v(\zeta)| \le Ca^{-1}(1-|\zeta|)^{-a-1},\ |\zeta| < 1.$$

□

Proof The proof follows from the fundamental theorem of calculus. For $|\zeta| < 1$, we have

$$v(\zeta) = \int_0^1 \frac{d}{dt}(v(t\zeta))\,dt = \int_0^1 \zeta v'(t\zeta)\,dt.$$

Therefore,

$$\begin{aligned} |v(\zeta)| &\le |\zeta| \int_0^1 |v'(t\zeta)|\,dt \le C|\zeta| \int_0^1 (1-t|\zeta|)^{-a-1}\,dt \\ &\le Ca^{-1}[(1-|\zeta|)^{-a} - 1] \le Ca^{-1}(1-|\zeta|)^{-a-1}. \end{aligned}$$

□

The above Lemma 11.3 easily gets extended to the case when bounds on the higher-order derivatives are given. It also extends to multi-variable case by working with one variable at a time, to obtain the following corollary:

Corollary 11.4. Suppose $v(z)$ is an analytic function in Ω satisfying

$$|D^{\beta}v(z)| \leq C\prod_{i=1}^{n}(1-|z_i|)^{-a-1}$$

for $z \in \Omega$ and some positive constants C, a and a multi-index $\beta > 0$ and the initial conditions

$$D_j^k v = 0, \text{ for } z_j = 0 \text{ if } 0 \leq k < \beta_j,\ j = 1, 2, \ldots, n.$$

Then,

$$|v(z)| \leq Ca^{-|\beta|}\prod_{i=1}^{n}(1-|z_i|)^{-a-1}$$

for $z \in \Omega$. □

Lemma 11.5. Let $v(\zeta)$ be an analytic function of one complex variable in $\{|\zeta| < 1\} \subset \mathbb{C}$ such that $|v(\zeta)| \leq C(1-|\zeta|)^{-a}$ for $|\zeta| < 1$ and some positive constants C and a. Then, for $k = 1, 2, \ldots$, we have

$$|v^{(k)}(\zeta)| \leq C(e(k+a))^k(1-|\zeta|)^{-a-k},\ |\zeta| < 1.$$

Here $v^{(k)}$ denotes the kth derivative of v. □

Proof Let $0 < \varepsilon < \rho \equiv 1 - |\zeta|$, $|\zeta| < 1$. If $|\xi - \zeta| \leq \varepsilon$, then $|\xi| \leq |\zeta| + \varepsilon < 1$. Hence, $|v(\xi)| \leq C(1 - |\zeta| - \varepsilon)^{-a} = (\rho - \varepsilon)^{-a}$. Next, by Cauchy formula, we have

$$v^{(k)}(\zeta) = \frac{(k-1)!}{2\pi i}\int_{|\xi-\zeta|=\varepsilon}\frac{v(\xi)}{(\xi-\zeta)^{k+1}}\,d\xi.$$

Therefore,

$$|v^{(k)}(\zeta)| \leq C(k-1)!\varepsilon^{-k}(\rho-\varepsilon)^{-a}.$$

Minimizing the right-hand side expression over ε, occurring at $\varepsilon = \frac{k\rho}{a+k}$, we obtain the estimate

$$|v^{(k)}(\zeta)| \leq C(k-1)!\frac{(a+k)^k}{k^k}\left(1+\frac{k}{a}\right)^a\rho^{-a-k}.$$

Using the trivial estimates

$$\frac{(k-1)!}{k^k} \leq 1 \text{ and } \left(1+\frac{k}{a}\right)^a \leq e^k,$$

the required estimate follows and the proof is complete. □

As in the case of Lemma 11.3, the multi-variable version of Lemma 11.5 is the following corollary:

Corollary 11.6. Suppose $v(z)$ is an analytic function in the polydisc Ω satisfying

$$|v(z)| \leq C\prod_{i=1}^{n}(1-|z_i|)^{-a},$$

for $z \in \Omega$ and some positive constants C and a. Then, for any multi-index β the estimate

$$|D^{\beta}v(z)| \leq C(e(a+|\beta|))^{|\beta|}\prod_{i=1}^{n}(1-|z_i|)^{-a-|\beta|},$$

holds for $z \in \Omega$. □

Proof of Theorem 11.1 Replacing u by $u-\varphi$ reduces the proof to the case $\varphi = 0$, with a possible change in the right-hand side term. We may thus assume $\varphi = 0$. We then solve equation (11.3) by iteration, using the recursion formula

$$D^{\beta}u_{m+1} = \sum_{|\alpha|\leq|\beta|} a_{\alpha}D^{\alpha}u_m + f, \tag{11.5}$$

for $m = 0, 1, \ldots$, with each u_m satisfying the initial conditions

$$D_j^k u = 0, \text{ when } z_j = 0,\ 0 \leq k < \beta_j,\ j = 1, 2, \ldots, n. \tag{11.6}$$

We take u_0 to be any analytic function satisfying conditions (11.6), for example $u_0 = 0$. The existence of the functions u_m, $m = 1, 2, \ldots$ then follows from Lemma 11.2. Now define $v_m = u_{m+1} - u_m$ for $m = 0, 1, \ldots$. Then, v_m's satisfy the recursion relations

$$D^{\beta}v_{m+1} = \sum_{|\alpha|\leq|\beta|} a_{\alpha}D^{\alpha}v_m. \tag{11.7}$$

We need to establish the convergence of u_m and v_m. We proceed as follows: First assume that the functions a_{α}, f and u_0 are analytic in a neighborhood of $\overline{\Omega}$ and let

$$\sum_{|\alpha|\leq|\beta|} |a_{\alpha}| \leq A \text{ and } |v_0| \leq M, \text{ in } \Omega.$$

We claim that

$$|v_m(z)| \leq C^m M\prod_{i=1}^{n}(1-|z_i|)^{-m|\beta|} \tag{11.8}$$

for $m = 0, 1, \dots$ and $z \in \Omega$. For $m = 0$, this is just the assumption on v_0. Assume that (11.8) holds for some m. Then, using (11.7) and applying Corollary 11.6, we obtain that

$$|D^\beta v_{m+1}(z)| \leq AC^m M(e|\beta|(m+1))^{|\beta|} \prod_{i=1}^{n} (1 - |z_i|)^{-(m+1)|\beta|}, \; z \in \Omega.$$

Since v_{m+1} satisfies the initial conditions in (11.6), it follows from Corollary 11.4 that

$$|v_{m+1}(z)| \leq AC^m M(e|\beta|(m+1)/m|\beta|)^{|\beta|} \prod_{i=1}^{n} (1 - |z_i|)^{-(m+1)|\beta|}, \; z \in \Omega.$$

Hence, (11.8) is valid with $C = A(2e)^{|\beta|}$.

Next we establish the existence of a solution to (11.3).

Step 1: Assume first that $C < 1$ and choose γ such that $C < \gamma < 1$. It follows from (11.8) that the series $\sum_{m=1}^{\infty} |v_m(z)|$ is uniformly convergent in the neighborhood $\widetilde{\Omega}$ of 0 defined by

$$\widetilde{\Omega} = \left\{ z \in \Omega : \prod_{j=1}^{n} (1 - |z_j|)^{|\beta|} > \gamma \right\}.$$

Since $u_m = v_1 + \cdots + v_{m-1}$, it follows that $\lim_{m\to\infty} u_m$ exists and call the limit u. Then u is analytic in $\widetilde{\Omega}$. Since $D^\alpha u = \lim_{m\to\infty} D^\alpha u_m$ for every multi-index α, letting $m \to \infty$ in (11.5) and (11.6), proves that u is a required solution in $\widetilde{\Omega}$.

To prove uniqueness, let u be an analytic solution of (11.1) with $f = 0$ and satisfying the conditions (11.2)(with $\varphi = 0$), in a neighborhood Ω_1 of 0. If we set $v_m = u$ for every m, then we have a solution of the recursive formula (11.7) satisfying the conditions (11.6). Now $u_{m+1} - u_m = v_m = u$ and since u_m converges as shown above, we see that $u = 0$ and uniqueness follows.

Step 2: We now prove the theorem removing the restriction on C introduced in Step 1. For this purpose, let $r > 0$ be a parameter and consider the following functions

$$U(z) = u(rz), \; F(z) = r^{|\beta|} f(rz).$$

If u and f satisfy (11.1) in a neighborhood of 0, then U satisfies

$$D^\beta U(z) = \sum_{|\alpha| \leq |\beta|} r^{|\beta|-|\alpha|} a_\alpha(rz) D^\alpha U(z) + F(z)$$

in a neighborhood of 0, while the conditions in (11.4) are the same for both u and U. Now writing

$$\sum_{|\alpha|\leq|\beta|} r^{|\beta|-|\alpha|}|a_\alpha(rz)| = \sum_{|\alpha|=|\beta|} a_\alpha(rz) + \sum_{|\alpha|<|\beta|} r^{|\beta|-|\alpha|}|a_\alpha(rz)|,$$

we see that the second term on the right-hand side can be made small by choosing r small. Thus, the restrictions made in the first part of the proof are fulfilled if we choose r sufficiently small and assume that

$$\sum_{|\alpha|=|\beta|} |a_\alpha(0)| < (2e)^{-|\beta|}. \tag{11.9}$$

This proves the existence of U using Step 1 and thus we obtain u. □

We now remove the restriction in (11.9) and prove the following:

Theorem 11.7. Let K be the set of all the multi-indices α in the sum on the right-hand side of (11.3) for which $a_\alpha \not\equiv 0$ and assume that the multi-index β is not in the convex hull of K. Then, the conclusion of Theorem 11.1 is still valid without the assumption (11.9). □

Note that the assumption on the multi-index β is automatically satisfied if the sum on the right-hand side of (11.1) contains only multi-indices α with $|\alpha| < |\beta|$ and in this case (11.9) is trivially satisfied.

Proof For $\rho = (\rho_1, \cdots, \rho_n) \in \mathbb{R}^n$, let $\rho(\alpha) = \sum_{i=1}^{n} \rho_i\alpha_i$, for any multi-index α. Using the hypothesis, it follows from Hahn–Banach separation theorem that there is a $\rho \in \mathbb{R}^n$ such that $\rho(\alpha) < \rho(\beta)$ for all $\alpha \in K$. By replacing $\rho(\alpha)$ by $\rho(\alpha) + t|\alpha|$, $t > 0$, we may assume that $\rho_j > 0$ for $j = 1, 2, \ldots, n$.

Next consider the change of variables

$$z \mapsto \tilde{z} = (e^{-\lambda\rho_1}z_1, \ldots, e^{-\lambda\rho_n}z_n),$$

where $\lambda > 0$ is a parameter and the functions

$$U(z) = u(\tilde{z}) \text{ and } F(z) = e^{-\lambda\rho(\beta)}f(\tilde{z}).$$

If u and f satisfy (11.3), then U satisfies

$$D^\beta U(z) = \sum_{\alpha\in K} a_\alpha(\tilde{z}) \exp\left(\lambda(\rho(\alpha) - \rho(\beta))\right) D^\alpha U(z) + F(z),$$

and the conditions (11.3) are the same for both u and U. For λ sufficiently large, the hypothesis of Theorem 11.1 is satisfied by the just derived equation satisfied by U and this completes the proof. □

We now come to the main result.

Theorem 11.8 (Cauchy–Kovalevsky). Consider a linear differential equation

$$\sum_{|\alpha|\leq m} a_\alpha(z)D^\alpha u(z) = f(z), \tag{11.10}$$

of order m, where the coefficients $a_\alpha(z)$ and $f(z)$ are analytic in a neighborhood of the origin 0 and the coefficient of D_n^m is non-zero when $z = 0$. Then, for every analytic function φ in a neighborhood of 0, there exists a unique solution u of (11.10) that is analytic in a neighborhood of 0 and satisfies the initial conditions

$$D_n^j(u-\varphi) = 0 \text{ when } z_n = 0, \text{ and } 0 \leq j < n.$$

□

Proof By the assumption that the coefficient of D_n^m is non-zero in a neighborhood of 0. Dividing (11.10) throughout by this coefficient and transferring the remaining terms to the right-hand side, equation (11.10) reduces to (11.1) with $\beta = (0, \ldots, 0, m)$. Then, we have $\alpha_n < m = \beta_n$ for all α occurring in the right-hand side of (the reduced equation)(11.1). Therefore, the theorem follows from Theorem 11.7. □

11.2.1 Real Analytic Functions

Much of the above discussion can also be done with real analytic functions in place of complex analytic functions.

Let Ω be an open set in $\mathbb{R}^n$. A function $f : \Omega \to \mathbb{R}$ is said to be *real analytic* at $x_0 \in \Omega$, if there exists $r > 0$ such that $B_r(x_0) \subset \Omega$ and f has the (convergent) power series representation

$$f(x) = \sum_\alpha a_\alpha(x-x_0)^\alpha, \tag{11.11}$$

for all $x \in B_r(x_0)$, where the sum extends over all multi-indices α and a_α are real constants. By the elementary properties of convergent power series, it follows that $f \in C^\infty(\Omega)$ and

$$D^\beta f(x) = \sum_\alpha \frac{(\beta+\alpha)!}{\alpha!} a_{\beta+\alpha}(x-x_0)^\alpha,$$

for all multi-indices β; here $\alpha! = \alpha_1! \cdots \alpha_n!$ for a multi-index α. In particular, $D^\beta f(x_0) = \beta! a_\beta$ for all multi-indices β.

Remark 11.9. As the radius of convergence of the power series (11.11) is determined by the coefficients a_α, we can replace the real variable x by a complex variable z. Thus, we infer that a real analytic function is also complex analytic. However, this is only a local result. For example, it is possible to construct a real analytic function on $\mathbb{R}$ that *cannot* be extended as a complex analytic function to the strip $\{z \in \mathbb{C} : |\Im z| < \delta\}$ for any $\delta > 0$. □

Remark 11.10. Consider the function $f : \mathbb{R} \to \mathbb{R}$ defined by

$$f(x) = \begin{cases} \exp(-1/x^2) & \text{if } x > 0, \\ 0 & \text{if } x \leq 0. \end{cases}$$

It is easily verified that $f \in C^\infty(\mathbb{R})$ and that $f^{(k)}(0) = 0$ for all $k \geq 0$. Thus, f cannot be analytic at $x = 0$. □

A natural question thus arises is that, which C^∞ functions are real analytic. We state the following theorem without proof:

Theorem 11.11. Let Ω be an open set in $\mathbb{R}^n$ and $f : \Omega \to \mathbb{R}$ be a C^∞ function. Then, f is real analytic at $x_0 \in \Omega$, if and only if there exist $r > 0$, $M > 0$ such that $B_r(x_0) \subset \Omega$ and the estimates

$$|D^\beta f(x)| \leq M \frac{\beta!}{r^{n|\beta|}} \tag{11.12}$$

are satisfied for all $x \in B_r(x_0)$ and all multi-indices β. □

11.2.2 Non-characteristic Cauchy Problem

In Cauchy–Kovalevsky Theorem 11.8, the Cauchy data was prescribed on the hypersurface $z_n = 0$, assuming it is non-characteristic with respect to equation (11.10). This result can be extended by prescribing the Cauchy data on a smooth hypersurface, which is non-characteristic with respect to the given differential equation. By a change of variables, this general case can be reduced to the one considered above, by *flattening* the given hypersurface to one of the co-ordinate hypersurface. This procedure has already been described in detail for a second-order linear equation in Chapter 6. Here we just describe the problem and the reader can consult the references cited at the beginning for details.

Thus, we consider a general quasilinear equation of order m:

$$\sum_{|\alpha| \leq m} a_\alpha D^\alpha u + a_0 = 0 \tag{11.13}$$

in an open set Ω in $\mathbb{R}^n$, where the coefficients

$$a_\alpha = a_\alpha(x, D^\beta u,\ |\beta| < m)$$

for $\alpha \geq 0$, are in general functions of x and the derivatives of u up to order $m - 1$.

Consider a smooth $n-1$-dimensional hypersurface $\Gamma \subset \Omega$, with a smoothly varying normal $\nu(x) = (\nu_1(x), \dots, \nu_n(x))$, $x \in \Gamma$. The jth *normal derivative* of u at $x \in \Gamma$ is defined by

$$\frac{\partial^j u}{\partial \nu^j}(x) = \sum_{|\alpha|=j} \nu^\alpha(x) D^\alpha u(x), \tag{11.14}$$

for $j = 0, 1, \dots, m-1$. Suppose $g_j : \Gamma \to \mathbb{R}$, $j = 0, 1, \dots, m-1$ are given smooth functions. The *Cauchy problem* for equation (11.13) consists of finding a function u satisfying (11.13) and the *boundary conditions*

$$\frac{\partial^j u}{\partial \nu^j}(x) = g_j(x),\ x \in \Gamma \text{ for } j = 0, 1, \dots, m-1. \tag{11.15}$$

The set of functions $\{g_0, \dots, g_{m-1}\}$ is referred to as a Cauchy data on Γ. We say the surface Γ is a *non-characteristic surface* with respect to equation (11.13) if

$$A(x) \equiv \sum_{|\alpha|=m} a_\alpha \nu^\alpha(x) \neq 0 \text{ for all } x \in \Gamma. \tag{11.16}$$

Note that from our assumption on the coefficients a_α, the right-hand side in (11.16) only depends on the Cauchy data on Γ. In seeking a solution u of (11.13), which is an analytic function in a neighborhood of Γ, first we need to find all derivatives of u on Γ. This is accomplished using the Cauchy data, equation (11.13) and the non-characteristic condition (11.16).

11.3 A GENERALIZATION: APPLICATION TO FIRST-ORDER SYSTEMS

Consider the Cauchy problem for a linear system of first-order equations:

$$\begin{aligned} \frac{\partial u}{\partial t} &= \sum_{j=1}^{n} B_j(t,z)\frac{\partial u}{\partial z_j} + B_0(t,z)u + f(t,z) \\ u(0,z) &= u_0(z). \end{aligned} \tag{11.17}$$

Here u and f are N vectors, B_0, B_j are $N \times N$ matrices. Assuming that the coefficient matrices, f and the initial function u_0 lie in a certain class of analytic functions of z in a strip of width ρ_1 in $\mathbb{C}^n$ for each $t \in [0, T]$, we seek a solution of (11.17) in the same class. Actually, the solution we will be obtaining is going to be analytic in a smaller strip of width $\rho_0 < \rho_1$. Before stating the main theorem, let us introduce some notations and function spaces we are going to work with. Fix $\rho_1 > 0$ and let $0 < \rho_0 < \rho_1$. Put $\delta = \rho_1 - \rho_0 > 0$. For $\rho \in [\rho_0, \rho_1]$, we write $\rho = \rho_0 + \sigma(\rho_1 - \rho_0)$ with $\sigma \in [0, 1]$. We write $\rho = \rho(\sigma)$ and it is convenient to use σ for labelling the domains and the function spaces. Note that if $\rho(\sigma)$ and $\rho(\sigma')$ are in $[\rho_0, \rho_1]$, then $\rho(\sigma') < \rho(\sigma)$ if and only if $\sigma' < \sigma$ and we have $\rho(\sigma) - \rho(\sigma') = \delta(\sigma - \sigma')$. Next, for $\sigma \in [0, 1]$, denote by Ω_σ the strip in $\mathbb{C}^n$:

$$\Omega_\sigma = \{z = (z_1, \dots, z_n) \in \mathbb{C}^n : |\Im z_j| < \rho(\sigma), j = 1, \dots, n\}.$$

Here $\Im z$ denotes the imaginary part of a complex number $z \in \mathbb{C}$. The Banach space of all the bounded analytic functions in Ω_σ is denoted by $\mathcal{A}_\sigma$, with the norm

$$\|u\|_\sigma = \sup_{z \in \Omega_\sigma} |u(z)|.$$

We next introduce the function space $X_\sigma = C^1([0, T]; \mathcal{A}_\sigma)$ defined by[1]

$$C^1([0, T]; \mathcal{A}_\sigma) = \left\{ u = u(t, z) : u(t, \cdot), \frac{\partial u}{\partial t}(t, \cdot) \in \mathcal{A}_\sigma \text{ for all } t \in [0, T] \text{ and are continuous} \right\}.$$

For brevity, we say that a vector-valued function or a matrix-valued function is in X_σ when each component of the vector or matrix is in X_σ. Assume that the coefficients B_0, B_j and f all lie in X_σ and let

$$M = \max_{t \in [0, T]} \{ \|B_0(t, \cdot)\|_1, \|B_j(t, \cdot)\|_1, \|f(t, \cdot)\|_1 \}.$$

Note that B_j, $j = 0, 1, \dots, n$ are $N \times N$ matrix-valued functions and f is a vector-valued ($\mathbb{C}^N$ valued) function.

Observe that the inclusion map $\mathcal{A}_\sigma \to \mathcal{A}_{\sigma'}$ is a continuous linear map with norm ≤ 1 for all $\sigma' < \sigma$. Further, if $u \in X_\sigma$ and $\|u\|_{\sigma'} = 0$, it follows from analyticity that $\|u\|_\sigma = 0$. On the other hand, the derivative map $D_j : \mathcal{A}_\sigma \to \mathcal{A}_{\sigma'}$ is a linear continuous map satisfying

$$\|D_j u\|_{\sigma'} \le \delta^{-1}(\sigma - \sigma')^{-1} \|u\|_{\sigma}, \tag{11.18}$$

for all $\sigma' < \sigma$ and $u \in \mathcal{A}_\sigma$. See the proof of Lemma 11.5. In particular, the map $D_j : \mathcal{A}_1 \to \mathcal{A}_0$ is a continuous linear map with norm $\le \delta^{-1}$. If we now put

$$B(t) = \sum_{j=1}^{n} B_j(t, z) \frac{\partial}{\partial z_j} + B_0(t, z),$$

it follows from (11.18) and the assumptions we have made on B_j that $B(t) : \mathcal{A}_\sigma \to \mathcal{A}_{\sigma'}$ is a linear continuous map with norm $\le (n\delta^{-1}(\sigma - \sigma')^{-1} + 1)M$, for all $\sigma' < \sigma$. If we put

$$C = (n\delta^{-1} + 1)M,$$

[1] In general, for a function $u : [0, T] \to X$, where X is a Banach space, its Fréchet derivative $u'(t)$ at $t \in [0, T]$ can be identified with an element in X.

then the norm of $B(t)$ is $\leq C/(\sigma - \sigma')$ for all $0 \leq \sigma' < \sigma \leq 1$, where C is independent of t, σ. By taking M larger, if necessary, we also assume that $(Ce)^{-1} < T$. We have the following theorem:

Theorem 11.12 (Cauchy–Kovalevsky). Let $u_0 \in \mathcal{A}_1$. Then, there exists $T_0 \in (0, T]$ such that the following statement is true: there is a unique $u \in C^1([0, T_0); \mathcal{A}_0)$ satisfying the equation and the initial condition in (11.17), for all $z \in \Omega_0$ and $t \in [0, T_0)$. □

It is easy to deduce that if the data is provided in a time interval $[-T, T]$ around the origin, then the solution also exists in an interval $(-T_0, T_0)$ for some $T_0 > 0$.

Proof The proof is carried out by writing (11.17) as an integral equation and then using the Picard iterations. Consider the system of integral equation

$$u(t, z) = u_0(z) + \int_0^t (Bu(s, z) + f(s, z))\, ds, \tag{11.19}$$

where

$$Bu(t, z) = \sum_{j=1}^{n} B_j(t, z)D_j u(t, z) + B_0(t, z)u(t, z).$$

If $u \in C([0, T]; \mathcal{A}_0)$ is a solution of (11.19), then a differentiation with respect to t shows that $u \in C^1([0, T]; \mathcal{A}_0)$ and satisfies the equation and the initial condition in (11.17), for $z \in \Omega_0$. The converse statement is also easy to verify. Thus, it suffices to prove the existence of a solution to (11.19) and this is done using Picard's iterations.

Existence: We will show that there exists a function u in $C^1([0, (Ce)^{-1}); \mathcal{A}_0)$ satisfying (11.17). We further show that, for any $0 \leq \sigma < 1$, this function u is also a C^1 function of $t \in [0, (Ce)^{-1}(1 - \sigma))$, valued in $\mathcal{A}_\sigma$.

Let $u_1(t, z) = u_0(z) + \int_0^t (Bu_0(z) + f(s, z))\, ds$ for $t \in [0, T]$ with the initial function $u_0 \in \mathcal{A}_1$. Define, inductively

$$u_{k+1}(t, z) = u_0(z) + \int_0^t (Bu_k(s, z) + f(s, z))\, ds, \tag{11.20}$$

for $k = 1, \ldots, t \in [0, T]$ and $z \in \Omega_\sigma$, $\sigma < 1$. It follows, from the observation we made above regarding the operator B, that for each $k = 1, 2, \ldots$, the function u_k is a continuous function of $t \in [0, T]$, valued in $\mathcal{A}_\sigma$, $\sigma < 1$.

Now, set $v_1 = u_1$ and $v_k(t, z) = u_{k+1}(t, z) - u_k(t, z)$ for $k = 1, 2, \ldots$ and $z \in \Omega_\sigma$. By linearity of B, we have

$$v_{k+1}(t, z) = \int_0^t Bv_k(s, z)\, ds,$$

for $k = 1, 2, \ldots$.

We now claim that

$$\|v_k(t,\cdot)\|_\sigma \le M_1\left(\frac{Cet}{1-\sigma}\right)^k, \quad t\in[0,T], \tag{11.21}$$

where

$$M_1 = \|u_0\|_0 + \max_{t\in[0,T]}\int_0^t \|f(s,\cdot)\|_0\, ds.$$

The estimate (11.21) trivially holds for $k = 1$. Assume it holds for some $k \ge 1$. Choose any $\tilde{\sigma} > \sigma$. Then, using the estimate of the norm of the operator B, we see that

$$\begin{aligned}\|v_{k+1}(t)\|_\sigma &\le \frac{C}{\tilde{\sigma}-\sigma}\int_0^t \|v_k(s)\|_{\tilde{\sigma}}\, ds\\ &\le M_1\frac{C}{\tilde{\sigma}-\sigma}\left(\frac{Ce}{1-\tilde{\sigma}}\right)^k\frac{t^{k+1}}{k+1}, \quad \text{using (11.21)}.\end{aligned}$$

Now choose $\tilde{\sigma} = \sigma + (1-\sigma)/(k+1)$, so that $1-\tilde{\sigma} = \frac{k}{k+1}(1-\sigma)$.
 Therefore,

$$\|v_{k+1}(t)\|_\sigma \le M_1\left(\frac{Ct}{1-\sigma}\right)^{k+1} e^k\left(1+\frac{1}{k}\right)^k.$$

Observing that $\left(1+\frac{1}{k}\right)^k \le e$, we complete the induction argument and thus proving the claim for all $k = 1, 2, \ldots$. From (11.21), we derive that the series

$$\sum_{k=1}^{\infty} v_k(t)$$

converges absolutely in $\mathcal{A}_\sigma$, uniformly in every closed interval of $[0,(Ce)^{-1}(1-\sigma))$. The sum u of the above series is therefore the solution of the integral equation (11.19) and hence that of (11.17). Of course u is also the limit of the sequence u_k as $k\to\infty$.

Uniqueness: We prove the statement: If u, v are two solutions of (11.17) in some interval $[0, T')$, valued in $\mathcal{A}_\sigma$, $\sigma\in(0,1]$, then they must be equal.

Put $w = u - v$. Then, by linearity w satisfies

$$w(t) = \int_0^t Bw(s)\, ds.$$

By hypothesis, $w(0) = 0$ and by continuity, the set of t at which w vanishes is closed in $[0, T')$. To complete the proof, we show that this set is also open in $[0, T')$. To this end, let $t_0 \in (0, T')$ and $w(t_0) = 0$.

Now take the initial time as $t = t_0$. Thus, we have

$$w(t) = \int_{t_0}^{t} Bw(s)\, ds. \tag{11.22}$$

Let $0 \leq \sigma' < \sigma$. We claim that equation (11.22) implies the following estimate for w:

$$\|w(t)\|_{\sigma'} \leq M_2(t)(\sigma - \sigma')^{-k}(Ce)^k |t - t_0|^k, \tag{11.23}$$

for $k = 0, 1, 2, \ldots$, where

$$M_2(t) = \sup \|w(s)\|_\sigma$$

with sup taken over all s over the line segment joining t and t_0. Here we are taking the values of t on both the sides of t_0. The estimate (11.23) trivially holds for $k = 0$. If it holds for some $k \geq 0$, then using the estimate of the norm of the operator B and (11.22), we have

$$\|w(t)\|_{\sigma'} \leq M_2(t) C \varepsilon^{-1} (Ce)^k (\sigma - \sigma' - \varepsilon)^{-k} \frac{|t - t_0|^{k+1}}{k+1},$$

where $\varepsilon = (\sigma - \sigma')/(k+1)$. This immediately shows that (11.23) holds with k replaced by $k+1$. Now choose t in a small open interval around t_0 such that $|t - t_0| < (Ce)^{-1}(\sigma - \sigma')/2$. Then, the estimate (11.23) implies that

$$\|w(t)\|_{\sigma'} \leq M_2(t)/2^k$$

for $k = 0, 1, 2, \ldots$. Letting $k \to \infty$, we conclude that $\|w(t)\|_{\sigma'} = 0$ for such t. But since the inclusion map from $\mathcal{A}_\sigma$ to $\mathcal{A}_{\sigma'}$ is continuous, we infer from analyticity that $\|w(t)\|_\sigma = 0$ for such t. This completes the proof. □

Remark 11.13. The CKT proved for the complex analytic case is equally valid for operators with real analytic coefficients with real analytic data prescribed on real analytic non-characteristic surface. This can be proved directly by using the method of majorants (see John, 1978; Folland, 1995). We may also use the fact that any real analytic function is also complex analytic in a small neighborhood of the real space and appeal to the theorem proved in the complex analytic case. □

11.4 HOLMGREN'S UNIQUENESS THEOREM

The CKT gives the existence and uniqueness result for a linear PDE (also for a system of PDE) with analytic coefficients, in the class of analytic solutions. This result, however, does not rule out the possibility of existence and/or uniqueness of a solution in the class of non-analytic functions. The Holmgren's uniqueness theorem therefore assumes significance as it asserts the uniqueness of any smooth solution to linear PDE with analytic coefficients; the assumption that the PDE has analytic coefficients is required to apply the CKT. We remark that the problem of existence of a solution, in the non-analytic class, of a non-characteristic Cauchy problem of a hyperbolic PDE or a system, is quite difficult. The interested reader may consult Courant and Hilbert (1989), Hörmander (1976, 1984), Benzoni-Gavage and Serre (2007).

The Holmgren's uniqueness theorem is more generally true if we assume the existence of a solution in the class of *distributions with compact support* (see Hörmander, 1976, 1984; Trèves, 2006). Here we merely sketch a proof of this theorem, for smooth solutions, based on the approach in Courant and Hilbert (1989) and John (1978). For a more detailed discussion, see Rauch (1992), Renardy and Rogers (2004), and Smoller (1994).

Theorem 11.14 (Holmgren's Uniqueness Theorem). Let Ω be an open set in $\mathbb{R}^n$ and $P(x, D) = \sum_{|\alpha| \leq m} a_\alpha(x) D^\alpha$ be a linear PDO of order m, with analytic coefficients. Suppose S is an analytic non-characteristic surface with respect to P. Then, the Cauchy problem

$$\begin{aligned} P(x, D)u &= f(x), \; x \in \Omega \\ D^\beta u(x) &= g_\beta(x), \; |\beta| \leq m-1, \; x \in S \end{aligned} \tag{11.24}$$

has at most one smooth solution in a neighborhood on S. □

Note that f and g_β are assumed to be smooth, but not necessarily analytic.

Proof We only sketch a proof. First note that by linearity it suffices to prove that any smooth solution of $P(x, D)u = 0$ with vanishing Cauchy data on S vanishes in a neighborhood of S. The idea of the proof is very simple: A linear operator T in a Hilbert space (or a Banach space) is one-one if and only if the adjoint T' of T has dense image or range. In the present case, it is still simpler. If, for a bounded open set $G \subset \Omega$, we can show that $\int_G uw \, dx = 0$ for a dense set of functions w, it follows that $u = 0$ in G, completing the proof.

We now explain how this can be achieved without going to technicalities. The (formal) adjoint of P is given by $P'(x, D)v = \sum_{|\alpha| \leq m} (-1)^{|\alpha|} D^\alpha (a_\alpha v)$. Note that the principal symbols of P and P' differ by a factor of ± 1. Thus, they have the same characteristic surfaces. Suppose S_0 is a compact subset of S. Assume that an analytic non-characteristic surface S_1 can be chosen so that the boundary of S_0 and S_1, which lie in $n-2$-dimensional space, are the same; in case of two dimensions, this means that the end points of S_0 and S_1 are the same. Thus, S_0 and S_1 are non-characteristic surfaces for both P and P'. Consider the non-characteristic Cauchy problem

$$\begin{aligned} P'(x, D)v &= w(x), \; x \in \Omega \\ D^{\beta}v(x) &= 0, \; |\beta| \leq m-1, \; x \in S_1, \end{aligned} \tag{11.25}$$

where w is an arbitrary polynomial. Let G be the region whose boundary $\partial G = S_0 \cup S_1$ and assume that a solution v of (11.25) exists in G. Then, we have

$$\int_G uw \, dx = \int_G uP'v \, dx = \int_G (Pu)v \, dx = 0.$$

Here the second equality follows by performing integration by parts; there are no boundary terms as $D^{\beta}u = 0$ on S_0 and $D^{\beta}v = 0$ on S_1 for all $|\beta| \leq m-1$. Therefore, we have accomplished the required statement on u, modulo some assumptions made above.

The first question is how to construct the surface S_1. This is done by *continuously deforming* the surface S_0. This is the first technicality. The second question is how to ensure that a solution v of (11.25) exists in G, that too for all polynomials w. The CKT certainly gives a solution in a neighborhood of S_1, but not necessarily in G. This is the second technicality. Here the linearity of P' comes into play and the existence of the solution v in G is done step-by-step. For a detailed discussion, we refer to Rauch (1992), Renardy and Rogers (2004), and Smoller (1994). □

11.5 NOTES

There are many interesting questions regarding the uniqueness of solutions in the non-analytic class of equations considered in Section 3.2. The result of Section 11.3 can be used to prove the existence and uniqueness of solutions to symmetric hyperbolic systems in some suitable class of Sobolev spaces, using smoothing operator techniques that are quite technical in nature.

CHAPTER 12

A Peep into Weak Derivatives, Sobolev Spaces and Weak Formulation

Throughout the earlier chapters, we have seen the necessity of defining solutions in a sense other than the well-known classical solutions. In other words, we need to go out of the realm of smooth solutions to capture the physically relevant solutions. In fact, we have established the existence of certain types of weak solutions in the context of Hamilton–Jacobi equations and Conservation Laws. Further, we have also indicated in Chapter 7, the existence of an integral formulation corresponding to a minimization problem given by the Dirichlet functional. Generically, we term such solutions as *weak solutions*. In fact, there are different concepts of weak solutions that have been developed in the last 100 years or so; like *distribution solutions, transposition solutions, entropy solutions, viscosity solutions*, and so on. Among them the notion of weak solutions in the sense of distributions, weak formulation and Sobolev spaces took centre stage in the first half of the last century, and subsequently changed the scenario of the study of partial differential equations. As remarked earlier, the necessity of introducing the above concepts was evident from the study of Dirichlet functionals from the second half of the eighteenth century and early part of the nineteenth century. Sobolev was quite successful in defining appropriate function spaces (the so-called Sobolev spaces) using generalized functions. A stable foundation was given later by the introduction of *distributions* by L. Schwartz in the 1940s (see Brezis, 2011; Kesavan, 1989; Schwartz, 1966). This new development was not only useful in the study of PDEs, but it could also rigorously establish the notions of Dirac δ function, its derivatives and the symbolic calculus developed by the physicists including Paul Dirac.

Without going much into the details, we would like to introduce here the notion of *weak derivative* of functions that are, otherwise, not classically differentiable and certain associated spaces, which are required to study solutions of weak formulation. In the process, we see some ideas about the modern theory of PDEs.

12.1 WEAK DERIVATIVES

Quite often, given a function f, the value $f(x)$ represents a physical quantity at the point x, say, temperature at x. Point being a mathematical concept, physically (or experimentally), it cannot determine any physical quantity exactly at a point, rather we can only obtain the average quantity in a neighborhood of the point. Thus, we actually measure $\int f\varphi$ by taking all possible functions φ having compact support. This is an idea that is behind the introduction of *distributions*: $T(\varphi)$, which are more general than the integral, requiring linearity and continuity of T (coming from the integral). Does this average, a good approximation to the exact quantity $f(x)$? Mathematically, we need to know that $\frac{1}{2h}\int_{x-h}^{x+h} f(t)\,dt \to f(x)$ as $h \to 0$. This is true if f is continuous. Thus, if f is not continuous, we need to work with the averages (actual experimental data) $\frac{1}{2h}\int_{x-h}^{x+h} f(t)\,dt$ for all $h > 0$ rather than the point-wise information of the function f. In other words, we need to go out of the comfort zone of the class of functions. The above localization procedure can be dealt with using the compactly supported functions that we describe now.

Localization and Linear Functionals: Recall the following result from basic analysis; If $f : (a, b) \to \mathbb{R}$ is continuous and $\int_a^b f(t)\phi(t)\,dt = 0$ for all continuous ϕ, then $f(t) = 0$ for all $t \in (a, b)$. in other words $f \equiv 0$ if and only if $\int_a^b f(t)\phi(t)\,dt = 0$ for all continuous ϕ. More generally

$$f \equiv g \text{ if and only if } \int_a^b f(t)\phi(t)\,dt = \int_a^b g(t)\phi(t)\,dt$$

for all $\phi \in \mathcal{D}(a, b)$, which is a smaller class than the class of continuous functions. The class $\mathcal{D}(a, b)$ is known as the class of test functions. This has the following localization effect; that is if $\int_a^b f(t)\phi(t)\,dt = 0$ for all $\phi \in \mathcal{D}(a, b)$ with $\operatorname{supp}\phi \subset (c, d) \subset (a, b)$, then $f \equiv 0$ in (c, d) if f is continuous. If f is any locally integrable function, then we get $f = 0$ a.e. in (c, d).

Thus, instead of viewing a function f as point-wise association $x \to f(x)$, it can also be viewed as a mapping $\phi \to \int f\phi$ from $\mathcal{D} \to \mathbb{R}$. This is more general in the sense that every locally integrable function can be viewed in the above sense, but converse need not be true that we will see later and thus providing a bigger class of objects. The impact of the above general viewing is more relevant when we consider the derivative of functions. Consider $f : (a, b) \to \mathbb{R}$ that is C^1. Then, f' can be interpreted as $\phi \to \int f'\phi$. But

$$\int_a^b f'(t)\phi(t)\,dt = -\int_a^b f(t)\phi'(t)\,dt. \tag{12.1}$$

The equality implies that f' can be realized from $\phi \to -\int f\phi'$ and this does not require the (classical) differentiability of f. This motivates us to define a notion of derivative in a generalized or weak sense of locally integrable function f, denoted by Df as

$$Df(\phi) = -\int_a^b f(t)\phi'(t)\,dt.$$

Difficulties and New Objects: Indeed (12.1) holds if f is C^1, but if f is not differentiable, the association $\phi \to -\int f\phi'$ may not represent a function leading to unknown new objects. Understanding these new objects, developing an analysis around them and putting the whole thing in a correct framework is essentially the study of distributions.

Example 12.1. Let $f(x) = |x|$ in $(-1, 1)$ which is not C^1. Now, for $\phi \in \mathcal{D}(-1, 1)$, we have

$$-\int_{-1}^{1} f(t)\phi'(t)\,dt = \int_{-1}^{0} x\phi' - \int_0^1 x\phi' = -\int_{-1}^{0} \phi + \int_0^1 \phi$$

since $\phi(-1) = \phi(1) = 0$. Now, define

$$g(x) = \begin{cases} -1 \text{ if } x \in (-1, 0) \\ 1 \text{ if } x \in (0, 1). \end{cases}$$

Then the weak derivative of f is $Df = g$ a.e. □

Example 12.2. Consider *the Heaviside function*

$$H(x) = \begin{cases} 0 \text{ if } x \in (-\infty, 0) \\ 1 \text{ if } x \in [0, \infty). \end{cases}$$

Then

$$-\int_{-\infty}^{\infty} H(t)\phi'(t)\,dt = -\int_0^{\infty} \phi'(t)\,dt = \phi(0)$$

for any $\phi \in \mathcal{D}(\mathbb{R})$. Thus

$$Df(\phi) = \phi(0).$$

Can we get a locally integrable function g so that

$$Df(\phi) = \phi(0) = \int_{-\infty}^{\infty} g(t)\phi(t)\,dt, \text{ for all } \phi \in \mathcal{D}(\mathbb{R})?$$

If so, we get $Df = g$ a.e. But such a g does not exist and can be seen as follows: Suppose such a g exists. Choose a sequence of test functions $\phi_k \in \mathcal{D}(\mathbb{R})$, $\text{supp}(\phi_k) \subset (-1/k, 1/k)$, $\phi_k(0) = 1$, $|\phi_k(x)| \le 1$. We can use the mollifiers to construct such a sequence. Then, clearly $1 \leq \int_{-1/k}^{1/k} |g(t)|\,dt$ for any k. This is a contradiction as the integral goes to 0 as $k \to \infty$ since g is locally integrable. □

For an open set $\Omega \subset \mathbb{R}^n$, denote by $\mathcal{D}(\Omega)$ the set of all real or complex-valued functions defined on Ω, having compact supports in Ω. The above examples indicate that we need to consider more general mappings $T : \mathcal{D}(\Omega) \to \mathbb{R}$ than the ones given by the locally integrable functions of the form $\phi \to \int f\phi$. An important observation of the earlier association is the linearity of the mappings that we retain for T. The major hurdle is the introduction of a suitable topology on $\mathcal{D}(\Omega)$, which is the real breakthrough in the development of distributions. We define the topology in terms of convergence. It is actually given by an *inductive limit topology*.

Definition 12.3 (Topology). A sequence $\{\phi_k\} \subset \mathcal{D}(\Omega)$ is said to converge to $\phi \in \mathcal{D}(\Omega)$, if there is a compact set $K \subset \Omega$ such that $\text{supp}(\phi_k)$, $\text{supp}(\phi) \subset K$ for all k and $D^\alpha \phi_k \to D^\alpha \phi$, *uniformly* on K, for all multi-indices α. □

The above-defined notion of convergence comes from introducing a *locally convex* topology in $C_c^\infty(\Omega)$. For details, see Rund (1973) and Kesavan (1989). The linear space $C_c^\infty(\Omega)$ with this topology is usually denoted by $\mathcal{D}(\Omega)$. If $\Omega = \mathbb{R}^n$, we write $\mathcal{D} = \mathcal{D}(\mathbb{R}^n)$, for convenience. It is not difficult to see that $\mathcal{D}(\Omega)$ is complete, that is, every Cauchy sequence in $\mathcal{D}(\Omega)$ converges to some function in $\mathcal{D}(\Omega)$. The only minor inconvenience is that $\mathcal{D}(\Omega)$ is *not metrizable*. The space $\mathcal{D}(\Omega)$ is referred to as the *space of test functions*. The mollifiers defined in Chapter 2 are interesting examples of test functions.

Definition 12.4 (Distributions). The space $\mathcal{D}(\Omega)$ being a locally convex space, possesses a topological vector space structure. Its *dual space*, denoted by $\mathcal{D}'(\Omega)$, is called the space of *distributions*.[1] To be more specific, if T is a distribution, that is, if $T \in \mathcal{D}'(\Omega)$, then

1. $T : \mathcal{D}(\Omega) \to \mathbb{R}$ or $\mathbb{C}$ is a linear map.
2. T is *continuous*: whenever $\phi_k \in \mathcal{D}(\Omega)$ and $\phi_k \to \phi$ in $\mathcal{D}(\Omega)$, then the numerical sequence $T(\phi_k) \to T(\phi)$. □

Observe that $\mathcal{D}'(\Omega)$ has a linear structure and inherits the topology from that in $\mathcal{D}(\Omega)$. This means that $T_k, T \in \mathcal{D}'(\Omega)$, then $T_k \to T$ in $\mathcal{D}'(\Omega)$, if $T_k(\phi) \to T(\phi)$ for every $\phi \in \mathcal{D}(\Omega)$.

Lemma 12.5. Suppose $T_k \in \mathcal{D}'(\Omega)$ is a sequence such that $T_k(\phi)$ converges for every $\phi \in \mathcal{D}(\Omega)$. Define T by

$$T(\phi) = \lim_{k\to\infty} T_k(\phi),$$

for every $\phi \in \mathcal{D}(\Omega)$. Then, $T \in \mathcal{D}'(\Omega)$. □

[1] Distributions are also called *generalized functions*, more so in the Russian literature.

This lemma shows that the space $\mathcal{D}'(\Omega)$ is also complete. The following result is of some interest:

Lemma 12.6. Suppose $T_k \in \mathcal{D}'(\Omega)$ is a sequence converging to 0 in $\mathcal{D}'(\Omega)$ and $\phi_k \in \mathcal{D}(\Omega)$ is a sequence converging to 0 in $\mathcal{D}(\Omega)$. Then,

$$T_k(\phi_k) \to 0 \text{ as } k \to \infty.$$

□

Example 12.7.

1. For any locally integrable function f defined on Ω, define $T_f(\phi) = \int_\Omega f\phi$, then $T_f \in \mathcal{D}'(\Omega)$. In particular, all $L^p(\Omega)$ functions can be viewed as distributions. From now onwards if a distribution is given by a function f, we use the notation f itself.
2. Define $\delta(\phi) = \phi(0)$, then δ is a distribution. This is known as the *Dirac δ function,* thus giving a rigorous interpretation to δ function. As already seen Dirac δ cannot be realized through a function.
3. More generally, for any Radon measure μ, define $T_\mu(\phi) = \int_\Omega \phi \, d\mu$, then $T_\mu \in \mathcal{D}'(\Omega)$.
4. Thus the class $\mathcal{M}(\Omega)$ of Radon measures is a subspace of $\mathcal{D}'(\Omega)$. However, the distribution T defined by $T(\phi) = \phi'(0)$, is not realized from any Radon measure. □

Definition 12.8 (Weak Derivative). For any $T \in \mathcal{D}'(\Omega)$, multi-index $\alpha = (\alpha_1, \dots \alpha_n)$, define the αth weak derivative denoted by $D^\alpha T$ of T as

$$(D^\alpha T)(\phi) = (-1)^{|\alpha|} \, T(D^\alpha \phi). \tag{12.2}$$

□

It is trivial to verify that $D^\alpha T \in \mathcal{D}'(\Omega)$. We have $D(|x|) = g$, where g is defined as in Example 12.1 and $DH = \delta$ from Example 12.2. Further, $D\delta(\phi) = -\phi'(0)$ that is not even given by a measure. The distribution δ, $D\delta$ are objects developed by Dirac in physics, but not with proper mathematical foundation and the introduction of distribution theory gave the mathematical rigor in the analysis of Dirac δ and other objects. This theory now can be used to define weak notions of solutions to PDEs. Proving existence, uniqueness, and so on, are a different issue and more sophisticated tools need be developed using functional analysis.

We remark that when a locally integrable function (of one variable) possesses the usual derivative f' a.e., this need not be its weak derivative. That is, it may not be true that $\int f\phi' = -\int f'\phi$ for all test functions ϕ. The Heaviside function H ($H' = 0$ a.e.) provides an example. An important result due to Lebesgue asserts that every monotonic function is differentiable a.e. The Cantor function provides another example of the assertion made above.

Example 12.9. We know that if $f' = 0$ in (a, b) in the classical sense, then f is a constant. This result holds in the sense of distributions as well, that is, if $Df = 0$, equivalently $\int_a^b f\psi' = 0$ for all test functions ψ, then f is constant a.e. To see this, let ϕ be any test function. Choose a non-negative test function η such that $\int_a^b \eta = 1$. We have already constructed such a η in Chapter 2. Define

$$\psi(x) = \int_a^x (\phi(y) - c\eta(y))\, dy,$$

where $c = \int_a^b \phi$. It is straightforward to check that ψ is also a test function and that $\psi' = \phi - c\eta$. Therefore, it follows that

$$0 = \int_a^b f\psi' = \int_a^b (f\phi - c \int_a^b f\eta).$$

We rewrite this as

$$\int_a^b (f - k)\phi = 0,$$

where the constant $k = \int_a^b f\eta$. Since ϕ is any arbitrary test function, it follows that f is the constant function k a.e. □

Example 12.10. As was shown in Chapter 9, Section 9.5, any continuous or even locally integrable function $u(x, t)$ of the form $u(x, t) = v(x \pm ct)$ is a weak solution of the 1D wave equation: $u_{tt} - c^2 u_{xx} = 0$. □

12.2 EXISTENCE OF AN L^2 WEAK SOLUTION

Here, we establish an existence result of an L^2 weak solution of a linear partial differential operator. This result requires only the Riesz representation theorem in the L^2 space and the Hahn–Banach theorem.

Consider an mth-order linear partial differential operator

$$P(x, D) \equiv \sum_{|\alpha| \le m} a_\alpha(x) D^\alpha, \tag{12.3}$$

defined in an open set $\Omega \subset \mathbb{R}^n$. Here the coefficients a_α are smooth (real or complex-valued) functions defined in Ω and $a_\alpha \neq 0$ for at least one multi-index α with $|\alpha| = m$.

We have, for a function u defined in Ω

$$P(x, D)u = \sum_{|\alpha|\leq m} a_\alpha(x)D^\alpha u. \tag{12.4}$$

Recall that the polynomials in $\xi \in \mathbb{R}^n$:

$$p(x, \xi) = \sum_{|\alpha|\leq m} a_\alpha(x)\xi^\alpha \text{ and } p_m(x, \xi) = \sum_{|\alpha|=m} a_\alpha(x)\xi^\alpha \tag{12.5}$$

are, respectively, the *complete* or *full symbol* and *principal symbol* of P.

The (formal) *adjoint* of P in (12.3) is the operator

$$P'(x, D) = \sum_{|\alpha|\leq m} D^\alpha \bar{a}_\alpha(x), \tag{12.6}$$

that is,

$$P'(x, D)u = \sum_{|\alpha|\leq m} D^\alpha(\bar{a}_\alpha(x)u(x)). \tag{12.7}$$

Given $f \in L^2(\Omega)$, by a *weak solution* u of $P(x, D)u = f$, we mean that $u \in L^2(\Omega)$ satisfying the condition

$$(u, P'\phi) = (f, \phi), \text{ for all } \phi \in \mathcal{D}(\Omega). \tag{12.8}$$

Here $(\cdot, \cdot)$ denotes the inner product in $L^2(\Omega)$. The corresponding norm is denoted by $\|\cdot\|$. We first derive a necessary and sufficient condition for the existence of a weak solution.

Theorem 12.11. The equation $P(x, D)u = f$ with $f \in L^2(\Omega)$, has a weak solution $u \in L^2(\Omega)$ if and only if

$$|(f, \phi)| \leq C\|P'\phi)\|, \tag{12.9}$$

for all $\phi \in \mathcal{D}(\Omega)$, for some positive constant C. □

Proof Suppose there is a weak solution u. Then, we have

$$|(f, \phi)| = |(u, P'\phi)| \leq \|u\| \, \|P'\phi)\| = C\|P'\phi)\|,$$

for all $\phi \in \mathcal{D}(\Omega)$. Thus, (12.9) holds. Conversely, suppose the condition in (12.9) holds. Let

$$W = \{P'\phi : \phi \in \mathcal{D}(\Omega)\}.$$

It is easy to check that W is a subspace of $L^2(\Omega)$. Now define $T : W \to \mathbb{C}$ by $T(P'\phi) = (f, \phi)$. We will first verify that T is well-defined. Suppose $P'\phi_1 = P'\phi_2$ for $\phi_1, \phi_2 \in \mathcal{D}(\Omega)$. Then, by the assumed necessary condition, we have

$$|(f, \phi_1 - \phi_2)| \leq C\|P'\phi_1 - P'\phi_2\| = 0,$$

using linearity. Hence, $(f, \phi_1) = (f, \phi_2)$.

Clearly T is linear and bounded:

$$|T(P'\phi)| = |(f, \phi)| \leq C\|P'\phi\|,$$

for all $\phi \in \mathcal{D}(\Omega)$, again, by the assumed necessary condition. By Hahn-Banach theorem, T can be extended to a bounded linear functional on $L^2(\Omega)$. Then, by Riesz representation theorem, there is a $u \in L^2(\Omega)$ satisfying $(u, P'\phi) = (f, \phi)$, for all $\phi \in \mathcal{D}(\Omega)$. Thus, u is a weak solution of (12.3). □

Before proceeding further, we make the following observations:

(1) The above result merely gives that a weak solution $u \in L^2(\Omega)$. We may ask the question whether u is a *classical solution*. That is, whether u is differentiable m times and the equation is satisfied at all the points in Ω. We may also ask the uniqueness question, both for weak and classical solutions. These are some of the deeper questions and the answers are not easy. Some of the answers are given by the regularity results.
(2) The next question concerns about operators satisfying the necessary and sufficient condition (12.9). There is one particularly simple class of operators, namely the constant-coefficient operators, that satisfy condition (12.9).

12.2.1 Constant Coefficient Operators

We assume that the coefficients a_α of the operator P in (12.3) are constants. Then, P' is also a constant coefficient operator. We have the following:

Theorem 12.12. Suppose $P = P(D)$ be a constant coefficient differential operator of order m in a bounded domain Ω. Then, there exists a positive constant C such that

$$\|\phi\| \leq C\|P\phi\|,$$

for all $\phi \in \mathcal{D}(\Omega)$. □

This immediately gives the existence of a weak solution.

Proof Let $p(\xi) = \sum_{|\alpha|\leq m} a_\alpha \xi^\alpha$ be the full symbol of P. We write the differentiation of a product as follows:

$$D_k(uv) = uD_kv + vD_ku \equiv (\overset{u}{D_k} + \overset{v}{D_k})uv,$$

where $\overset{u}{D}_k$ means that u is considered as a constant during the differentiation. Hence,

$$P(D)(uv) = P(\overset{u}{D} + \overset{v}{D})uv.$$

By Taylor's formula, we have

$$p(\eta + \xi) = \sum_{|\alpha| \le m} \frac{1}{\alpha!} p^{(\alpha)}(\eta)\xi^{\alpha},$$

where $p^{(\alpha)}(\eta) = \frac{\partial^{|\alpha|} p(\eta)}{\partial \eta_1^{\alpha_1} \cdots \eta_n^{\alpha_n}}$, with $p^{(0)}(\eta) = p(\eta)$. Therefore,

$$P(uv) = P(D)(uv) = \sum_{|\alpha| \le m} \frac{1}{\alpha!} (P^{(\alpha)}(D)u) D^{\alpha} v. \tag{12.10}$$

If $\bar{p}(\xi) = \sum_{|\alpha| \le m} \bar{a}_{\alpha} \xi^{\alpha}$ denotes the full symbol of the adjoint operator P', it is easy to check that

$$\|P\phi\| = \|P'\phi\|, \quad \text{for all } \phi \in \mathcal{D}(\Omega),$$

The theorem now follows from the following lemma and its corollary: □

Lemma 12.13. Fix k, $1 \le k \le n$ and suppose that

$$\Omega \subset \{x : |x_k - a| \le M/2\}.$$

Let $p^{(k)}(\xi) = \dfrac{\partial p(\xi)}{\partial \xi_k}$. Then,

$$\|P^{(k)}(D)\phi\| \le mM \|P(D)\phi\|, \quad \text{for all } \phi \in \mathcal{D}(\Omega).$$

Proof By induction on m, the order of the given differential operator. Suppose □

$$\|Q^{(k)}(D)\phi\| \le mM \|Q(D)\phi\|, \quad \text{for all } \phi \in \mathcal{D}(\Omega),$$

where $Q(D) = \sum_{|\alpha| < m} c_{\alpha} D^{\alpha}$. By a translation, we may assume that $a = 0$. Then, using (12.10), we have

$$P(D)(x_k\phi) = x_k P(D)\phi + P^{(k)}(D)\phi.$$

Hence, using the induction hypothesis, we have

$$
\begin{aligned}
\|P^{(k)}(D)\phi\|^2 &= (P^{(k)}(D)\phi, P^{(k)}(D)\phi) \\
&= (P(D)(x_k\phi) - x_k P(D)\phi, P^{(k)}(D)\phi) \\
&= (P(D)(x_k\phi), P^{(k)}(D)\phi) - (x_k P(D)\phi, P^{(k)}(D)\phi) \\
&= (\overline{P^{(k)}(D)}(x_k\phi), \overline{P(D)}\phi) - (P(D)\phi, x_k P^{(k)}(D)\phi) \\
&= (x_k \overline{P^{(k)}(D)}\phi + \overline{P^{(kk)}(D)}\phi, \overline{P(D)}\phi) - (P(D)\phi, x_k P^{(k)}(D)\phi) \\
&\leq \|P(D)\phi\| \left(M\|P^{(k)}(D)\phi\| + \|P^{(kk)}(D)\phi\|\right),
\end{aligned}
$$

where we have used $|x_k| \leq M$ and $\|\bar{P}(D)\phi\| = \|P(D)\phi\|$ and

$$p^{(kk)}(\xi) = \frac{\partial^2}{\partial \xi_k^2} p(\xi).$$

Since, $p^{(k)}(\xi)$ is of degree $< m$, we have

$$\|P^{(kk)}(D)\phi\| \leq (m-1)M\|P^{(k)}(D)\phi\|.$$

This completes the proof of the induction step. For $m = 1$, $p^{(kk)}(\xi) = 0$, so the lemma holds true for $m = 1$. This completes the proof. □

Corollary 12.14. Suppose

$$\Omega \subset \{x : |x_k - a_k| \leq \frac{1}{2} M_k,\ 1 \leq k \leq n\}.$$

Then, for any multi-index α, we have

$$\|P^{\alpha}(D)\phi\| \leq \frac{m!}{(m - |\alpha|)!} M^{\alpha} \|P(D)\phi\|,$$

where $M^{\alpha} = M_1^{\alpha_1} \cdots M_n^{\alpha_n}$. □

To see how Theorem 12.12 follows from Corollary 12.14, choose a multi-index α such that $p^{(\alpha)}(\xi) = \text{constant} \neq 0$. Necessarily $|\alpha| = m$ and such an α always exists.

We now make a few remarks regarding the case of variable coefficients. The analysis is much harder and deeper. Hörmander (1976) has given a necessary condition for the existence of a solution; a strengthened version of this condition is also sufficient provided that there are *no multiple characteristics*. Nirenberg and Trèves have given a much more complete analysis of the first-order case.

Consider the operator P given by (12.3) with smooth coefficients and let $p_m(x, \xi)$ be its principle symbol. Put $\bar{p}_m(x, \xi) = \sum_{|\alpha|=m} \bar{a}_\alpha \xi^\alpha$. Let

$$p_m^{(j)}(x, \xi) = \frac{\partial}{\partial \xi_j} p_m(x, \xi), \; p_{m,j}(x, \xi) = \frac{\partial}{\partial x_j} p_m(x, \xi).$$

Set

$$C_{2m-1}(x, \xi) = \sum_{j=1}^{n} i \left(p_m^{(j)}(x, \xi) \bar{p}_{m,j}(x, \xi) - p_{m,j}(x, \xi) \bar{p}_m^{(j)}(x, \xi) \right).$$

Note that C_{2m-1} is a polynomial (in ξ) of degree $2m - 1$. Further, if the coefficients in P are real or constants, then $C_{2m-1} \equiv 0$. Here is the necessary condition given by Hörmander.

Theorem 12.15. Suppose $P(x, D)u = f$ has a solution $u \in \mathcal{D}'(\Omega)$ for every $f \in \mathcal{D}(\Omega)$, then

$$C_{2m-1}(x, \xi) = 0 \text{ if } p_m(x, \xi) = 0. \tag{12.11}$$

□

12.3 SOBOLEV SPACES

Recall Section 7.5 of Chapter 7 of the Laplace equation $-\Delta u = f$ on the discussion on weak formulation. There, we encountered the problem of identifying the completion X of the space $C^1(\overline{\Omega})$ with respect to the norm

$$\|u\|^2 := \|u\|^2_{L^2(\Omega)} + \|\nabla u\|^2_{L^2(\Omega)}. \tag{12.12}$$

This is due to the consequence of the fact that we had to take minimizing sequences with respect to the above norm and it may happen that the limits need not differentiable or not even continuous.

For the PDE

$$-\Delta u + u = f \text{ in } \Omega \quad u = 0 \text{ on } \partial\Omega, \tag{12.13}$$

we have an integral formulation

$$\int \nabla u \cdot \nabla v + \int u\, v = \int fv. \tag{12.14}$$

We have included lower-order term in (12.13) to avoid certain other technicalities. The relation (12.14) is referred to as a weak formulation of (12.13) and is obtained by multiplying (12.13) by $v \in C_0^1(\overline{\Omega})$ and performing an integration by parts. However, if $u \in C^2(\overline{\Omega})$ with zero condition on the boundary, and (12.14) holds for $v \in C_0^1(\overline{\Omega})$, then (12.13) can be recovered, that is u is a solution of (12.13).

In order to show the existence of u satisfying (12.14) for all v, we need to work in a complete space as we can then apply the Riesz representation theorem in a Hilbert space. Thus, we need to know the completion of $C^1(\overline{\Omega})$ and/or $C_0^1(\overline{\Omega})$ with respect to the norm given by (12.12). Observe that all the terms in the integral formulation hold if u, v and its first-order derivatives are L^2 functions. In fact, it is sufficient to have weak derivatives, but the weak derivatives should exist as functions. This motivates the introduction of the following function spaces:

Let $u \in L^2(\Omega)$, where $\Omega \subset \mathbb{R}^n$ is a smooth domain. As we have seen earlier, this may not imply that the weak derivative $D_i u$ with respect to x_i is given by a function; it is simply an element in $\mathcal{D}'(\Omega)$. We say u has an ith weak derivative in $L^2(\Omega)$ if there is a function $v_i \in L^2(\Omega)$ such that

$$D_i u(\phi) = -\int_{\Omega} u(x)\frac{\partial \phi}{\partial x_i} = \int_{\Omega} v_i(x)\phi(x)dx.$$

Now, define

$$H^1(\Omega) = \{u \in L^2(\Omega) : D_i u \in L^2(\Omega),\ 1 \le i \le m\} \tag{12.15}$$

and

$$H_0^1(\Omega) = \{u \in H^1(\Omega) : u = 0 \text{ on } \partial\Omega\}. \tag{12.16}$$

Interpretation: The above spaces are Hilbert spaces with respect to the norm

$$\|u\|^2 \equiv \|u\|^2_{L^2(\Omega)} + \|\nabla u\|^2_{L^2(\Omega)}. \tag{12.17}$$

The space $H_0^1(\Omega)$ is introduced to take care of the boundary condition. The definition of $H_0^1(\Omega)$ is much more delicate due to the following reason: Given a smooth domain $\Omega \subset \mathbb{R}^n$, its boundary $\partial\Omega$ is a measure zero set. Hence a function in $L^2(\Omega)$, which is defined a.e. does not have any meaning on sets of measure zero. This statement in particular applies to the functions in $H^1(\Omega)$. But, we can rigorously interpret through delicate analysis, the boundary values (known as traces) of $H^1(\Omega)$ functions. This is a non-trivial result and is known as *trace theorem*. Interpreting boundary values of non-smooth functions has paramount importance in the study of boundary value problems.

Together with trace results, the other important issues are *prolongation, density and compactness results*. These four issues constitute the initial study of Sobolev spaces before undertaking the study of existence, uniqueness and regularity results of weak formulation. We are not planning to elaborate on these points, but quickly explain what it means.

Prolongation: Given $u \in L^2(\Omega)$, we know that the trivial extension $\tilde{u} \in L^2(\mathbb{R}^n)$, where

$$\tilde{u}(x) = \begin{cases} u(x) \text{ if } x \in \Omega \\ 0 \text{ if } x \in \mathbb{R}^n \setminus \Omega. \end{cases}$$

But $u \in H^1(\Omega)$ does not necessarily imply that $\tilde{u} \in H^1(\mathbb{R}^n)$ as can be seen from the following example: Take u as the constant function 1 in $(0, 1)$. Then, $\frac{d\tilde{u}}{dx} = \delta_0 - \delta_1$, where δ_0 and δ_1 are the Dirac delta concentrated at 0 and 1 respectively given by $\delta_0(\phi) = \phi(0)$ and $\delta_1(\phi) = \phi(1)$. However, it may be possible to have non-trivial extensions of u so that the extended function is in $H^1(\mathbb{R}^n)$. It is important to have such space-preserving extensions in the analysis of PDE, since working with full space has several advantages like applying Fourier transform, convolution, and so on. We remark that the smoothness of the boundary plays an important role in the analysis of such an extension.

Density: We know that the test function space $\mathcal{D}(\Omega)$ is dense in $L^2(\Omega)$, However, in general, $\mathcal{D}(\Omega)$ is not dense in $H^1(\Omega)$. It holds true if $\Omega = \mathbb{R}^n$ and further, $\mathcal{D}(\Omega)$ is dense in $H^1_0(\Omega)$. Again approximating $H^1(\Omega)$ functions via smooth functions is important in proving many results as quite often, such results are first proved for smooth functions and derive those results for non-smooth functions via density arguments. There are many density results for $H^1(\Omega)$ functions.

Compactness Theorem: This is heavily used in a-postori analysis. After obtaining a weak solution of a PDE in an appropriate Sobolev space, we might ask: Is the weak solution, a classical solution? More precisely, this leads to the abstract question of imbedding (continuous, compact) of a Sobolev space into a classical space. We remark at this stage that we can also define higher-order H^k spaces requiring that all the weak derivatives up to order k are L^2 functions. We state that higher-order Sobolev spaces are always imbedded compactly in a lower-order Sobolev space. Further, it is possible to imbed a higher-order H^k space into a lower-order smooth space C^l with certain relations connecting k, l and the dimension n. The regularity results play an important role here. After obtaining a solution in a suitable H^m space, we try to establish the weak solution is in a higher-order H^k space (these are called regularity results). If k is sufficiently large, we may use imbedding theorems to establish that the weak solution is also a classical solution.

Sobolev Spaces via Fourier Transform (FT): In $\mathbb{R}^n$, we can define $H^1(\mathbb{R}^n)$ or more generally $H^s(\mathbb{R}^n)$ for any[2] $s > 0$ without appealing to the theory of distributions and weak derivatives. Observe that $u \in L^2(\mathbb{R}^n)$ if and only if the FT, $\hat{u} \in L^2(\mathbb{R}^n)$ and the norm equality holds. Thus, $u \in H^1(\mathbb{R}^n)$ if and only if $\hat{u}, \widehat{D_i u} \in L^2(\mathbb{R}^n)$. Hence, we can recast $H^1(\mathbb{R}^n)$ as

$$H^1(\mathbb{R}^n) = \{u \in L^2(\mathbb{R}^n) : \left(1 + |\xi|^2\right)^{1/2} \hat{u}(\xi) \in L^2(\mathbb{R}^n)\}$$

[2]The definition may also be made for $s < 0$, using the tempered distributions in place of L^2 functions.

and the norm is given by

$$\|u\|_{H^1(\mathbb{R}^n)} = \| \left(1 + |\xi|^2\right)^{1/2} \hat{u}\|_{L^2(\mathbb{R}^n)}.$$

Interestingly, $H^1(\mathbb{R}^n) = H^1_0(\mathbb{R}^n)$ and hence $H^1(\mathbb{R}^n)$ functions can be approximated by $\mathcal{D}(\mathbb{R}^n)$ functions.

If s is not an integer, the spaces $H^s(\mathbb{R}^n)$ are known as Sobolev spaces of fractional order. Defining $H^s(\Omega)$ for proper subsets Ω of $\mathbb{R}^n$ is more delicate and, in fact, there is no unique way of defining these spaces. It is also possible to define negative-order Sobolev spaces $H^{-s}(\Omega), s > 0$ as the dual space of $H^s_0(\Omega)$. Here, we remark that $H^{-s}(\Omega), s > 0$ are not function spaces. Further, there are extensions using $L^p(\Omega)$ in place of $L^2(\Omega)$ for $p \neq 2$. Indeed, all these spaces are extremely useful in the study of different classes of PDE.

12.4 NOTES

The presentation in this chapter is just like scratching a surface without going anywhere deep. A good understanding requires advanced topics from modern functional analysis. A second course in PDE begins from here. There are many books in this direction, see for example, Kesavan (1989), Brezis (2011), Hörmander (1976, 1984), and Trèves (2006).

References

Abramowitz, M. and Stegun, A. (1972). *Handbook of Mathematical Functions*, New York: Dover.

Apostol, T. M. (2002). *Mathematical Analysis*, 2nd ed., New Delhi: Narosa.

———. (2011). *Calculus,* vols. 1 and 2. New Delhi: Wiley India.

Bardi, M. and Capuzzo Dolcetta, I. (1997). *Optimal Control and Viscosity Solution of Hamilton–Jacobi–Bellman Equations*, Boston: Birkhäuser.

Barták, J., Herrman, L., Lovicar, V. and Vejvoda, O. (1991). *Partial Differential Equations of Evolution*, New York: Ellis Horwood.

Benton, S. (1977). *The Hamilton–Jacobi Equation: A Global Approach*, New York: Academic Press.

Benzoni-Gavage, S. and Serre, D. (2007). *Multidimensional Hyperbolic Partial Differential Equations*, Oxford: Clarendon Press.

Brezis, H. (2011). *Functional Analysis, Sobolev Spaces and Partial Differential Equations*, New York: Springer.

Burgers, J. M. (1974). *The Nonlinear Diffusion Equation*, Dordrecht-Holland: D. Reidel Publishing Company.

Caflisch, R. E. (1990). A simplified version of the abstract Cauchy–Kowalewski theorem with weak singularities. *Bulletin of the American Mathematical Society*, **23**(2), 495–500.

Cole, J. D. (1951). On a quasi-linear parabolic equation occurring in aerodynamics. *Quarterly of Applied Mathematics*, **9**, 225–236.

Copson, E. T. (1975). *Partial Differential Equations*, Cambridge: Cambridge University Press.

Courant, R. and Friedrichs, K. O. (1976). *Supersonic Flow and Shock Waves*, reprint ed., New York: Springer-Verlag.

Courant, R. and Hilbert, D. (1989). *Methods of Mathematical Physics: Vol II Partial Differential Equations*, New York: John Wiley and Sons.

Crandall, M. G., Evans, L. C. and Lions, P. L. (1984). Some properties of viscosity solutions of Hamilton–Jacobi equations. *Transactions of the American Mathematical Society*, **281**, 487–502.

Crandall, M. G., Ishii, H. and Lions, P. L. (1992). Users guide to viscosity solutions of second order HJEs. *Bulletin of the American Mathematical Society*, **27**, 1–67.

Crandall, M. G. and Lions P. L. (1981). Conditions d'unicite pour les solutions generalises des equations d'Hamilton–Jacobi du premier ordre. *Comptes rendus de l'Académie des Sciences Paris Serie I: Mathematique*, **292**, 487–502.

——. (1983). Viscosity solutions of Hamilton–Jacobi equations. *Transactions of the American Mathematical Society*, **292**, 1–42.

Dafermos, C. M. (2009). *Hyperbolic Conservation Laws in Continuum Physics*, 2nd ed., New York: Springer.

DiBenedetto, E. (2010). *Partial Differential Equations*, 2nd ed., New Delhi: Springer International.

Evans, L. C. (1998). *Partial Differential Equations*, Rhode Island: AMS, Providence.

Folland, G. B. (1992). *Fourier Analysis and Its Applications*, Hyderabad: American Mathematical Society, University Press (India).

——. (1995). *Introduction to Partial Differential Equations*, 2nd ed., Princeton: Princeton University Press.

Forsyth, A. R. (1906). *Theory of Differential Equations, Part IV, Partial Differential Equations*, vol. VI. Cambridge: Cambridge University Press.

Friedland, S., Robbins, J. W. and Sylvester, J. H. (1984). On the crossing rule. *Communications on Pure and Applied Mathematics*, **37**, 19–37.

Gilberg, D. and Trudinger, N. S. (2001). *Elliptic Partial Differential Equations of Second Order*, Berlin-Heidelberg: Springer.

Gilkey, P. B. (1984). *Invariance Theory, the Heat Equation and the Atiyah-Singer Index Theorem*, Houston: Publish or Perish Inc.

Glimm, J. (1965). Solutions in the large for nonlinear hyperbolic systems of equations. *Communications on Pure and Applied Mathematics*, **18**, 697–715.

Goldstein, J. A. (1985). *Semigroups of Linear Operators and Applications*, Oxford: Oxford University Press.

Hadamard, J. (1902). Sur les problmes aux drives partielles et leur signification physique. *Princeton University Bulletin*, **13**, 49–52.

Hopf, E. (1950). The partial differential equation $u_t + uu_x = \mu u_{xx}$. *Communications on Pure and Applied Mathematics*, **3**, 201–230.

Hörmander, L. (1976). *Linear Partial Differential Operators, Fourth Printing*, New York: Springer-Verlag.

——. (1984). *Linear Partial Differential Operators*, vols. I and II. New York: Springer-Verlag.

——. (1988). *Non-linear Hyperbolic Differential Equations*, Lund: University of Lund and Lund Institute of Technology.

——. (1997). *Lectures on Nonlinear Hyperbolic Differential Equations, Mathèmatiques and Applications*, vol. 26. Heidelberg: Springer.

John, F. (1971). *Partial Differential Equations*, 1st ed., New York: Springer Verlag.

——. (1975). *Partial Differential Equations*, 2nd ed., New York: Springer Verlag.

——. (1978). *Partial Differential Equations*, 3rd ed., New York: Springer Verlag.

——. (1983). Lower bounds for the life span of solutions of nonlinear wave equations in three space dimensions. *Communications on Pure and Applied Mathematics*, **36**, 1–35.

John, F. and Klainerman, S. (1983). Almost global existence to nonlinear wave equations in three space dimensions. *Communications on Pure and Applied Mathematics*, **37**, 443–455.

Kesavan, S. (1989). *Topics in Functional Analysis and Applications*, New Delhi: Wiley Eastern.

Klainerman, S. (1980). Global existence for nonlinear wave equations. *Communications on Pure and Applied Mathematics*, **33**, 43–101.

———. (1985). Uniform decay estimates and the Lorentz invariance of the classical wave equation. *Communications on Pure and Applied Mathematics*, **38**, 321–332.

Klainerman, S. and Ponce, G. (1983). Global small amplitude solutions to nonlinear evolution equations. *Communications on Pure and Applied Mathematics*, **36**, 133–141.

Koshlyakov, N. S., Smirnov, M. M. and Gliner, E. B. (1964). *Differential Equations of Mathematical Physics*, Amsterdam: North-Holland.

Kreiss, H. O. and Lorenz, J. (2004). *Initial-Boundary Value Problems and the Navier–Stokes Equations*, Philadelphia: SIAM.

Ladyzhenskaya, O. A. (1985). *Boundary Value Problems of Mathematical Physics*, New York: Springer.

Ladyzhenskaya, O. A., Solonnikov, V. A. and Ural'ceva, N. N. (1968). *Linear and Quasilinear Equations of Parabolic Type*, Rhode Island: American Mathematical Society, Providence.

Lax, P. (1973). *Hyperbolic Systems of Conservation Laws and Mathematical Theory of Shock Waves*, Philadelphia: SIAM.

———. (1982). Multiplicity of eigenvalues. *Bulletin of the American Mathematical Society*, **6**(2), 213–214.

Lewy, H. (1957). An example of a smooth partial differential equation without solution. *Annals of Mathematics*, **66**, 155–158.

Liberzon, D. (2012). *Calculus of Variations and Optimal Control Theory: A Concise Introduction*, Princeton: Princeton University Press.

Lighthill, M. J. and Whitham, G. B. (1955). *On Kinematic Waves: I. Flood movement in long rivers; II. Theory of traffic flow on long crowded roads*. Proceedings of the Royal Society London, **A229**, 281–345.

Lions, P. L. (1982). *Generalized Solutions of Hamilton–Jacobi Equations*, London: Pitman.

Liu, T. (1977). The deterministic version of the Glimm scheme. *Communications in Mathematical Physics*, **55**, 163–177.

Majda, A. (1985). Compressible fluid flow and systems of conservation laws in several space variables, New York: Springer-Verlag.

Markowitz, P. (2005). *Partial Differential Equations*, New York: Springer.

McOwen, R. C. (2005). *Partial Differential Equations: Methods and Applications*, 2nd ed., Delhi: Pearson Education.

Mikhailov, V. P. (1978). *Partial Differential Equations*, Moscow: Mir Publishers.

Mitrea, D. (2013). *Distributions, Partial Differential Equations and Harmonic Analysis*, New York: Springer.

Morawetz, C. S. (1981). *Lectures on Non-linear Waves and Shocks, TIFR Lecture Notes*, New York: Springer-Verlag.
Munkres, J. R. (1991). *Analysis on Manifolds*, Redwood City: Addison-Wesley.
Murray, J. D. (2003). *Mathematical Biology*, 3rd ed., New York: Springer.
Nandakumaran, A. K., Datti, P. S. and George, R. K. (2017). *Ordinary Differential Equations: Principles and Applications*, Cambridge: Cambridge University Press.
Nirenberg, L. (1976). *Lectures on Linear Partial Differential Equations, Regional Conference Series in Mathematics*, vol. 17, second printing ed., American Mathematical Society, Providence.
Oleinik, O. A. (1959). Uniqueness and stability of the generalised solution of the Cauchy problem for a quasi-linear equation. *Uspekhi Matematicheskikh Nauk*, **14**, 165–70.
Pazy, A. (1983). Semigroups of linear operators and applications to partial differential equations, New York: Springer-Verlag.
Peetre, J. (1960). Réctifications à l'article "Une caracterérisation abstraite des opéteurs différentials". *Mathematica Scandinavica*, **8**, 116–120.
Pinchover, Y. and Rubinstein, J. (2005). *An Introduction to Partial Differential Equations*, Cambridge: Cambridge University Press.
Prasad, P. and Ravindran, R. (1996). *Partial Differential Equations*, 3rd ed., New Delhi: New Age International.
Rauch, J. (1992). *Partial Differential Equations*, New Delhi: Narosa.
Renardy, M. and Rogers, R. C. (2004). *Partial Differential Equations*, New York: Springer.
Rhee, H.-K., Aris, R. and Amundson, N. R. (1986). *First-Order Partial Differential Equations*, vols. I and II. Mineola: Dover Publications.
Rubinstein, I. and Rubinstein, L. (1998). *Partial Differential Equations in Classical Mathematical Physics*, Cambridge: Cambridge University Press.
Rudin, W. (1976). *Principles of Mathematical Analysis*, London: McGraw-Hill.
Rund, H. (1973). *The Hamilton–Jacobi Theory in the Calculus of Variations*, Huntington, NY: Krieger.
Salsa, S. (2008). *Partial Differential Equations in Action: From Modelling to Theory*, Milano: Springer.
Schwartz, L. (1966). *Théorie des Distributions*, Paris: Hermann.
Simmons, G. F. (1991). *Differential Equations with Applications and Historical Notes*, New York: McGraw-Hill International.
Smoller, J. (1994). *Shock Waves and Reaction–Diffusion Equations*, 2nd ed., New York: Springer-Verlag.
Spivak, M. (1965). *Calculus on Manifolds*, Reading, MA: Addison-Wesley Publishing Company.
Stroock, D. W. and Varadhan, S. R. S. (1979). *Multidimensional Diffusion Processes*, New York: Springer-Verlag.
Taylor, A. E. and Mann, W. R. (1983). *Advanced Calculus*, New York: John Wiley and Sons.
Trèves, F. (2006). *Basic Linear Partial Differential Equations*, New York: Dover.

Vladimirov, V. S. (1979). *Generalized Functions of Mathematical Physics*, Moscow: Mir Publishers.
———. (1984). *Equations of Mathematical Physics*, Moscow: Mir Publishers.
von Wahl, W. (1971). L^p decay rates for homogeneous wave equations. *Mathematische Zeitschrift*, **120**, 93–106.
Whitham, G. B. (1974). *Linear and Non-linear Waves*, New York: Wiley-Interscience.
Widder, D. V. (1961). *Advanced Calculus*, Delhi: PHI Learning.
———. (1975). *The Heat Equation*, New York: Academic Press.
Wilcox, C. H. (1962). Initial-boundary value problems for linear partial differential equations of the second order. *Archive for Rational Mechanics and Analysis*, **10**, 361–400.
Yosida, K. (1974). *Functional Analysis*, 4th ed., New Delhi: Narosa Publishing House.

Index